Nonlinear Photonics Devices

Nonlinear Photonics Devices

Editors

Luigi Sirleto
Giancarlo C. Righini

MDPI • Basel • Beijing • Wuhan • Barcelona • Belgrade • Manchester • Tokyo • Cluj • Tianjin

Editors
Luigi Sirleto
National Research Council,
Institute of Applied Sciences
and Intelligent Systems
Italy

Giancarlo C. Righini
"Nello Carrara" Institute of
Applied Physics (IFAC),
National Research Council
Italy

Editorial Office
MDPI
St. Alban-Anlage 66
4052 Basel, Switzerland

This is a reprint of articles from the Special Issue published online in the open access journal *Micromachines* (ISSN 2072-666X) (available at: https://www.mdpi.com/journal/micromachines/special_issues/Nonlinear_Photonics_Devices).

For citation purposes, cite each article independently as indicated on the article page online and as indicated below:

LastName, A.A.; LastName, B.B.; LastName, C.C. Article Title. *Journal Name* **Year**, *Volume Number*, Page Range.

ISBN 978-3-03943-721-4 (Hbk)
ISBN 978-3-03943-722-1 (PDF)

© 2020 by the authors. Articles in this book are Open Access and distributed under the Creative Commons Attribution (CC BY) license, which allows users to download, copy and build upon published articles, as long as the author and publisher are properly credited, which ensures maximum dissemination and a wider impact of our publications.

The book as a whole is distributed by MDPI under the terms and conditions of the Creative Commons license CC BY-NC-ND.

Contents

About the Editors vii

Luigi Sirleto and Giancarlo C. Righini
Editorial for the Special Issue on Nonlinear Photonics Devices
Reprinted from: *Micromachines* 2020, 11, 760, doi:10.3390/mi11080760 1

Iännis Roland, Marco Ravaro, Stéphan Suffit, Pascal Filloux, Aristide Lemaître, Ivan Favero and Giuseppe Leo
Second-Harmonic Generation in Suspended AlGaAs Waveguides: A Comparative Study
Reprinted from: *Micromachines* 2020, 11, 229, doi:10.3390/mi11020229 7

Francesco De Lucia and Pier J. A. Sazio
Thermal Poling of Optical Fibers: A Numerical History
Reprinted from: *Micromachines* 2020, 11, 139, doi:10.3390/mi11020139 15

Luigi Sirleto and Maria Antonietta Ferrara
Fiber Amplifiers and Fiber Lasers Based on Stimulated Raman Scattering: A Review
Reprinted from: *Micromachines* 2020, 11, 247, doi:10.3390/mi11030247 33

Maria Antonietta Ferrara and Luigi Sirleto
Integrated Raman Laser: A Review of the Last Two Decades
Reprinted from: *Micromachines* 2020, 11, 330, doi:10.3390/mi11030330 53

Iolanda Ricciardi, Simona Mosca, Maria Parisi, François Leo, Tobias Hansson, Miro Erkintalo, Pasquale Maddaloni, Paolo De Natale, Stefan Wabnitz and Maurizio De Rosa
Optical Frequency Combs in Quadratically Nonlinear Resonators
Reprinted from: *Micromachines* 2020, 11, 230, doi:10.3390/mi11020230 73

Gabriele Frigenti, Daniele Farnesi, Gualtiero Nunzi Conti and Silvia Soria
Nonlinear Optics in Microspherical Resonators
Reprinted from: *Micromachines* 2020, 11, 303, doi:10.3390/mi11030303 95

Varun Raghunathan, Jayanta Deka, Sruti Menon, Rabindra Biswas and Lal Krishna A.S
Nonlinear Optics in Dielectric Guided-Mode Resonant Structures and Resonant Metasurfaces
Reprinted from: *Micromachines* 2020, 11, 449, doi:10.3390/mi11040449 117

Vincenzo Bruno, Stefano Vezzoli, Clayton DeVault, Thomas Roger, Marcello Ferrera, Alexandra Boltasseva, Vladimir M. Shalaev and Daniele Faccio
Dynamical Control of Broadband Coherent Absorption in ENZ Films
Reprinted from: *Micromachines* 2020, 11, 110, doi:10.3390/mi11010110 147

Mourad Baira, Bassem Salem, Niyaz Ahamad Madhar and Bouraoui Ilahi
Intersubband Optical Nonlinearity of GeSn Quantum Dots under Vertical Electric Field
Reprinted from: *Micromachines* 2019, 10, 243, doi:10.3390/mi10040243 157

Alessandro Belardini, Grigore Leahu, Emilija Petronijevic, Teemu Hakkarainen, Eero Koivusalo, Marcelo Rizzo Piton, Soile Talmila, Mircea Guina and Concita Sibilia
Circular Dichroism in the Second Harmonic Field Evidenced by Asymmetric Au Coated GaAs Nanowires
Reprinted from: *Micromachines* 2020, 11, 225, doi:10.3390/mi11020225 167

Xi-Rong Su, Yi-Wen Huang, Tong Xiang, Yuan-Hua Li and Xian-Feng Chen
Generation of Pure State Photon Triplets in the C-Band
Reprinted from: *Micromachines* **2019**, *10*, 775, doi:10.3390/mi10110775 . . . . **175**

Juan S. Totero Gongora, Luana Olivieri, Luke Peters, Jacob Tunesi, Vittorio Cecconi, Antonio Cutrona, Robyn Tucker, Vivek Kumar, Alessia Pasquazi and Marco Peccianti
Route to Intelligent Imaging Reconstruction via Terahertz Nonlinear Ghost Imaging
Reprinted from: *Micromachines* **2020**, *11*, 521, doi:10.3390/mi11050521 . . . . **187**

About the Editors

Luigi Sirleto is a research scientist, working at National Research Council of Italy. In september 2001, he spent 2 months at Institute of Nanotechnology, University of Twente, (NL). In July 2003, he spent 9 months, as a visiting scientist, at Electrical Engineering Department of UCLA (University of California, Los Angeles)-USA. In september 2006 he founded the Ultrafast and Nonlinear Optics Lab at Institute of Applied Sciences and Intelligent Systems (ISASI) of CNR and he has led the research activities of the same, until now. In April 2018, he received his qualification as Full Professor of Experimental Matter Physics. He is co-author of over 180 papers, mostly on nonlinear optics and photonics devices. He has served as a reviewer of many international journals and as a committee member of many national and international conferences.

Giancarlo C. Righini is a physicist; fellow of EOS, OSA, SIOF, and SPIE; and meritorious member of the Italian Physical Society (SIF). He worked for almost 40 years at CNR, the National Research Council of Italy, in Florence and Rome, acting as director of various structures. After his retirement from CNR, he was director of the Enrico Fermi Centre in Rome. He is author or co-author of over 500 papers, mostly on photonic glasses, integrated optics, and microresonators (ORCID ID: 0000-0002-6081-6971). He is editor of a book on glass micro- and nanospheres, and he is co-editor of other books. He was vice-president of IUPAP and of ICO, co-founder and president of the Italian Society of Optics and Photonics (SIOF), secretary of EOS, and member of the Board of Directors of SPIE. Currently, he is chair of the TC20 Committee on Glasses for Optoelectronics of ICG and honorary chair of the series of PRE (Photoluminescence in Rare Earths) Workshops.

Editorial

Editorial for the Special Issue on Nonlinear Photonics Devices

Luigi Sirleto [1],* and Giancarlo C. Righini [2],*

1. National Research Council (CNR), Institute of Applied Sciences and Intelligent Systems (ISASI), Via Pietro Castellino 111, 80131 Napoli, Italy
2. National Research Council (CNR), Institute of Applied Physics (IFAC) "Nello Carrara", Via Madonna del Piano 10, 50019 Sesto Fiorentino, Florence, Italy
* Correspondence: luigi.sirleto@cnr.it (L.S.); righini@ifac.cnr.it (G.C.R.)

Received: 13 July 2020; Accepted: 5 August 2020; Published: 7 August 2020

There is some incertitude on the creation of the term "photonics" and some ambiguity about its frontiers (and differences with respect to optoelectronics and electro-optics). Many authors consider the French scientist Pierre Agrain as the "father" of photonics, as of 1967, even if it would be more correct to refer to an almost simultaneous invention of the word by a group of French physicitsts working in lasers and fiber optics and by a Dutch group of high speed photography specialists. The first appearance of this word was apparently in 1952 [1]. A very interesting analysis of the use of the term photonics, embracing history, philosophy, and sociology of science, was published recently [2].

What is sure is that "photonics" was increasingly used after the broadening of laser applications, and nowadays there is a rather general consensus on the definition given in the web page of UNESCO 2015 International Year of Light and Light-Based Technologies: "Photonics is the science and technology of generating, controlling, and detecting photons, which are particles of light" [3].

In most cases, the response of a material to an optical field is linear (i.e., the strength of the response is proportional to the strength of the optical field), but all the way back in the second half of the XIX century, John Kerr, in Glasgow, observed effects that were proportional to the square of the applied field. The field of nonlinear optics, however, started to grow up only after the invention of the laser, when intense light sources became easily available. The seminal studies by Peter Franken [4] and Nicolaas Bloembergen [5], in the 1960s, paved the way to the development of today's nonlinear photonics, the field of research which encompasses all the studies, designs, and implementations of nonlinear optical devices which can be used for the generation, communication, and processing of information.

Ten years after Franken's paper, Anderson and Boyd performed the first nonlinear optics experiment in waveguides: as for bulk nonlinear optics, it dealt with frequency conversion, namely, second harmonic generation (SHG) in gallium arsenide (GaAs) waveguides [6]. It became soon clear that waveguides would offer fundamental advantages for nonlinear optics due to the intrinsic radiation confinement, leading to high optical power densities over long propagation distances.

Of course, the general trend of science towards the nano-world has also influenced the development of photonics, which started from waveguides at micrometer scale, going through microphotonics structures, and finally coming to nanophotonics. In the last few decades, with the development of integrated and nano-optics, biophotonics, quantum, and free-space optical communication, the concept of "photonics" acquired a broader sense. Nowadays, "photonics" is used almost synonymously with the term "optics," referring equally to both science and applications, while nonlinear optical phenomena, and devices based on them, play a key role both in the knowledge of the matter and in many applications of photonics. This justifies the continuation of fundamental studies, and the search for new or advanced materials—with higher nonlinear coefficients and/or better overall properties.

The goal of identifying an efficient device integration platform is another hot issue: it would enable the development of low-cost and reliable devices and systems, wherein nonlinear phenomena may find new or more effective applications in areas such as all-optical switching, all-optical signal processing, and quantum photonics. The use of nonlinear effects in optical waveguides and microcavities is also at the forefront of this research.

This field attracts huge attention, as confirmed by a search made by using the Clarivate Web of Science: almost 200,000 papers were published which refer to the topic "nonlinear optic*". Over 36,000 papers with the same keyword were published in the last four years (2015–2018), and over 17,000 used the keyword "nonlinear photonic*".

The present Special Issue (SI) of *Micromachines* journal, titled "Nonlinear Photonics Devices," aims at highlighting the current state of the art, some recent advances, and some perspectives for further development. Fundamental and applicative aspects have been considered, with special attention to the hot topics that could lead to technological and scientific breakthroughs. Contributions were solicited from both leading researchers and emerging investigators. As a result, this SI contains six reviews and six research articles.

The first group of articles has to do with nonlinear optical phenomena in optical waveguides, of fiber and integrated optical types. Going to the nanoscale level, the paper by Roland et al. [7] investigates the nonlinear properties of nanowire and nanorib waveguides in AlGaAs, which have the advantage of exhibiting an adjustable modal birefringence and supporting phase-matched frequency mixing in the whole AlGaAs transparency range, even close to the gap. In particular, the experimental performances and drawbacks of two different designs (a nanowire in straight or snake-shaped configurations, and a nanorib waveguide) of AlGaAs suspended nonlinear waveguides are compared. The authors conclude that, while the optical performances are almost identical for the two designs, the nanorib exhibits far better mechanical properties.

Optical fibers are often exploited for non-linear photonic devices due to their higher order intrinsic non-linear susceptibility $\chi(3)$: third harmonic generation (THG), self-focusing, and four-wave mixing (FWM) are some examples of the studied effects. SHG, on the contrary, would not be allowed in silica fibers, due to the absence of intrinsic second-order properties in centrosymmetric materials. This limitation has been overcome by the introduction, almost 30 years ago, of the technique of thermal poling. The article by De Lucia and Sazio [8] focuses on the logical and chronological development of 2D numerical models, with the aim of explaining in the best possible dynamics of evolution of the poling process. The authors have also identified the single-anode configuration as the most effective method for thermal poling, in terms of both the absolute value of the created quadratic non-linearity and of simplification of the fabrication constraints.

Another important nonlinear effect in optical fibers is due to inelastic-scattering, in which the optical field transfers part of its energy to the nonlinear medium, thereby inducing stimulated effects such as stimulated Brillouin scattering (SBS) and stimulated Raman scattering (SRS). Either of those types of stimulated scattering process can be used as a source of gain in the fiber. The article by Sirleto and Ferrara [9] reviews the state of the art, achievements, challenges, and perspectives of fiber Raman amplifiers (FRAs) and lasers (FRLs). FRAs are now widely used in fiber optic communications, in order to respond to the growing demand in terms of transmission capacity: the dramatic increase in bandwidth requirement has ruled out the use of erbium-doped fiber amplifiers (EDFAs), leaving fiber Raman amplifiers as the key devices for future ultra-high-capacity systems. FRLs, on the other hand, provide a very attractive option in the field of high-power fiber lasers. Nowadays, commercially available fiber-based Raman lasers can deliver output powers in the range of a few tens of Watts in continuous-wave operation, with high efficiency and broad gain bandwidth, covering almost the entire near-infrared region. The development of integrated RLs is reviewed in another paper, wherein Ferrara and Sirleto [10] describe the transition from the all-silicon Raman laser realized in 2005, based on a single-mode rib waveguide containing a reverse-biased p-i-n diode structure and fabricated on a standard silicon-on-insulator (SOI) substrate, to the current interest toward Si microphotonic structures based on photon confinement effects (nanocrystal waveguides, nanowires, and nanocavities).

Resonating structures, especially at microscale and nanoscale, are very attractive, due to the small volume and consequent high power density of the optical field, which gives higher strength to the nonlinear phenomena. Nonlinear photonics in resonators are the subject of another group of papers in the present SI. Ricciardi et al. [11] discussed the advances that occurred since it was shown that quadratic χ(2) processes can lead to direct generation of optical frequency combs in cw-pumped quadratic nonlinear resonators. Recently, direct generation of quadratic frequency combs has been demonstrated also in chip-scale, lithium niobite, periodically-poled, linear waveguide resonators and in whispering-gallery-mode (WGM) resonators. In this study, the authors analyzed and experimentally demonstrated comb generation in two configurations: a SHG cavity, where combs were generated both around the pump frequency and its second harmonic, and a degenerate optical parametric oscillator, where combs were generated around the pump frequency and its subharmonic. It may be worth noting that optical frequency combs are now attracting interest as sources of complex quantum states of light for high-dimensional quantum computation.

Nonlinear effects in solid and hollow microspherical WGM resonators (WGMRs) are reviewed in the paper by Frigenti et al. [12]. These structures are easy to fabricate and exhibit a very high quality factor Q; they are excellent platforms to understand how light, sound, and matter interact. Nonlinear photonic effects can be easily generated, and their very dense mode spectra allow one to efficiently fulfill the phase-matching conditions required for parametric and hyper-parametric interactions. This review describes Kerr effects in silica and hybrid (silica sphere with organic coating) WGMRs, including third-harmonic generation, third-order sum-frequency generation, frequency combs, Kerr switching, and two-photon fluorescence. Stimulated Raman scattering and stimulated Brillouin scattering, and combinations of other nonlinear phenomena, such as four-wave-mixing, are also discussed.

With the emergence of accurate nanofabrication techniques, there is interest in exploring nonlinear optical effects at a scale comparable to, or much less than, the incident light wavelength. At the nanoscale, interesting regimes for nonlinear optics emerge, in which the resonant optical interaction, due to frequency-selective light scattering or light coupling into and out of the structures, becomes significant. The resonant effects lead to a build-up of electric field inside or in the vicinity of the structure, resulting in enhancement of the nonlinear optical effects. The review article by Raghunathan et al. [13] provides an overview of this emerging field in dielectric-based sub-wavelength periodic structures to realize efficient harmonic generations, wavelength mixers, optical switches, etc. The structures considered here are broadly classified into guided-mode resonant structures and resonant metasurfaces; reference is made, for instance, to 1D gratings, 2D arrays of nanodisks, bar-nanodisk structures, asymmetric bar dimers, asymmetric rectangular unit-cells, and disordered nanodisk arrays. The basic physical mechanisms, the various nonlinear phenomena, and their applications are discussed too.

Exploiting at the best the photonic nonlinear effects requires a careful choice of structures and materials. Thus, some papers in this SI present a detailed analysis of these aspects. Bruno et al. [14] have studied thin films of epsilon-near-zero (ENZ) materials, such as transparent conductive oxides, including aluminum-doped zinc oxide (AZO) and indium tin oxide (ITO). In their paper, they demonstrate, both theoretically and experimentally, that a broadband coherent perfect absorption (CPA) based on light-with-light modulation may be achieved in these films. By using Kerr optical nonlinearities, the visibility and the peak wavelength of the total energy modulation can be dynamically tuned. The coherent control of the absorption in ENZ media may open a route towards technologies such as optical data processing or devices that require efficient light absorption and dynamical tunability.

The investigation of linear and nonlinear intersubband optical properties of quantum dots (QDs), which are of a great interest for integrated quantum photonic technologies, is the subject of the paper by Baira et al. [15]. Recently, GeSn has been shown to have comparable properties to III–V materials, while being compatible with complementary metal-oxide semiconductor (CMOS) technology. In this paper, the effects of an applied electric field on the electron-related linear and third-order nonlinear optical properties are evaluated numerically, with the aim of helping future realizations of CMOS-compatible, nonlinear optical devices. Pyramidal GeSn quantum dots with different sizes are considered. The results show that the transition energies and the transition dipole moment, particularly for larger dot sizes, are altered by the electric-field-induced electron confining potential profile's modification.

Gallium arsenide has been widely used in photonic applications. Recently, it has also been proven that, due to its very high refractive index, nanostructures, such as GaAs nanowires, are able to effectively guide light by using leaky waves; this may lead to different applications as emitters or even as laser sources. Belardini et al. [16] have shown that glancing angle deposition of gold on GaAs nanowires induces a symmetry breaking that leads to an optical circular dichroism (CD) response that mimics chiral behavior. The presence of extrinsic chirality can have applications in different fields, including the ability to generate photons in a second-harmonic field, while selective pumping with circular polarized light could boost the processes of circular polarized photon generation or absorption. Geometric resonance that can be finely tuned by changing the diameter of the nanowires, is an essential feature in this extrinsic chiral behavior.

Periodically-poled lithium niobate (PPLN) is a material widely exploited for the implementation of nonlinear optical devices, both in bulk and in integrated optical format. Su et al. [17] used PPLN and MgO-doped PPLN to generate pure state photon triplets by cascaded second-order spontaneous parametric down-conversion (SPDC). Through numerical simulation, the most suitable parameters, in terms of pump duration and crystal length, were identified to eliminate the frequency correlation between the photon pairs in each SPDC process. Quantum interference is vital for quantum information science, since it is not only the basis of quantum manipulation technology, but is also an important tool for implementing quantum computing and quantum communication. The preparation of three photons with hyperspectral purity in the telecommunication C band is critical for research into quantum information processes and for applications.

Finally, another application of nonlinear phenomena, concerning the development of imaging techniques that are capable of reconstructing the full-wave properties (amplitude and phase) of arbitrary electromagnetic field distributions, is discussed in the paper by Gongora et al. [18]. Interestingly, the direct detection of the field evolution is achievable at terahertz (THz) frequencies thanks to the availability of the time-domain spectroscopy (TDS) technique. Such a capability, coupled with the existence of specific and distinctive spectral fingerprints in the terahertz frequency range, are critical enabling tools for advanced applications; a promising alternative to TDS imaging arrays is single-pixel imaging, or ghost imaging (GI). In this paper, the key advantages and practical challenges in the implementation of time-resolved nonlinear ghost imaging (TIMING) are discussed. TIMING combines nonlinear THz generation with time-resolved time-domain spectroscopy detection. The reported results establish a comprehensive theoretical and experimental framework for the development of a new generation of terahertz hyperspectral imaging devices.

Overall, this collection of scientific articles presents and discusses some interesting research topics in nonlinear photonics. It is our wish that this Special Issue will serve as a stimulus for students and researchers to further expand the potential of nonlinear photonics devices, via fundamental investigations and practical applications.

We would like to thank all the authors for their submissions to this special issue; we really have appreciated their contributions. We also thank all the reviewers for dedicating their time and helping to ensure the quality of the submitted papers. Last but not least, we are grateful to the staff at the editorial office of *Micromachines*—in particular to Mr. Dikies Zhang—for their efficient assistance.

Conflicts of Interest: The authors declare no conflict of interest.

References

1. "Photonics.", Merriam-Webster Dictionary. Available online: https://www.merriam-webster.com/dictionary/photonics (accessed on 21 April 2020).
2. Krasnodębski, M. Throwing light on photonics: The genealogy of a technological paradigm. *Centaurus* **2018**. [CrossRef]
3. Why Light Matters. Available online: http://www.light2015.org/Home/WhyLightMatters.html (accessed on 21 April 2020).
4. Franken, P.A.; Hill, A.E.; Peters, C.W.; Weinreich, G. Generation of optical harmonics. *Phys. Rev. Lett.* **1961**, *7*, 118. [CrossRef]
5. Armstrong, J.A.; Bloembergen, N.; Ducuing, J.; Pershan, P.S. Interactions between light waves in a nonlinear dielectric. *Phys. Rev.* **1962**, *127*, 1918. [CrossRef]
6. Anderson, D.B.; Boyd, T.J. Wideband CO_2 laser second harmonic generation phase matched in gaas thin-film waveguides. *Appl. Phys. Lett.* **1971**, *19*, 266. [CrossRef]
7. Roland, I.; Ravaro, M.; Suffit, S.; Filloux, P.; Lemaître, A.; Favero, I.; Leo, G. Second-Harmonic Generation in Suspended AlGaAs Waveguides: A Comparative Study. *Micromachines* **2020**, *11*, 229. [CrossRef] [PubMed]
8. De Lucia, F.; Sazio, P.J.A. Thermal Poling of Optical Fibers: A Numerical History. *Micromachines* **2020**, *11*, 139. [CrossRef] [PubMed]
9. Sirleto, L.; Ferrara, M.A. Fiber Amplifiers and Fiber Lasers Based on Stimulated Raman Scattering: A Review. *Micromachines* **2020**, *11*, 247. [CrossRef]
10. Ferrara, M.A.; Sirleto, L. Integrated Raman Laser: A Review of the Last Two Decades. *Micromachines* **2020**, *11*, 330. [CrossRef]
11. Ricciardi, I.; Mosca, S.; Parisi, M.; Leo, F.; Hansson, T.; Erkintalo, M.; Maddaloni, P.; De Natale, P.; Wabnitz, S.; De Rosa, M. Optical Frequency Combs in Quadratically Nonlinear Resonators. *Micromachines* **2020**, *11*, 230. [CrossRef] [PubMed]
12. Frigenti, G.; Farnesi, D.; Nunzi Conti, G.; Soria, S. Nonlinear Optics in Microspherical Resonators. *Micromachines* **2020**, *11*, 303. [CrossRef] [PubMed]
13. Raghunathan, V.; Deka, J.; Menon, S.; Biswas, R.; Lal Krishna, A.S. Nonlinear Optics in Dielectric Guided-Mode Resonant Structures and Resonant Metasurfaces. *Micromachines* **2020**, *11*, 449. [CrossRef] [PubMed]
14. Bruno, V.; Vezzoli, S.; DeVault, C.; Roger, T.; Ferrera, M.; Boltasseva, A.; Shalaev, V.M.; Faccio, D. Dynamical Control of Broadband Coherent Absorption in ENZ Films. *Micromachines* **2020**, *11*, 110. [CrossRef]
15. Baira, M.; Salem, B.; Ahamad Madhar, N.; Ilahi, B. Intersubband Optical Nonlinearity of GeSn Quantum Dots under Vertical Electric Field. *Micromachines* **2019**, *10*, 243. [CrossRef]
16. Belardini, A.; Leahu, G.; Petronijevic, E.; Hakkarainen, T.; Koivusalo, E.; Rizzo Piton, M.; Talmila, S.; Guina, M.; Sibilia, C. Circular Dichroism in the Second Harmonic Field Evidenced by Asymmetric Au Coated GaAs Nanowires. *Micromachines* **2020**, *11*, 225. [CrossRef] [PubMed]
17. Su, X.-R.; Huang, Y.-W.; Xiang, T.; Li, Y.-H.; Chen, X.-F. Generation of Pure State Photon Triplets in the C-Band. *Micromachines* **2019**, *10*, 775. [CrossRef] [PubMed]
18. Totero Gongora, J.S.; Olivieri, L.; Peters, L.; Tunesi, J.; Cecconi, V.; Cutrona, A.; Tucker, R.; Kumar, V.; Pasquazi, A.; Peccianti, M. Route to Intelligent Imaging Reconstruction via Terahertz Nonlinear Ghost Imaging. *Micromachines* **2020**, *11*, 521. [CrossRef] [PubMed]

© 2020 by the authors. Licensee MDPI, Basel, Switzerland. This article is an open access article distributed under the terms and conditions of the Creative Commons Attribution (CC BY) license (http://creativecommons.org/licenses/by/4.0/).

Article

Second-Harmonic Generation in Suspended AlGaAs Waveguides: A Comparative Study

Iännis Roland [1], Marco Ravaro [1], Stéphan Suffit [1], Pascal Filloux [1], Aristide Lemaître [2], Ivan Favero [1] and Giuseppe Leo [1,*]

[1] MPQ, Université de Paris & CNRS, 10 rue A. Domon et L. Duquet, 75013 Paris, France; iannis.roland@univ-paris-diderot.fr (I.R.); marco.ravaro@univ-paris-diderot.fr (M.R.); stephan.suffit@univ-paris-diderot.fr (S.S.); pascal.filloux@univ-paris-diderot.fr (P.F.); ivan.favero@univ-paris-diderot.fr (I.F.)
[2] C2N, CNRS, Université Paris-Saclay, 10 boulevard T. Gobert, 91120 Palaiseau, France; aristide.lemaitre@c2n.upsaclay.fr
* Correspondence: giuseppe.leo@u-paris.fr

Received: 30 December 2019; Accepted: 18 February 2020; Published: 23 February 2020

Abstract: Due to adjustable modal birefringence, suspended AlGaAs optical waveguides with submicron transverse sections can support phase-matched frequency mixing in the whole material transparency range, even close to the material bandgap, by tuning the width-to-height ratio. Furthermore, their single-pass conversion efficiency is potentially huge, thanks to the extreme confinement of the interacting modes in the highly nonlinear and high-refractive-index core, with scattering losses lower than in selectively oxidized or quasi-phase-matched AlGaAs waveguides. Here we compare the performances of two types of suspended waveguides made of this material, designed for second-harmonic generation (SHG) in the telecom range: (a) a nanowire suspended in air by lateral tethers and (b) an ultrathin nanorib, made of a strip lying on a suspended membrane of the same material. Both devices have been fabricated from a 123 nm thick AlGaAs epitaxial layer and tested in terms of SHG efficiency, injection and propagation losses. Our results point out that the nanorib waveguide, which benefits from a far better mechanical robustness, performs comparably to the fully suspended nanowire and is well-suited for liquid sensing applications.

Keywords: second-harmonic generation; waveguide; AlGaAs

1. Introduction

Recent technological advances have allowed reducing the size of semiconductor photonic devices to the sub-micrometer scale, with a remarkable impact in several research domains like integrated optofluidics [1] and nonlinear photonics [2]. Because of the high-refractive-index contrast and subwavelength size, the normal field component can be very strong at the semiconductor–air interface. This makes nanophotonic devices very sensitive to the complex refractive index of the surrounding medium and thus promising candidates for chemical or biological sensing in liquid or gaseous environments with lab-on-chip integrated photonic sensors [3]. This is all the more true for resonators and waveguides operating in the mid-infrared, where many absorption resonances of important analytes occur [4]. For these reasons, suspended silicon structures operating in the linear regime have been recently proposed as an alternative to their silicon-on-insulator counterparts [5,6], where the SiO_2 substrate exhibits nonnegligible losses around 2.8 µm and beyond 4 µm, while the transparency of silicon itself ends beyond 8.5 µm [7]. A further asset of nanoscale high-contrast photonics in respect to µm-sized devices is the combination of strong nonlinear light–matter interaction with higher flexibility in dispersion and mode coupling engineering [8].

In this context, $Al_xGa_{1-x}As$ is an attractive material for its high second-and third-order nonlinear coefficients ($d_{14} \approx 100$ pm·V^{-1} [9], $n_2 \approx 10^{-17}$ m^2·W^{-1} [10]), well-established processing technology, direct bandgap (for x < 0.45) that increases with Al molar fraction x and its broad transparency spectral region ranging from near- to mid-IR. The exploitation of AlGaAs nonlinearity for frequency mixing was once challenging because of its optical isotropy, which hinders birefringent phase-matching (PM), and its optical losses associated with the implementation of quasi-PM in the near-IR. In the last two decades, however, efficient guided-wave frequency mixing has been reported, based on form birefringence [11,12], modal PM [13] and counterpropagating PM [14]. In each of those cases, the nonlinear waveguides relied on total internal reflection between an aluminum-poor AlGaAs core and aluminum-rich claddings with a relatively low refractive-index step ($\Delta n \approx 0.2$), which was also the case for the demonstration of $\chi^{(3)}$ guided-wave devices [15].

In the last years, high-contrast AlGaAs nonlinear photonic structures have been reported at the nanoscale level, based on either selective oxidation of an AlAs substrate [16,17] or epitaxial liftoff followed by bonding on glass [18], for both second-harmonic generation (SHG) [16–18] and spontaneous parametric down-conversion (SPDC) [19]. Their higher refractive-index step ($\Delta n \approx 1.5$) made them suitable for shallow etching fabrication, with a huge impact on integration up until the demonstration of the first $\chi^{(2)}$ metasurfaces [20,21]. Similar AlGaAs-on-oxide structures have also been demonstrated for waveguides and microresonators fabricated by wafer bonding, both in $\chi^{(3)}$ [22] and $\chi^{(2)}$ devices [23,24]. However, the potential of AlGaAs-on-oxide guided-wave devices is still affected by either the intrinsic limits of wafer bonding technology in terms of homogeneity and throughput or by the intrinsic scattering loss of devices based on native AlAs oxide [25,26].

Within this context, an alternative approach to high-contrast AlGaAs photonics was pioneered more than a decade ago with substrate-removed electrooptic modulators [27,28]; then, suspended microdisk resonators were used both in optomechanics [29] and nonlinear optics [30–32]. Finally, suspended nonlinear nanowires [33] and nanorib waveguides [34] have been reported, and a suspended nonlinear photonic integrated circuit has been demonstrated for both SHG and SPDC in a microdisk coupled with two distinct waveguides at ω and 2ω [35].

Both nanowire and nanorib waveguides naturally lend themselves to mode birefringence phase-matching with a few advantages over multilayered form birefringent waveguides: (a) the attainable modal birefringence is sufficient to compensate dispersion in the whole AlGaAs transparency range, even close to the gap; (b) the modal areas of the fields are extremely small and tightly confined within the GaAs core, resulting in high conversion efficiency; and (c) the absence of aluminum oxide layers and the smoothness of top and bottom surfaces, which is defined by epitaxial growth, result in low scattering losses.

Here we compare the experimental performances and drawbacks of two different designs for AlGaAs suspended nonlinear waveguides (Figure 1): (a) a nanowire that recently allowed the demonstration of phase-matched SHG in both straight and snake-shaped configurations [33] and (b) a nanorib waveguide developed for frequency down-conversion towards the mid-IR range [34].

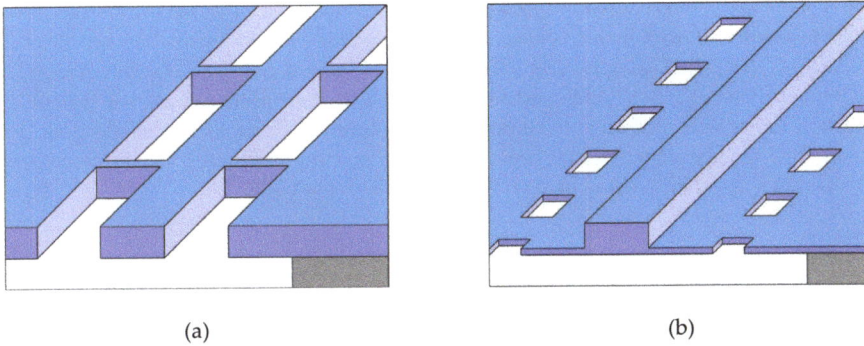

Figure 1. Suspended waveguide schemes: (**a**) nanowire anchored by tethers; (**b**) nanorib bounded by etch windows. Tethers and windows have no impact on optical propagation.

2. Materials and Methods

Both the above devices were processed from a planar AlGaAs heterostructure consisting of a 123 nm thick film of $Al_{0.19}Ga_{0.81}As$ on top of a 4 µm thick $Al_{0.8}Ga_{0.2}As$ layer, grown on a GaAs {001} substrate by molecular-beam epitaxy.

Suspended nanowires 1 µm wide and 1 mm long (Figure 2a) were patterned along with their anchoring points by e-beam lithography followed by $Ar/SiCl_4$-assisted inductively coupled plasma reactive-ion etching (ICP-RIE). The anchoring points were pairs of 100 nm wide and 1 µm long lateral tethers placed every 50 µm along the wire. A 1 mm wide, 100 µm deep mesa was then defined in the GaAs substrate by optical lithography and wet etching, giving access to the input and output ends for butt coupling. Finally, the $Al_{0.8}Ga_{0.2}As$ layer was underetched with 1% HF solution at 4 °C for 6 minutes without stirring before sample CO_2 critical point drying.

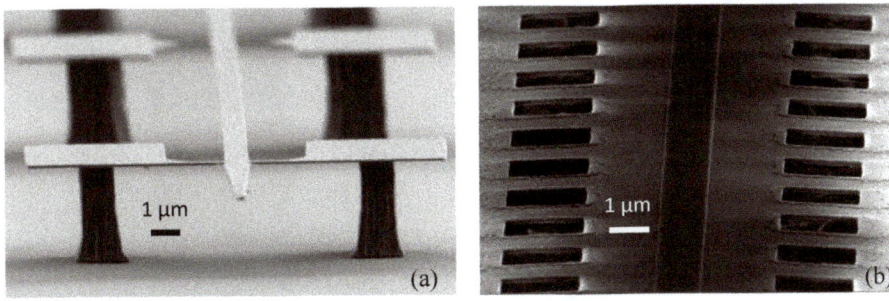

Figure 2. Scanning electron microscope (SEM) images of the suspended nanowire (**a**) and nanorib (**b**) waveguides.

Suspended nanorib waveguides (Figure 2b) were patterned by means of a two-step e-beam lithography plus ICP-RIE process: the former defined a 1 µm wide, 200 µm long and 80 nm thick rib in the $Al_{0.19}Ga_{0.81}As$ layer, while the latter opened two lines of 2 µm × 2 µm square windows through the same layer, 2 µm away from the strip. The windows allowed wet isotropic underetching (10 min in 1% HF at room temperature with moderate stirring) of the underlying $Al_{0.8}Ga_{0.2}As$ layer, which thus liberated a suspended 40 nm thick, 15 µm wide and 200 µm long $Al_{0.19}Ga_{0.81}As$ membrane supporting the guiding rib. It is worth noticing that rib waveguides, due to intrinsic robustness, do not require critical point drying at the end of processing but can be simply flash dried (isopropanol evaporation on a hot plate at 270 °C).

Both types of waveguides were terminated with inverted tapers designed for efficient input/output coupling at fundamental frequency ω and second-harmonic 2ω.

All devices were tested using two continuously tunable laser sources: aCW external cavity laser diode emitting between 1.5 and 1.6 µm and a single mode CW Ti:sapphire tunable between 0.7 and 1 µm. Both laser beams butt coupled at the input and the output with microlensed, single mode optical fibers. Linear and nonlinear spectra have been recorded by injecting and tuning the laser sources while detecting the outcoupled light either by an InGaAs or an Si photodiode.

3. Results

The transverse section of the waveguides was designed for Type-I phase-matched SHG from the TE00 mode at ω ($\lambda \approx 1.6$ µm) to the TM00 mode at 2ω ($\lambda \approx 800$ nm): (a) the thickness of the $Al_{0.19}Ga_{0.81}As$ film was chosen so as to ensure strong modal birefringence while keeping the interacting modes well-confined; (b) the wire/rib width was then adjusted in order to precisely set the phase-matching wavelength [33]. The electric field amplitude profiles of both modes are shown in Figure 3. It can be observed that the 40 nm thick membrane does not significantly affect the lateral confinement for both modes. Accordingly, phase-matching is obtained for an almost identical width (≈ 1 µm), and the SHG efficiency expected from numerical simulations (not shown) is also very similar for the two devices: $\eta = 300\%$ $W^{-1}mm^{-2}$ (nanowire) and $\eta = 401\%$ $W^{-1}mm^{-2}$ (nanorib).

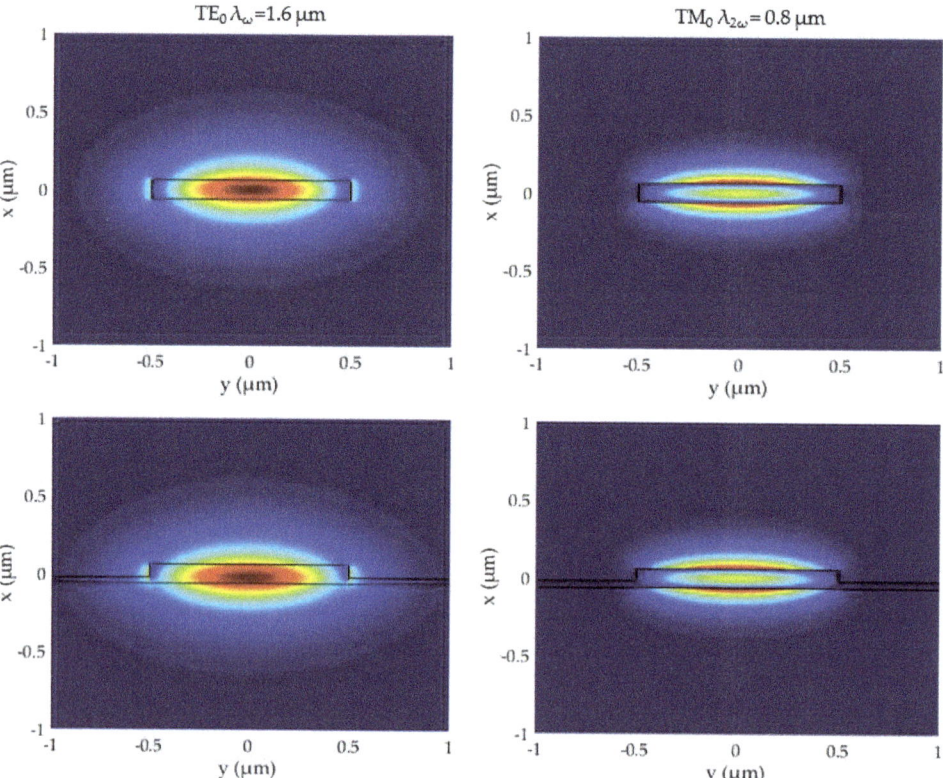

Figure 3. TE$_{00}$ amplitude at ω (E$_y$, left) and TM$_{00}$ amplitude at 2ω (E$_x$, right) in the suspended nanowire (top) and rib waveguide (bottom).

Propagation losses at ω and 2ω were measured by acquiring Fabry–Perot transmission interference fringes in on-purpose processed 200 μm long waveguides terminated by flat ICP etched facets (which have higher reflectivity than the tapered counterparts). The combined loss–reflection coefficient R' = R exp(-αL) can be extracted by the contrast K of the transmission fringes as follows:

$$K = \frac{T_{max} - T_{min}}{T_{max} + T_{min}} \quad (1)$$

$$R' = \frac{1 - \sqrt{1 - K^2}}{K} \quad (2)$$

where L is the length of the waveguide and T_{max} and T_{min} are the maximum and minimum transmission values, respectively. The modal reflectivity R was calculated at both ω and 2ω via 3D FDTD modeling, and the propagation loss coefficient was then found as:

$$\alpha = \frac{1}{L} \ln\left(\frac{R}{R'}\right) \quad (3)$$

The coupling efficiency κ of a waveguide terminated by inverted tapers, assumed to be equal at the input and at the output, was finally obtained by measuring its overall transmission and dividing it by the propagation loss exp(-αL):

$$\kappa = \sqrt{\frac{T}{e^{-\alpha L}}} \quad (4)$$

The results of the above linear characterization are summarized in Table 1. For both designs, for L = 1 mm, the propagation loss at ω is quite limited ($e^{-\alpha L} \approx 70\%$), while at 2ω it turned out to be one order of magnitude higher, due to the proximity between photon energy at 2ω and the forbidden band (740 nm) and to stronger scattering at the waveguide sidewalls at shorter wavelength. As for the input/output coupling, we estimate that the efficiency at ω for the rib waveguides can reach the same level as in nanowires after further optimization of design and processing. The low coupling efficiency at 2ω is to be ascribed to the multimode nature of the waveguide at this wavelength.

Table 1. Measured linear optical features.

Design	λ (nm)	R (%)	L_{trans} (μm)	K (%)	A (cm^{-1})	T (%)	κ (%)
Wire	1600	16.7	200	30.3 ± 0.3	3.7 ± 0.5	34.0 ± 0.2	60.7 ± 0.3
	800	24.3	200	23 ± 5	38 ± 12	0.50 ± 0.02	10.0 ± 1.0
Rib	1600	16.7	200	30.1 ± 0.6	4.0 ± 1.0	5.0 ± 0.2	23.3 ± 0.7
	800	24.3	200	22 ± 3	39 ± 7	0.50 ± 0.02	10.4 ± 0.7

Figure 4 shows the SHG efficiency spectra acquired by injecting and tuning the TE polarized telecom-range laser into a 1 mm nanowire (black trace) and a 200 μm long nanorib (red trace) waveguide, collecting the outcoupled TM mode at 2ω. The internal efficiency η was calculated by normalizing the overall efficiency P_{SHG}/P_{in}^2 to the coupling efficiency at ω and 2ω:

$$\eta = \frac{1}{\kappa_\omega^2 \kappa_{2\omega}} \frac{P_{SHG}}{P_{in}^2} \quad (5)$$

with peak values of 16% W^{-1} (wire) and 3% W^{-1} (rib). The normalized efficiency equations ($\eta_{norm} = \eta/L^2$) of the two devices are expected to be very similar; nevertheless, the ratio (η_{wire}/η_{rib}) does not scale as the square of the ratio of the lengths $(L_{wire}/L_{rib})^2$. This is due to propagation loss at

2ω, which limits the interaction length to $<< L_{wire}$. By taking into account the effect of propagation loss on the efficiency η, we can calculate the normalized SHG efficiency η_{norm}, defined as follows: [36]

$$\eta = \eta_{norm} L^2 \exp[-(\alpha_\omega + \alpha_{2\omega}/2)L] \frac{\sinh^2\left[(\alpha_\omega - \alpha_{2\omega}/2)\frac{L}{2}\right]}{\left[(\alpha_\omega - \alpha_{2\omega}/2)\frac{L}{2}\right]^2} \qquad (6)$$

obtaining 128% $W^{-1}mm^{-2}$ (wire) and 119% $W^{-1}mm^{-2}$ (rib). The results are summarized in Table 2.

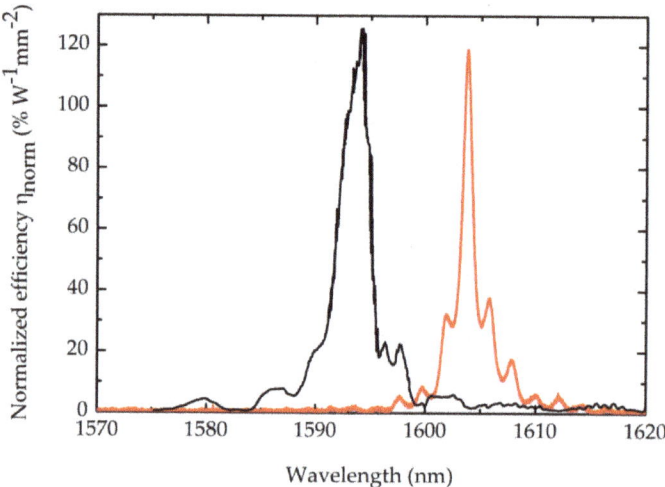

Figure 4. Nonlinear second-harmonic generation (SHG) efficiency spectra for the nanowire (black) and nanorib (red) waveguides.

Table 2. Measured nonlinear optical features.

Design	L_{SHG} (μm)	P_{in} (W)	P_{SHG} (W)	η (% W^{-1})	η_{norm} (% W^{-1} mm^{-2})
Wire	1000	8.0×10^{-4}	3.9×10^{-9}	16 ± 2	128 ± 20
Rib	200	4.0×10^{-4}	2.7×10^{-11}	3.0 ± 0.5	119 ± 20

4. Discussion

We demonstrated phase-matched optical SHG from the telecom range in suspended submicron AlGaAs waveguides with two different designs: a nanowire and a nanorib. The two approaches exhibit similar performances in terms of injection and propagation loss at ω, which are low enough to fabricate 1 mm long devices and in terms of nonlinear efficiency. Propagation loss at 2ω is intrinsically higher due to scattering and residual absorption, and it limits the SHG efficiency with respect to the expected values. Yet, the experimental conversion efficiency is higher than in oxidized form birefringent AlGaAs nonlinear waveguides ($\approx$ 10% $W^{-1}mm^{-2}$), [37,38] and comparable to state-of-the-art SiO_2 cladded submicron GaAs waveguides ($\approx$ 130% $W^{-1}mm^{-2}$) [23]. While the optical performances are almost identical for the two designs, the nanorib exhibits far better mechanical properties. Its mechanical robustness makes its processing easier, not requiring CO_2 supercritical drying, with a higher fabrication yield and less delicate handling. In addition, its ability to withstand several wetting and flash drying cycles without any damage makes the nanorib perfectly adapted to chemical and biological sensing in liquids, which could be easily injected through the etch windows.

Author Contributions: Conceptualization, G.L.; Investigation, M.R., S.S., P.F.; Technology contributions, A.L., I.F.; Supervision, G.L.; Writing and editing, I.R., M.R., G.L. All authors have read and agreed to the published version of the manuscript.

Funding: I.R.'s post-doc grant was co-funded by the Université de Paris and the SEAM Labex.

Conflicts of Interest: The authors declare no conflict of interest.

References

1. Monat, C.; Domachuk, P.; Eggleton, B. Integrated optofluidics: A new river of light. *Nat. Photonics* **2007**, *1*, 106–114. [CrossRef]
2. Helmy, A.S.; Abolghasem, P.; Stewart Aitchison, J.; Bijlani, B.J.; Han, J.; Holmes, B.M.; Hutchings, D.C.; Younis, U.; Wagner, S.J. Recent advances in phase matching of second-order nonlinearities in monolithic semiconductor waveguides. *Laser Photon. Rev.* **2011**, *5*, 272. [CrossRef]
3. Rodriguez-Ruiz, I.; Ackermann, T.N.; Muñoz-Berbel, X.; Llobera, A. Photonic Lab-on-a-Chip: Integration of Optical Spectroscopy in Microfluidic Systems. *Anal. Chem.* **2016**, *88*, 6630–6637. [CrossRef] [PubMed]
4. Popa, D.; Udrea, F. Towards integrated Mid-Infrared Gas Sensors. *Sensors* **2019**, *19*, 2076. [CrossRef] [PubMed]
5. Soler Penades, J.; Ortega-Moñux, A.; Nedeljkovic, M.; Wangüemert-Pérez, J.G.; Halir, R.; Khokhar, A.Z.; Alonso-Ramos, C.; Qu, Z.; Molina-Fernández, I.; Cheben, P.; et al. Suspended silicon mid-infrared waveguide devices with subwavelength grating metamaterial cladding. *Opt. Express* **2016**, *24*, 22908–22916. [CrossRef]
6. Chiles, J.; Khan, S.; Ma, J.; Fathpour, S. High-contrast, all-silicon waveguiding platform for ultra-broadband mid-infrared photonics. *Appl. Phys. Lett.* **2013**, *103*, 151106. [CrossRef]
7. Nedeljkovic, M.; Khokhar, A.Z.; Hu, Y.; Chen, X.; Soler Penadés, J.; Stankovic, S.; Chong, H.M.H.; Thomson, D.J.; Gardes, F.Y.; Reed, G.T.; et al. Silicon photonic devices and platforms for the mid-infrared. *Opt. Mater. Express* **2013**, *3*, 1205–1214. [CrossRef]
8. Li, X.; Zhou, P.; He, S.; Gao, S. Dispersion engineering of suspended silicon photonic waveguides for broadband mid-infrared wavelength conversion. *J. Opt. Soc. Am. B* **2014**, *31*, 2295–2301. [CrossRef]
9. Shoji, I.; Kondo, T.; Kitamoto, A.; Shirane, M.; Ito, R. Absolute scale of second-order nonlinear-optical coefficients. *J. Opt. Soc. Am. B* **1997**, *14*, 2268–2294. [CrossRef]
10. Dolgaleva, K.; Ng, W.C.; Qian, L.; Aitchison, J.S. Compact highly-nonlinear AlGaAs waveguides for efficient wavelength conversion. *Opt. Express* **2011**, *19*, 12440–12455. [CrossRef]
11. De Rossi, A.; Berger, V.; Calligaro, M.; Leo, G.; Ortiz, V.; Marcadet, X. Parametric fluorescence in oxidized aluminum gallium arsenide waveguides. *Appl. Phys. Lett.* **2001**, *79*, 3758–3760. [CrossRef]
12. Savanier, M.; Ozanam, C.; Lanco, L.; Lafosse, X.; Andronico, A.; Favero, I.; Ducci, S.; Leo, G. Near-infrared optical parametric oscillator in a III-V semiconductor waveguide. *Appl. Phys. Lett.* **2013**, *103*, 261105. [CrossRef]
13. Boitier, F.; Orieux, A.; Autebert, C.; Lemaître, A.; Galopin, E.; Manquest, C.; Sirtori, C.; Favero, I.; Leo, G.; Ducci, S. An electrically injected photon-pair source at room temperature. *Phys. Rev. Lett.* **2014**, *112*, 183901. [CrossRef] [PubMed]
14. Caillet, X.; Berger, V.; Leo, G.; Ducci, S. A semiconductor source of counterpropagating twin photons: a versatile device allowing the control of the two-photon state. *J. Mod. Opt.* **2009**, *56*, 232–239. [CrossRef]
15. Wathen, J.J.; Apiratikul, P.; Richardson, C.J.; Porkolab, G.A.; Carter, G.M.; Murphy, T.E. Efficient continuous-wave four-wave mixing in bandgap-engineered AlGaAs waveguides. *Opt. Lett.* **2014**, *39*, 3161–3164. [CrossRef] [PubMed]
16. Gili, V.F.; Carletti, L.; Locatelli, A.; Rocco, D.; Finazzi, M.; Ghirardini, L.; Favero, I.; Gomez, C.; Lemaître, A.; Celebrano, M.; et al. Monolithic AlGaAs second-harmonic nanoantennas. *Opt. Expr.* **2016**, *24*, 15965–15971. [CrossRef]
17. Liu, S.; Saravi, S.; Keeler, G.A.; Sinclair, M.B.; Yang, Y.; Reno, J.; Pertsch, T.; Brener, I. Resonantly Enhanced Second-Harmonic Generation Using III–V Semiconductor All-Dielectric Metasurfaces. *Nano Lett.* **2016**, *16*, 5426–5432. [CrossRef]
18. Camacho-Morales, R.; Rahmani, M.; Kruk, S.; Wang, L.; Xu, L.; Smirnova, D.A.; Solntsev, A.S.; Miroshnichenko, A.; Tan, H.H.; Karouta, F.; et al. Nonlinear Generation of Vector Beams From AlGaAs Nanoantennas. *ACS Nano Lett.* **2016**, *16*, 7191–7197. [CrossRef]

19. Marino, G.; Solntsev, A.S.; Xu, L.; Gili, V.F.; Carletti, L.; Poddubny, A.N.; Rahmani, M.; Smirnova, D.A.; Chen, H.; Lemaître, A.; et al. Spontaneous photon-pair generation from a dielectric nanoantenna. *Optica* **2019**, *6*, 1416–1422. [CrossRef]
20. Marino, G.; Gigli, C.; Rocco, D.; Lemaître, A.; Favero, I.; De Angelis, C.; Leo, G. Zero-Order Second Harmonic Generation from AlGaAs-on-Insulator Metasurfaces. *ACS Photonics* **2019**, *65*, 1226–1232. [CrossRef]
21. Gigli, C.; Marino, G.; Suffit, S.; Patriarche, G.; Beaudoin, G.; Pantzas, K.; Sagnes, I.; Favero, I.; Leo, G. Polarization- and diffraction-controlled second harmonic generation from semiconductor metasurfaces. *J. Opt. Soc. Am. B* **2019**, *36*, E55–E64. [CrossRef]
22. Pu, M.; Ottaviano, L.; Semenova, E.; Yvind, K. Efficient frequency comb generation in AlGaAs-on-insulator. *Optica* **2016**, *3*, 823–826. [CrossRef]
23. Chang, L.; Boes, A.; Guo, X.; Spencer, D.T.; Kennedy, M.J.; Peters, J.D.; Volet, N.; Chiles, J.; Kowligy, A.; Nader, N.; et al. Heterogeneously Integrated GaAs Waveguides on Insulator for Efficient Frequency Conversion. *Laser Photonics Rev.* **2018**, *12*, 1800149. [CrossRef]
24. Chang, L.; Xie, W.; Shu, H.; Yang, Q.; Shen, B.; Boes, A.; Peters, J.D.; Jin, W.; Liu, S.; Moille, G.; et al. Ultra-efficient frequency comb generation in AlGaAs-on-insulator microresonators. *Phys. Opt.* **2019**, submitted. arXiv:1909.09778.
25. Guillotel, E.; Ravaro, M.; Ghiglieno, F.; Langlois, C.; Ricolleau, C.; Ducci, S.; Favero, I.; Leo, G. Parametric amplification in GaAs/AlOx waveguide. *Appl. Phys. Lett.* **2009**, *94*, 171110–171113. [CrossRef]
26. Ozanam, C.; Savanier, M.; Lanco, L.; Lafosse, X.; Almuneau, G.; Andronico, A.; Favero, I.; Ducci, S.; Leo, G. Towards an AlGaAs/AlOx near-IR integrated optical parametric oscillator. *J. Opt. Soc. Am. B* **2014**, *31*, 542–550. [CrossRef]
27. Shin, J.; Wu, S.; Dagli, N. Bulk Undoped GaAs–AlGaAs Substrate-Removed Electrooptic Modulators With 3.7-V-cm Drive Voltage at 1.55 μm. *IEEE Photon. Technol. Lett.* **2006**, *18*, 2251–2253. [CrossRef]
28. Shin, J.; Chang, Y.-C.; Dagli, N. 0.3 V drive voltage GaAs/AlGaAs substrate removed Mach–Zehnder intensity modulators. *Appl. Phys. Lett.* **2008**, *92*, 201103. [CrossRef]
29. Nguyen, D.T.; Baker, C.; Hease, W.; Sejil, S.; Senellart, P.; Lemaître, A.; Ducci, S.; Leo, G.; Favero, I. Ultrahigh Q-Frequency product for optomechanical disk resonators with a mechanical shield. *Appl. Phys. Lett.* **2013**, *103*, 241112. [CrossRef]
30. Mariani, S.; Andronico, A.; Mauguin, O.; Lemaître, A.; Favero, I.; Ducci, S.; Leo, G. AlGaAs microdisk cavities for second-harmonic generation. *Opt. Lett.* **2013**, *38*, 3965–3968. [CrossRef]
31. Kuo, P.S.; Bravo-Abad, J.; Solomon, G.S. Second-harmonic generation using 4bar-quasi-phasematching in a GaAs whispering-gallery-mode microcavity. *Nat. Commun.* **2014**, *5*, 3109–3115. [CrossRef] [PubMed]
32. Mariani, S.; Andronico, A.; Lemaître, A.; Favero, I.; Ducci, S.; Leo, G. Second-harmonic generation in AlGaAs microdisks in the telecom range. *Opt. Lett.* **2014**, *39*, 3062–3065. [CrossRef] [PubMed]
33. Morais, N.; Roland, I.; Ravaro, M.; Hease, W.; Lemaître, A.; Gomez, C.; Wabnitz, S.; De Rosa, M.; Favero, I.; Leo, G. Directionally induced quasi-phase matching in homogeneous AlGaAs waveguides. *Opt. Lett.* **2017**, *42*, 4287–4290. [CrossRef] [PubMed]
34. Stievater, T.H.; Mahon, R.; Park, D.; Rabinovich, W.S.; Pruessner, M.W.; Khurgin, J.B.; Richardson, C.J.K. Mid-infrared difference-frequency generation in suspended GaAs waveguides. *Opt. Lett.* **2014**, *39*, 945–948. [CrossRef]
35. Roland, I.; Borne, A.; Ravaro, M.; De Oliveira, R.; Lemaître, A.; Favero, I.; Leo, G. Frequency doubling and parametric fluorescence in a 4-port photonic chip. *Opt. Lett.* **2019**. submitted.
36. Sutherland, R.L. *Handbook of Nonlinear Optics*; Marcel Dekker: New York, NY, USA, 1996; p. 96.
37. Scaccabarozzi, L.; Fejer, M.; Huo, Y.; Fan, S.; Yu, X.; Harris, J. Enhanced second-harmonic generation in AlGaAs/Al$_x$O$_y$ tightly confining waveguides and resonant cavities. *Opt. Lett.* **2006**, *31*, 3626–3628. [CrossRef]
38. Savanier, M.; Andronico, A.; Lemaître, A.; Galopin, E.; Manquest, C.; Favero, I.; Ducci, S.; Leo, G. Large Second-Harmonic Generation at 1.55 μm in oxidized AlGaAs waveguides. *Opt. Lett.* **2011**, *36*, 2955–2957. [CrossRef]

© 2020 by the authors. Licensee MDPI, Basel, Switzerland. This article is an open access article distributed under the terms and conditions of the Creative Commons Attribution (CC BY) license (http://creativecommons.org/licenses/by/4.0/).

Review

Thermal Poling of Optical Fibers: A Numerical History

Francesco De Lucia and Pier J. A. Sazio *

Optoelectronics Research Centre, University of Southampton, Southampton SO17 1BJ, UK; fdl1c13@soton.ac.uk
* Correspondence: pjas@soton.ac.uk

Received: 26 November 2019; Accepted: 7 January 2020; Published: 27 January 2020

Abstract: This review gives a perspective of the thermal poling technique throughout its chronological evolution, starting in the early 1990s when the first observation of the permanent creation of a second order non-linearity inside a bulk piece of glass was reported. We then discuss a number of significant developments in this field, focusing particular attention on working principles, numerical analysis and theoretical advances in thermal poling of optical fibers, and conclude with the most recent studies and publications by the authors. Our latest works show how in principle, optical fibers of any geometry (conventional step-index, solid core microstructured, etc) and of any length can be poled, thus creating an advanced technological platform for the realization of all-fiber quadratic non-linear photonics.

Keywords: non-linear photonics; optical fibers; thermal poling; numerical analysis

1. Introduction

Since their early implementation in the 1920s [1,2] and subsequent optimization in the 1970s [3] optical fibers have become the most widespread technological platform for telecommunications, mainly due to their relatively low losses and huge bandwidths which greatly exceed the performances of any other system for the transmission of information [4]. For example, it is possible to dope optical fibers with rare earth ions such as erbium to obtain optical amplifiers [5], with ytterbium or neodymium to create fiber lasers, or to embed Bragg grating mirrors and filters into them [6].

Optical fibers are typically exploited as a reliable technology for non-linear photonic devices based on their higher order intrinsic non-linear susceptibility $\chi^{(3)}$. This, by definition, requires high laser pump intensities and appropriate phase matching conditions to operate efficiently. Third harmonic generation (THG), optical Kerr effect, self-focusing, intensity dependent refractive index, four-wave mixing (FWM) are some of these $\chi^{(3)}$-related effects exploited in all-fiber non-linear devices such as, for example, supercontinuum sources [7]. Nevertheless, the absence of intrinsic second order properties in centrosymmetric materials, such as silicate glasses, does not in the first instance allow for their exploitation in creating parametric effects related to this lower order optical non-linearity [8].

However, in 1991 Myers et al. developed a technique, called thermal poling [9], to permanently create effective second-order susceptibility $\chi^{(2)}_{eff}$ inside glasses. The method consists in the concomitant heating process of a piece of glass and application of a relatively high static electric field through it. When the glass reaches the temperature where some alkali impurity ions (already included inside the glass matrix) have a non-negligible diffusion and drift mobility, the alkali ions start to electromigrate consequently forming a static electric field which is later frozen-in the glass after it is cooled down and the external electric field is removed. The thermal poling technique, at first adopted for bulk glasses, was later used for optical fibers [10] with the main motivation of overcoming some of the issues typical of the classical approach for the realization of non-linear optical devices, based on the interaction between intense light beams and non-linear crystals (such as for example lithium triborate (LBO), beta-barium borate (BBO) or lithium niobate (LiNbO$_3$)). These issues can include thermal instabilities of non-linear crystals when illuminated by very high pump powers [11,12], relatively short interaction lengths between light waves involved in the non-linear process, high costs and low damage thresholds of the non-linear crystals and coupling losses due to the presence of air/non-linear crystal interfaces as well as the onerous requirement for continuous optical alignment necessary in free-space optical setup. The appeal represented by the idea of a new technological platform for the realization of efficient and all-fiber non-linear devices produced a significant scientific effort towards the complete exploitation of the thermal poling technique. Since its first appearance, many papers have been published where continuous improvements of the experimental thermal poling technique are presented. In this work we focus our attention on the chronological development of the theoretical models implemented in the last 25 years to explain the glass chemistry and physics behind thermal poling, with the final aim of shedding more light on the mechanisms involved in the creation of the second order non-linearity and ideally understanding how eventually to push the features of the technique beyond its current limits.

2. Early Evidence of Second-Order Non-Linearities in Silica Fibers and Thermal Poling

An amorphous dielectric medium can be considered macroscopically isotropic and centrosymmetric and consequently invariant by parity inversion [13]. This means that a glass, as an amorphous medium, lacks any second order non-linear susceptibility $\chi^{(2)}$ in the electric dipole approximation, because of the parity invariance [8]. For example, silica optical fibers possess a zero $\chi^{(2)}$ as evidenced by the absence of any quadratic non-linear effect. However, in the 1980s some quadratic nonlinear phenomena were observed in silica optical fibers excited by the radiation generated by high power lasers, for example, the generation of wavelengths corresponding to sum-frequency radiation. The source adopted was a Q-switched and mode-locked neodymium-doped yttrum aluminum garnet (Nd:YAG) laser at 1.064 µm while the sum-frequency light was generated mixing the light of fundamental wavelength and the Stokes wavelengths generated via Raman inelastic scattering [14–16]. It is due to the work of Gabriagues et al. the first ever reported observation of second harmonic generation (SHG) in optical fibers [17], while a few years later Osterberg et al. studied the SHG process produced in a silica fiber with laser pulses characterized by a time duration of 100-130 psec and peak power of 70 kW. They observed that, after constantly illuminating a silica fiber, some SH light was collected at its output and the intensity of the light generated grew after a certain time [18,19]. In a later work, it was reported the possibility of reducing this "preparation" time from hours to minutes by illuminating the fiber not only with fundamental wavelength, but also with the SH one [20].

The SHG produced in optical fibers was explained in two different ways. Farries et al. considered that the existence of a non-linear electric quadrupole susceptibility causes the generation of a feeble SH radiation when elevated intensities of the pump light are used [21]. This process produces the formation of color centers (created where fundamental and SH radiation are in phase) in an axially periodic arrangement [22]. Stolen et al., instead, attribute the SHG to a sort of photoinduced phenomenon forming the $\chi^{(2)}$. Basically, they assume that the origin of the SHG process is the creation of a DC polarization due to the mix of the fundamental and the SH wavelength (already present inside the fiber or even fed from outside). The polarization is characterized by a certain periodicity and is capable of orienting defects and consequently create a phasematched $\chi^{(2)}$ [20].

Finally, Kashyap created an experimental setup to produce phase-matched electric field-induced second harmonic (EFISH) in single-mode Germania-doped silica fibers [23] by applying a periodic electric field across the core of an optical fiber. The static field created in the fiber's core generates a periodic $\chi^{(2)} \propto \chi^{(3)}E$. It was possible to tune the period of the electric field simply rotating the electrode by an appropriate angle.

As previously discussed, significant permanent effective second order non-linear susceptibility ($\approx$1 pm/V) in centrosymmetric media such as bulk silica glass was demonstrated by Myers et al. [9]. The technique is defined thermal poling and consists in the application of high electric potentials (3–5 kV) through a piece (thickness of 1.6 mm) of fused silica kept at a temperature between 250 and 325 °C for a temporal interval in the range 15–120 min. After the heating phase, the glass is cooled down to room temperature while the voltage is still maintained. The non-linearity is created permanently only in the first few microns of the sample close to the surface where the anodic potential is applied. The $\chi_{33}^{(2)}$ value for fused silica was found to be 20% of the typical value of the $\chi_{22}^{(2)}$ measured for LiNbO$_3$. A relevant experimental result obtained by Myers et al. is the strict relationship between the value of $\chi^{(2)}$ obtained and the concentration of the impurities present in the glass the fiber is made of. This observation suggested that the presence of the impurities is of critical importance to make thermal poling an efficient process.

2.1. First Theoretical Explanation of Thermal Poling: Single-Carrier Model

In 1994 Mukherjee et al. presented a first model to explain the thermal poling process dynamics [24]. The model is based on transport of ionic species together with bonds reorientation and states that, by applying an external electric field, the impurity ions already included in the glass matrix create locally static electric fields capable to orient the bonds (related to impurities or Si-O bonds). The induced $\chi^{(2)}$ is expressed by:

$$\chi^{(2)} \approx \chi^{(3)} E_{DC} + \frac{np\beta}{5k_B T} E_{DC} \qquad (1)$$

where $\chi^{(3)} E_{DC}$ is the term representing the optical rectification process of third order and E_{DC} is the local field due to the non-uniform charge distribution. The other term of Equation (1) represents the electric-field-induced orientation of the molecular second order hyperpolarizability β, with k_B the Boltzmann constant, T the absolute temperature of the sample, p the permanent dipole moment associated with the bond, N the number of dipoles involved in the process and a uniaxial molecular system is assumed for the sake of simplicity (the direction of E_{DC} is fixed). Mukherjee et al. introduced for the first time the concept of depletion region formation. The latter consists in the creation of a space-charge zone, situated in proximity of the anodic electrode, which is emptied of impurities. This portion of the glass includes the negatively charged non-bridging oxygen (NBO$^-$) centers. By applying high electric fields at high temperature it is possible to move away the ions originally electrostatically linked to them. Applying the Poisson's equation it is possible to obtain the electric field in the depletion region, which is given by [24]:

$$E_{DC} = \frac{qn}{\varepsilon}(a-x), \ 0 < x < a, \ a = \left(\frac{2\varepsilon V}{qn}\right)^{\frac{1}{2}} \qquad (2)$$

where a is the depletion width, n is the concentration of ionized impurities, q is the magnitude of the electronic charge, ε the dielectric constant and V the difference of potential externally applied between the two electrodes. This result has been obtained assuming that the depletion layer width on the anodic side is much greater than the corresponding cathode accumulation layer. This model is based on the assumption that there is only one type of carrier involved in the formation of the depletion region, but a later work of Alley et al. [25] highlighted a series of experimental observations which are incompatible with the single-carrier model, including in particular the observation of multiple time scales for the poling, and the dependence of the non-linearity on the sample thermal poling history.

2.2. Multiple-Carrier Model for Space-Charge Region Formation

Although the early experimental results seemed to confirm the formation of a negatively charged region underneath the anodic surface of the bulk silica sample [26], as predicted by the single carrier model described in the Section 2.1, other observations indicated that in a thermal poling process of silica there is something more complex than a simple uniformly negatively charged region. In particular, Kazansky et al. found regions of alternating charge below the anode [27] while Myers et al. found that the depth of the non-linearity generated in poled bulk samples was greater for samples poled for 2 h than for 15 min [9]. If the depletion region was a uniformly negatively charged region and the electric field frozen into the glass was expressed by Equation (2), according to the Equation (1), the $\chi^{(2)}$ induced would be peaked in the region closest to the anode and not at a certain distance as commonly observed in the early poling experiments [28].

After the work of Alley et al. [25], published in 1998, where the important experimental observation of multiple time scales for the formation of the SH signal was reported, in 2005 Kudlinski et al. realized a more exhaustive description of space-charge region formation and induced second order non-linearity in bulk silica glasses [29]. The samples used in their work were disks of fused silica (Infrasil^{TM}) characterized by the presence of some types of impurity carriers (typically Na$^+$, Li$^+$, K$^+$, ...) located in the glass matrix with a concentration value of 1 ppm. Disks of different thickness, sandwiched between two Si electrodes, were heated at 250 °C, and poled at 4 kV for different temporal durations. When the impurity carriers become mobile, a high electric field is applied through them, producing their electromigration toward the cathode of the system. As a consequence, a negative space charge is created underneath the anodic surface, due to the fact that negative charges are motionless in the glass matrix (NBO$^-$ centers). A huge electric field similar to the dielectric breakdown field is consequently established within the depletion region and a second order susceptibility is then created:

$$\chi^{(2)} = 3\chi^{(3)} E_{DC} \quad (3)$$

where we are assuming a system unidimensional with the electric fields involved all linearly polarized along the same direction for the sake of simplicity. For the first few seconds of the electromigration process the single-carrier model can be still used to describe the time evolution of the depletion region formation [30], while after a certain time, defined optimal time (t$_{opt}$) [31], it is necessary to use a multiple carrier model to describe the temporal evolution of the poling process. If we consider the fast carriers (impurity charges) and the slow carriers (hydrogenated species) and both the migration and the diffusion phenomena, the equation of continuity and the Poisson's equation can be written as [29]:

$$\frac{\partial p_i}{\partial t} = -\mu_i \frac{\partial (p_i E)}{\partial x} + D_i \frac{\partial^2 p_i}{\partial x^2} \quad (4)$$

$$\frac{\partial E}{\partial x} = \frac{q}{\varepsilon} \left[\sum_i (p_i - p_{0,i}) \right] \quad (5)$$

where p_i, $p_{0,i}$ and μ_i are respectively the instantaneous concentration (ions/m^3), the initial (at t = 0) concentration and the mobility (at the temperature where the poling experiment is realized) of the ith species, q is the electron charge, $\varepsilon = 3.8\varepsilon_0$ is the permittivity of the medium and $D_i = k_B T \mu_i / q$ is the diffusion constant of the ith species, with k_B the Boltzmann constant and T the temperature of the medium. The system of Equations (4) and (5) gives the spatial distribution of the electric field in the sample as function of the poling duration. The assumptions related to the voltage applied are that the potential at the anodic surface ($x = 0$) is V_{app}, while the potential at the cathodic surface ($x = l$) is zero. Therefore, the first boundary condition is:

$$\int_0^l E dx = V_{app} \quad (6)$$

While the impurity charges (such as Na$^+$) are already present into the sample with the initial uniform concentration p_{0,Na^+}, the hydrogenated species possess an initial density $p_{0,H^+} = 0$ and are injected into the glass with an injection rate which depends on the electric field strength at the anodic surface. Therefore, the second boundary condition can be written as:

$$\left(\frac{\partial p_{H^+}}{\partial t} \right)_{x=0} = \sigma_{H^+} E(x=0) \quad (7)$$

where σ_{H^+} is an adjustable parameter used to describe the charge injection into the glass of the hydrogenated species.

In order to describe the dynamical evolution of the space-charge region, we can assume that, as a consequence of the application of the voltage (V_{app}) throughout the whole sample of length l, an electric field is created equal to V_{app}/l and, because $\mu_{Na^+} \gg \mu_{H^+}$, at first a depleted layer close to the anodic surface of the glass is formed due to the Na ions migrating toward the cathode. The induced electric field at the surface increases and screens the external electric field in the part of the sample placed outside the depletion region. When the space charge region is completely created, the maximum value of $E_{DC} \approx 10^9 V/m$ is obtained. At this time the concentration of the injected carriers per second increases rapidly to the value of $7.5 \times 10^{-22} m^{-3} s^{-1}$ (according to the equation that governs the injection into the glass of the hydrogenated species, which affirms that the concentration of those species at the anodic surface is linearly proportional to the value of the electric field at the same surface). At the same time, the drift velocity of the injected hydrogenated species $v_{H^+} = \mu_{H^+} E_{DC}$ reaches the same order of magnitude of to the velocity of the Na ions, which are outside the depletion region, where the external electric field is reduced because it is screened by the formation of the space charge. For poling durations longer than few minutes, these injected ions move deeper and deeper into the glass replacing slowly the Na ions removed previously, consequently neutralizing the NBO$^-$ centers (refer to [29] and figures in that paper).

3. From Poling of Bulk Glasses to Silica Optical Fibers

The first experiment of thermal poling of a silica fiber was reported by Kazansky et al. in 1994 [10], when a D-shaped fused silica Germania-doped step-index fiber was poled using the setup reported in Figure 3a of ref [10].

This poling configuration was adopted until 1995, when a twin-hole step-index silica fiber was poled for the first time by applying a voltage between the electrodes embedded respectively into the two cladding channels of the fiber [32]. This "twin-hole" fiber became the most adopted geometry for thermal poling of optical fibers [33,34].

After the early works on the poling of silica twin-hole fibers, many other works were published on this topic, such as for example Wong et al. [35], who revealed for the first time the existence, in a poled fiber, of the frozen-in electric field E_{DC}, using a Mach–Zehnder interferometer. The technique adopted allowed them to measure both the magnitude and the direction of the frozen-in field. They also measured the third-order non-linearity $\chi^{(3)}$ of unpoled and poled fibers, concluding that the $\chi^{(3)}$ has increased by a factor of 2 after the thermal poling process.

A work of Blazkiewicz et al., published in 2001, shows the effects on the dynamics of the poling process of the inclusion of a deposited doped silicate glass ring or of a borosilicate glass ring inside the anode hole of a poled twin-hole fiber [36]. In particular, they observed that in the case of a doped silicate glass ring there is a rapid saturation of the electro-optic coefficient, while the borosilicate glass ring instead acts as a trap layer that retards the evolution of the growth of the electro-optic coefficient. This work demonstrates that by tailoring the structure of the optical fiber to be poled it is possible to modify significantly the characteristics of the poling process and consequently the properties of the poled fiber.

3.1. From Conventional Poling to Cathode-Less Poling

The conventional anode–cathode configuration for thermal poling of silica fibers, shown in Figure 2 of [35], generates a space-charge region exclusively in the region surrounding the channel where the anodic electrode is inserted. The space-charge region can be visually observed quite simply by etching the cleaved end of the poled fiber in Hydrofluoric (HF) acid (diluted at 50% in deionized (DI) water) for 1 min and is reported in Figure 5 of [33]. The anode-cathode configuration has the drawback of the tiny distance ($\approx$10–20 µm) between the two channels, which greatly increases the risk of unwanted electric arcing discharge through the glass as a consequence of the application of elevated voltages.

However, in 2009 Margulis et al. showed that it is possible to make a depletion region develop around both the embedded electrodes by connecting them to the same anodic potential [37]. Figure 1 of reference [37] shows the schematics of the cathode-less poling configuration. The first advantage of this new poling configuration is the possibility of reduction of the risk of electrical breakdown through the fiber. Margulis et al. demonstrated also that the $\chi^{(2)}$ created via the cathode-less method is larger and more stable than the one created via conventional poling.

In 2012, An et al. reported in [38] a study where four different electrodes configurations were adopted to thermally pole a twin-hole optical fiber, including having only one anode wire inserted in one of the two cladding channels, two anode wires embedded inside both the channels, one cathode wire in one channel, and two cathode wires in the channels, in comparison to the conventional one where each one of the two wires embedded in the two channels was respectively connected to the anode and the cathode. The technique of second harmonic microscopy (SHM) was used to visualize the spatial distribution of the second order non-linearity created inside the poled fibers and to measure their magnitude. The results of this work consisted mainly in the observation that both one- and two-anode configurations gave a strong non-linearity compared with the conventional anode-cathode one. At the same time An et al. observed that the two-anode configuration was more reproducible than the one-anode one; for the one cathode-wire and two-cathode-wire configuration, strong non-linearity in a ring shape concentric with the fiber outer surface was induced as if the cathode metal wire were in the center of the twin-hole fiber rather than substantially offset. Figure 1 shows second harmonic (SH) micrographs for the fibers poled in the five different configurations.

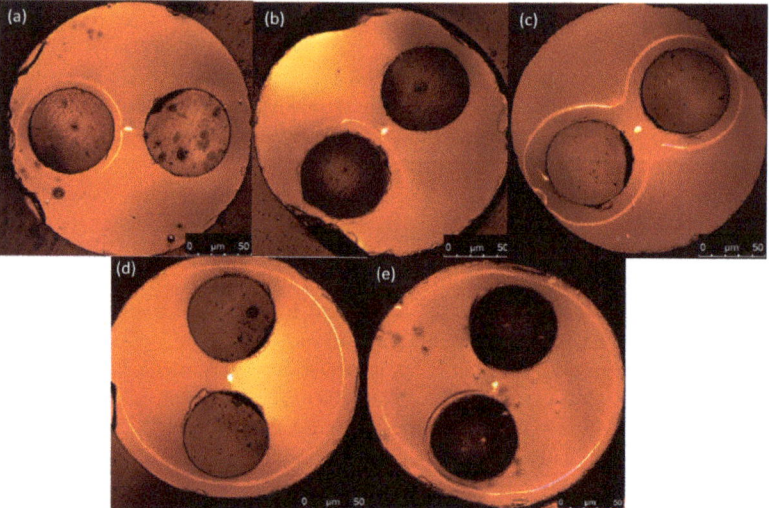

Figure 1. Second harmonic (SH) micrographs of twin-hole silica fibers poled in different electrical configurations, namely (**a**) conventional anode–cathode, (**b**) single anode, (**c**) anode–anode, (**d**) single cathode and (**e**) cathode–cathode. The figures are extracted from the work of An et al. [38].

In 2014 Camara et al. presented for the first time 2D numerical model of the cathode-less poling technique applied to optical fibers [39]. Their numerical simulations are based on a 2D implementation of the ion-exchange model (the one developed by Kudlinski et al. [29]), applied to poled fibers by using COMSOL™ Multiphysics, and consider the presence in the glass matrix of a faster cation (Na^+) and a slower cation (Li^+). Both the ions are assumed to be uniformly distributed in the glass matrix before the poling process starts, while a hydrogenated species (H_3O^+) is assumed to be injected from the surfaces in contact with the anodic electrodes. The physics of the 2D model is based on the transport of diluted species and assumes that ions characterized by a low concentration (1 ppm) move in consequence of processes of diffusion and drift due to an electric field [25,29]. The cladding holes of the twin-hole fiber are completely filled by metal [33], providing a perfect equipotential. The equation solved in x, y and t for the concentration of the ith ion (Na$^+$, Li$^+$ and the hydrogen species, such as H$_3$O$^+$) is [39]:

$$\frac{\partial c_i}{\partial t} + \nabla \cdot (-D_i \nabla c_i - z_i \mu_i F c_i \nabla V) = R_i \qquad (8)$$

where the first term in brackets represents the diffusion while the second term the drift in the electric field E, c is the concentration, D is the diffusivity, z is the charge, μ is the ionic mobility, F the Faraday constant, V the electric potential and R the consumption or production rate. The electric field and electric potential distribution are obtained from Maxwell's equations in the electrostatics regime (magnetic fields are neglected). The boundary conditions assumed in the model of Camara et al. are the initial electrical neutrality of the fiber (the mobile ions and the motionless NBO$^-$ centers are characterized by the same concentration inside the fiber), the potential at the surfaces of the holes is the applied voltage during the poling process and zero when the voltage is removed. Furthermore, the external surface of the fiber is at zero volts and that the cations exit it and do not come back. The hydrogenated species is injected from the surface of the cladding holes and move, pushed by the applied electric field. Two possible situations are studied; firstly, where the injection rate of H$_3$O$^+$ constant, which assumes the presence of ions already at the surface of the hole (in the glass) [25], and secondly where an injection proportional to the electric field on the surface of the hole, as implemented in [29]. The initial carrier concentrations are: c(Na$^+$) = 1 ppm uniformly distributed in the glass at t = 0 sec; c(Li$^+$) = 1 ppm uniformly distributed in the glass at t = 0 sec; c(H$_3$O$^+$) = up to 2 ppm injectable from the holes, initially zero inside the entire fiber, with a rate that is either constant, linearly dependent on the field at the electrode edge, or decaying exponentially as the ion supply is exhausted; c(NBO$^-$) = 2 ppm uniformly distributed in the glass at t = 0 sec for guaranteeing the initial charge neutrality. It is worth highlighting that the types of charges involved in the poling process, their initial concentration, and their mobility at the desired temperature represent all sources of error in the absolute determination of the precise dynamical evolution of the depletion region. Nevertheless, the results obtained represent a global trend which was strongly validated. Indeed, for the first time in the work of Alley et al. [40] and later in many other papers including the work of Camara et al. [39], the shape of depletion region developed around the anodic electrodes in a thermal poling process is revealed via a process of etching in hydrofluoric acid of the cross section of the fiber. An example of this shape is also reported in Figure 3b in Section 3.2 of this paper. In Figure 2, it is possible to gain an idea of the temporal evolution of the concentrations of the two impurity species already present into the glass matrix (Na$^+$ and Li$^+$) and of the hydrogenated species injected after the application of the external electric field (H$_3$O$^+$).

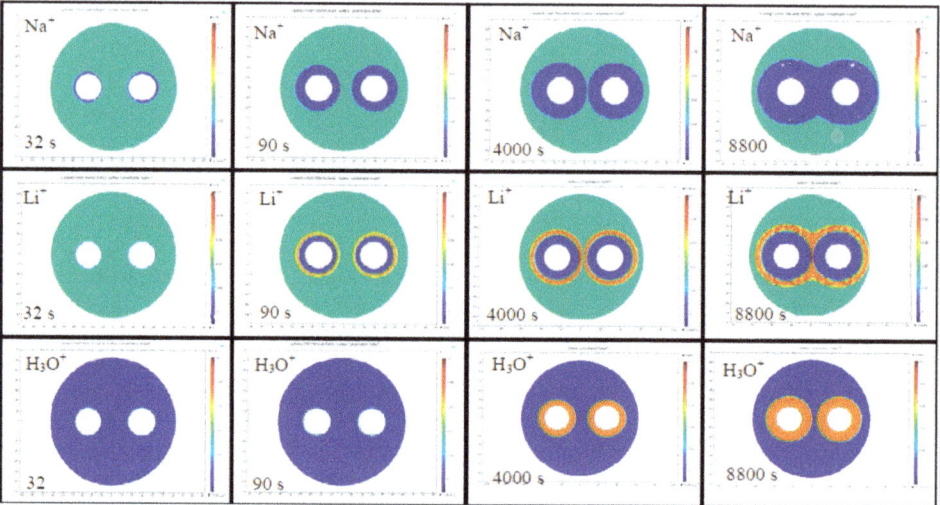

Figure 2. Temporal evolution of the mobile cations for a Germania doped twin-hole fused silica fiber poled in a cathode-less configuration (the two electrodes inserted in the two cladding channels are connected to the same potential of 5 kV). The injection of the H_3O^+ ions is considered inexhaustible and capable to neutralize the non-bridging oxygen (NBO^-) centres depleted of the impurity positive ions moved because of the application of the external electric field [39].

In 2009 another interesting contribution to the understanding of the dynamics of the thermal poling process in silica glass, was given by Zhang et al. [41], who studied multiple poling processes. In particular they demonstrated that the first poling process, in case of a thermal erasure of the non-linearity and subsequent re-poling process, has a strong effect on the latter. Using a two carriers model (the same introduced by Alley et al. [25] and improved by Kudlinski et al. [29]), they quantitatively show that the difference in the evolution of the $\chi^{(2)}$ is due to the different initial charge distributions before each poling process. The extra hydrogenated species injected during the initial poling process modifies the dynamic of the second poling process; in contrast to the first poling (where the $\chi^{(2)}$ increases in time), the $\chi^{(2)}$ tends to decrease in time after reaching a maximum value.

3.2. Induction Poling

The cathode-less configuration for poling optical fibers, presented by Margulis et al. in 2009, was adopted until 2014. At that time, De Lucia et al. presented a new technique of thermal poling of silica fibers, called "electrostatic induction" [42,43]. The setup to realize the induction poling process is reported in Figure 3a. Two samples of a twin-hole fused silica fiber, both equipped with solid electrodes embedded in both the cladding channels, are utilized. One of them (≈5 cm of length) is used as electrostatic inductor, while the other one (≈40 cm of length) is the fiber to be poled. The two fibers are kept (on top of a microscope slide placed inside a Petri dish in turn located on top of a heater) adjacent only along the 2.5 cm of the short side of the slide, while the rest of the longer fiber (the one to be poled) is fixed on top of the Petri dish surface with some Kapton tape to facilitate its thermalization. The rear surface of the microscope slide is coated with gold and represents the ground plane of the system. The electrodes embedded in the inductor are both connected to the anodic potential, while the two electrodes embedded in the cladding channels of the fiber to be poled by induction are left floating.

The HF etching, shown in Figure 3b, demonstrated that a depletion region was created all along the whole length of the floating electrodes embedded into the "induced" fiber. The creation of a $\chi^{(2)}_{eff}$ by induction poling has been also proven measuring a SHG signal produced by pumping the poled fiber with a 1550 nm laser. In order to observe a significant SHG signal it is necessary to periodically erase the non-linearity previously created by thermal poling. The erasure process is obtained by exposing the poled fiber to the light generated by an ultraviolet (UV) source. The periodic erasure allows for obtaining a quasi-phase-matching (QPM) condition between the pump at 1550 nm and the SH light at 775 nm. This induction poling technique thus allows for poling long fibers without any physical contact between the power supply and the embedded electrodes.

Figure 4a,b show the setup used to erase the non-linearity via UV light exposure and then to subsequently characterize the second harmonic signal and the SHG signal peaks obtained for the two different periods of erasure of the non-linearity created in two identical fibers poled under the same experimental conditions. The clear dependence of the wavelength doubled from the period of the erasure confirms that the signal measured is due to a SHG signal created by a quadratic non-linear process via induction poling.

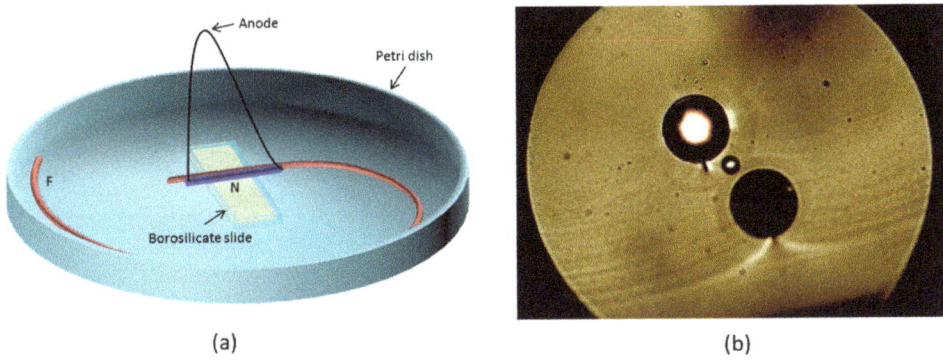

Figure 3. (**a**) Schematic of the setup to pole twin-hole silica fibers via electrostatic induction. The inductor is represented by the blue fiber. Two metallic wires are inserted in the two cladding holes and connected to the identical anodic potential. This fiber is basically used as a layer of dielectric material simply to avoid unwanted electrical arcing discharge in air. The fiber represented in red (whose embedded electrodes are left floating) represents the fiber to be poled. Inductor and sample are attached to a microscope slide by means of some Kapton tape and maintained adjacent along 2.5 cm of the short side of the microscope slide. A gold coating on the backside of the slide (created via e-beam evaporation) works as the ground plane. The Petri dish is maintained at a temperature of ≈ 300 °C during the duration of the process. (**b**) Cross section of a twin-hole silica fiber poled via induction poling technique. The depletion regions are visualized by means of a process of decorative etching in HF acid for 1 min [42,43].

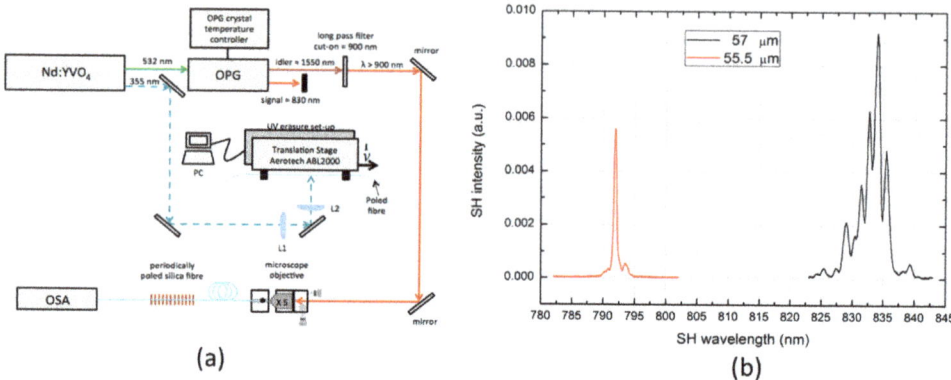

Figure 4. (**a**) Schematic of the setup used to realize the periodic erasure of the quadratic non-linearity created via induction poling. The fiber previously poled via the setup shown in Figure 3a is periodically exposed to ultraviolet (UV) light with the objective of obtaining the quasi-phase matching (QPM) condition between the pump (1550 nm) and the second harmonic generation (SHG) light (775 nm). The wavelength of the laser source used to erase the non-linearity is 355 nm. L1 and L2 stand for the cylindrical lenses of focal lengths f 500 mm and 85 mm, respectively, used to focalize the laser beam in a spot of area of 10 µm × 100 µm at the fiber's core. Also shown is the setup for the characterization of the SHG signal generated by the periodically poled fiber. (**b**) SHG spectra of induction poled fibers characterized by two different QPM periods of UV erasure [43].

In 2016, De Lucia et al. published the 2D numerical model of the process of creation of a depletion region inside a twin-hole fiber poled via electrostatic induction [44]. The model was inspired by Camara et al. [39], even if it shows some important differences. First of all, the model of Camara et al. assumes that the external surface of the fiber to be poled is always kept at ground potential. While this assumption is suitable for the setup presented by Margulis et al. in [37], the same is not reasonable for a situation where an external field is applied by an inductor to floating electrodes embedded inside the fibers to be poled. If, indeed, the external surface of the fiber to be poled by induction was assumed to be grounded, it would consequently screen the electric field created by the inductor, thus suppressing completely the process of electrostatic induction. Another difference consists in the fact that while in the model of Camara et al. the injection rate of the H_3O^+ ions can be always assumed to be constant, in the thermal poling process via electrostatic induction, the variable floating potentials intrinsic to this process require a field-dependent charge injection. Furthermore, it is necessary to consider the ion recombination process at the cladding–air interface and consequently to modify the field dependency.

The model for the induction poling scheme (whose setup is reported in Figure 3a) is obtained separating the setup in two distinguished parts (indicated in Figure 3a with the letters N and F). In the N part of the setup (the one where inductor and poled fiber are adjacent) the fiber to be poled is immersed in the electric field generated by the inductor, while in the part of the setup where inductor and sample are far from each other (F in Figure 3a) the fiber poled is not immersed in the field lines created by the inductor. This "double" model needs the assumption that the electrodes embedded inside the fiber to be poled are electrically continuous. When the two floating electrodes inserted in the cladding holes of the fiber to be poled are immersed in the external electric field, provided by the inductor in the N region, they become charged as a result of a process of electrostatic induction and reach a specific electric potential. If we assume that there is no drop of electric potential along the floating embedded electrodes, they will be characterized by an equipotential surface along the whole fiber. In other words, whatever is the potential picked up by the floating electrodes in the region N, will be transferred efficiently to any location along the whole fiber.

In the 2D model the two fibers are assumed to be made of two different types of glass. Specifically, the glass the inductor is made of is assumed to be pure silica, which lacks charge impurities, while the fiber to be poled is made of fused silica and is characterized by an initial concentration of 1 ppm of Na^+ uniformly distributed through its cross section. At the same time, the fiber has up to 1 ppm of H_3O^+ ions that can be injected at the cladding holes, while there are no H_3O^+ ions inside the fiber at t = 0 s. To initially fulfil the charge neutrality, it is necessary to assume that NBO^- centers (characterized by very low mobility) are uniformly distributed inside the fiber with the same concentration of the Na^+ ions at t = 0 s. The cladding channels of the fiber are considered equipotential. The H_3O^+ ions can be injected through the electrode-cladding interface if located at electric potentials higher than the surrounding cladding. A variable parameter σ_2 (whose value is chosen to be identical to that chosen in the model of Kudlinski et al. [29]) is used to describe the charge injection into the sample. The induction poling model considers also the particular case where the electric field is less than zero. In this case, H_3O^+ ions close to the cladding (either previously injected or diffused from other regions of the fiber) possess a negative injection rate, which substantially means an outflow. However, if the concentration of H_3O^+ is zero at the electrode-cladding interface, the injection rate is zero, even in the case of a "negative" electric field. Therefore, the variation of the injected H_3O^+ density per unit of time at the electrode-cladding interface can be expressed by:

$$\left(\frac{\partial c_2}{\partial t}\right)_{surface} = \sigma_2 E, \ E \geq 0 \ \text{or} \ E < 0 \ \text{and} \ c_2 > 0 \tag{9}$$

$$\left(\frac{\partial c_2}{\partial t}\right)_{surface} = 0, \ \text{otherwise,} \tag{10}$$

where c_2 is the concentration of the H_3O^+ species and σ_2 the parameter chosen.

In the near model, the two fibers lie adjacent each other on top of a microscope slide (1 mm thick), with the back face coated with gold and grounded. The far model, on the other hand, consists in the model of Camara [39] (modified according to the considerations reported in the initial part of this section) where the values of the electric potential applied to the two embedded floating electrodes are not constant, but are the values of potential (changing in time) calculated via the near model. Moreover, the far model assumes that the fiber lacks a ground plane. Figure 5 shows the time dynamics of the concentrations of both the "fast" (Na^+) and "slow" (H_3O^+) carriers calculated at three different times of the induction poling process in the near model, while in Figure 6 the concentrations obtained using the far model are reported. It is possible to note that the two depletion regions develop in a different way according to the location where they develop. The reason for this different behaviour is the fact that in the N area the external electric field created by the inductor affects the distribution of the total electric field developed in the region surrounding each electrode, while in the F region the electric field created around each electrode is not modified by the presence of any external electric field. Consequently, even the evolution of the depletion region will be different in each different region of the setup.

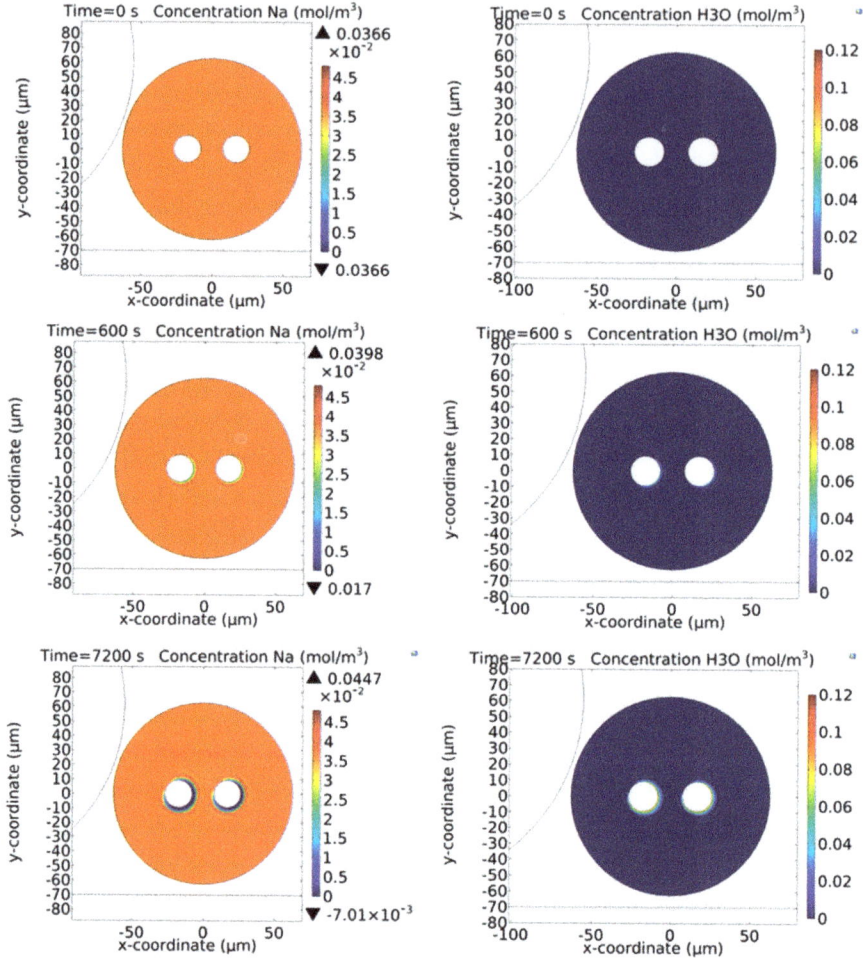

Figure 5. Time dynamics of the Na$^+$ ions calculated by means of the near model for induction poling. The Na$^+$ ions are considered mobile only in the fiber to be poled, while they are assumed to be motionless inside the inductor. Both the outer surfaces of the two fibers are not assumed to be grounded. The ground of the system is placed at a distance of 1 mm below the microscope slide surface. The concentration of the Na$^+$ ions is 1 ppm before the start of the poling process. Both the fibers are considered at a temperature of 300 °C and the injection of the H$_3$O$^+$ is assumed to be inexhaustible. The H$_3$O$^+$ ions can consequently neutralize the NBO$^-$ centres previously depleted of Na$^+$ ions migrated as a result of the application of the external electric field. The two electrodes inserted in the cladding channels of the inductor are both connected to the same electric potential of +5 kV [44].

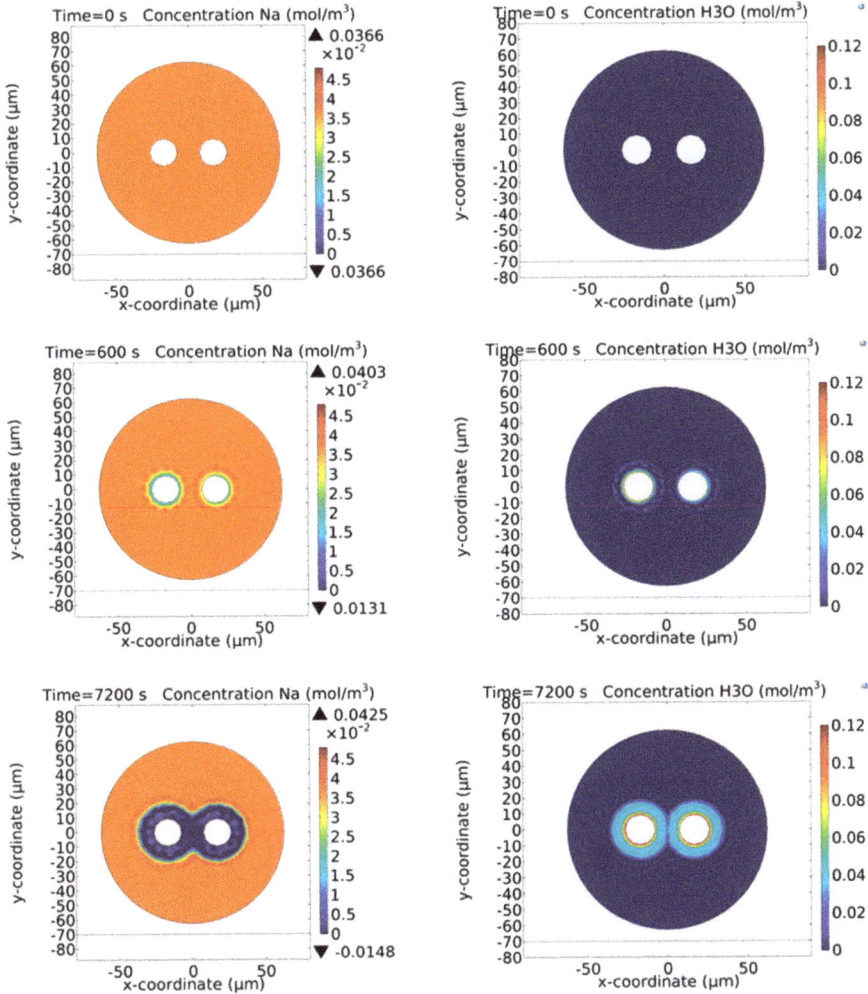

Figure 6. Time dynamics of the Na$^+$ ions calculated by means of the far model for induction poling. The two floating electrodes inserted in the cladding channels of the fiber to be poled assume values calculated via the near model in the conditions reported in the caption of Figure 5. The concentration of the Na$^+$ ions is 1 ppm before the start of the poling process. The fiber is considered at a temperature of 300 °C and the injection of the H$_3$O$^+$ is assumed to be inexhaustible [44].

3.3. Single Anode Poling

The most recent results obtained by De Lucia et al. [45] demonstrated both experimentally and theoretically that the single-anode (S-A) configuration, characterized by the fact that only one electrode is embedded inside one of the two cladding channels of the optical fiber and connected to a certain electrical potential, is superior (in terms of quadratic non-linearity created in the fiber core) to the double-anode (D-A) configuration, introduced for the first time by Margulis et al. [37]. Starting from the theoretical result for a fiber of symmetric geometry (the two cladding channels are at the same distance from the fiber core) and poled in D-A configuration the value of $\chi^{(2)}_{eff}$ at the center of the fiber is almost zero for long time poling ($\approx 2h$). However, it was observed that if only one electrode was connected to the high positive potential while the other electrode was completely removed, the value of the quadratic non-linearity was not null any more at the center of the fiber core. The hypothesis is that the behavior of the D-A configuration is due to the concomitant and competitive evolution of the space-charge formation around the two anodes. In contrast, the S-A poling scheme does not exhibit the same limitation. The two configurations (D-A and S-A) have been then theoretically studied for a fiber of asymmetric geometry (different distance from each cladding hole and the fiber core). Figure 7 shows the trend (simulated by means of COMSOL™ Multiphysics, Edition 5.1, COMSOL, Inc., Burlington, MA, USA) of $\chi_{eff}^{(2)}$ with the temporal duration of the poling process for both electrode configurations and at five different locations in the fiber core region for an asymmetric geometry of the fiber. The most significant outcome of the numerical simulations consists in the fact that the ultimate value (for long poling times) of $\chi_{eff}^{(2)}$ in S-A configuration is approximately double if compared to the one calculated in the D-A. The value of the non-linear susceptibility $\chi_{eff}^{(2)}$ has also been experimentally measured in a process of second harmonic generation (SHG) at 1550 nm in a fiber periodically poled in S-A configuration. The $\chi_{eff}^{(2)}$ has been periodically erased by exposing the poled fiber to a UV light emitted by a frequency doubled argon-ion laser (CW, 244 nm). The results obtained proved that the S-A scheme for poling silica fibers is preferable to the D-A one in terms of absolute value of $\chi_{eff}^{(2)}$. Indeed, in their 2019 paper [45], De Lucia et al. also demonstrated that the theoretical result shown in Figure 7 is verified experimentally. Two identical fibers poled in the same experimental conditions and for long time (2 h) but in the two different configurations (D-A and S-A) have been characterized in terms of the $\chi_{eff}^{(2)}$ obtained in a second harmonic generation process. The value of effective second order non-linear susceptibility obtained for the fiber poled in S-A configuration is double with respect to that obtained for the fiber poled in the D-A configuration. Furthermore, the S-A scheme allows for a significant simplification of the fiber fabrication scheme, as only a single cladding channel will be required for the electrode, thus allowing for with considerably relaxed tolerances on the fiber's core position relative to the single electrode.

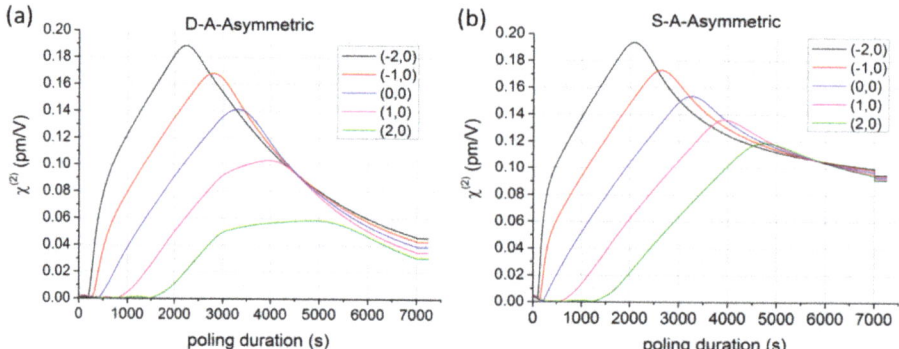

Figure 7. Time evolutions of the $\chi_{eff}^{(2)}$ numerically obtained for a twin-hole fiber of asymmetric geometry poled in (**a**) D-A and (**b**) S-A configurations. The values of $\chi_{eff}^{(2)}$ have been calculated via Equation (3) at five different locations in the fiber's core region (4 μm diameter) and in the plane Y = 0. The legend shows the (x,y) coordinates (in μm) of the points where the values of $\chi_{eff}^{(2)}$ have been calculated [45].

4. Conclusions

Thermal poling, a technique invented more than 25 years ago, nowadays still represents an important tool in the area of quadratic nonlinear photonics. More recently, two new developments have been presented. The adoption of liquid electrodes [46] and the demonstration of the induction poling technique [42,44], are potentially useful to create nonlinear quadratic all-fiber devices exploiting different types of waveguides, in terms of length and geometry. In this paper, we have mainly focused our attention on the logical and chronological development of 2D numerical models to explain as deeply as possible the dynamics of evolution of the poling process. In particular, we started from the early theoretical interpretation of the process as based on the electromigration of impurity ions immersed in high electric fields. Later, we presented the step towards a full understanding of the phenomenon, as represented by the work of Kudlinski et al. [29], then further refined by Camara et al. [39], and by the work of De Lucia et al. [44], who applied the 2D model to the induction poling technique, explaining its evolution. Finally, we presented our most recent theoretical work which allowed us to identify in the single-anode configuration the most effective method in terms of absolute value of quadratic non-linearity created inside the glass fiber and also in terms of simplification of the fabrication constraints.

Funding: This research was funded by Engineering and Physical Sciences Research Council (EPSRC), grant number EP/I035307/1).

Conflicts of Interest: The authors declare no conflict of interest.

References

1. Baird, J.L. An Improved Method of and Means for Producing Optical Images. British Patent No. 285 738, 1928.
2. Hansell, C.W. Picture Transmission, 1751584. 1930. Available online: https://patentimages.storage.googleapis.com/07/66/87/78111772a09892/US1751584.pdf (accessed on 20 November 2018).
3. Kapron, F.P.; Keck, D.B.; Maurer, R.D. Radiation Losses in Glass Optical Waveguides. *Appl. Phys. Lett.* **1970**, 423–425. [CrossRef]
4. Agrawal, G.P. *Fiber-Optic Communication Systems*; John Wiley & Sons: Hoboken, NJ, USA, 2012.
5. Ainslie, B.J. A Review of the Fabrication and Properties of Erbium-Doped Fibers for Optical Amplifiers. *J. Lightwave Technol.* **1991**, *9*, 220–227. [CrossRef]
6. Chen, J.; Liu, B.; Zhang, H. Review of Fiber Bragg Grating Sensor Technology. *Front. Optoelectron. China* **2011**, *4*, 204–212. [CrossRef]
7. Dudley, J.M.; Taylor, J.R. *Supercontinuum Generation in Optical Fibers*; Cambridge University Press: Cambridge, UK, 2010.
8. Butcher, P.N.; Cotter, D. *The Elements of Nonlinear Optics, volume 9*; Cambridge Studies in Modern Optics: Cambridge, UK, 1990.
9. Myers, R.A.; Mukherjee, N.; Brueck, S.R. Large Second-Order Nonlinearity in Poled Fused Silica. *Opt. Lett.* **1991**, *16*, 1732–1734. [CrossRef]
10. Kazansky, P.G.; Dong, L.; Russell, P.S. High Second-Order Nonlinearities in Poled Silicate Fibers. *Opt. Lett.* **1994**, *19*, 701–703. [CrossRef]
11. Batchko, R.G.; Miller, G.D.; Alexandrovski, A.; Fejer, M.M.; Byer, R.L. Limitations of High-Power Visible Wavelength Periodically Poled Lithium Niobite Devices Due to Green-Induced Infrared Absorption and Thermal Lensing. In Proceedings of the Lasers and Electro-Optics, CTuD6G, San Francisco, CA, USA, 3–8 May 1998.
12. Furukawa, Y.; Kitamura, K.; Alexandrovski, A.; Route, R.K.; Fejer, M.M.; Foulon, G. Green-Induced Infrared Absorption in MgO Doped LiNbO3. *Appl. Phys. Lett.* **2001**, *78*, 1970–1972. [CrossRef]
13. Yamane, M.; Asahara, Y. *Glasses for Photonics*; Cambridge University Press: Cambridge, UK, 2000.
14. Fujii, Y.; Kawasaki, B.S.; Hill, K.O.; Johnson, D.C. Sum-Frequency Light Generation in Optical Fibers. *Opt. Lett.* **1980**, *5*, 48–50. [CrossRef]
15. Sasaki, Y.; Ohmori, Y. Phasematched Sumfrequency Light Generation in Optical Fibers. *Appl. Phys. Lett.* **1981**, *39*, 466–468. [CrossRef]
16. Ohmori, Y.; Sasaki, Y. Two-Wave Sum-Frequency Light Generation in Optical Fibers. *IEEE Trans. Microw. Theory Tech. MTT* **1982**, *30*, 604–608. [CrossRef]
17. Gabriagues, J.M.; Fersing, L. Second and Third Harmonic Generation in Optical Fibers. In *Conference on Lasers and Electro-Optics*; Optical Society of America: Washington, DC, USA, 1984.
18. Osterberg, U.; Margulis, W. Dye Laser Pumped by Nd:YAG Laser Pulses Frequency Doubled in a Glass Optical Fibers. *Opt. Lett.* **1986**, *11*, 516–518. [CrossRef]

19. Osterberg, U.; Margulis, W. Experimental Studies on Efficient Frequency Doubling in Glass Optical Fibers. *Opt. Lett.* **1987**, *12*, 57–59. [CrossRef] [PubMed]
20. Stolen, R.H.; Tom, H.W.K. Self-Organized Phase-Matched Harmonic Generation in Optical Fibers. *Opt. Lett.* **1987**, *12*, 585–587. [CrossRef] [PubMed]
21. Bethune, D.S. Quadrupole Second-Harmonic Generation for a Focused Beam of Arbitrary Transverse Structure and Polarization. *Opt. Lett.* **1981**, *6*, 287–289. [CrossRef] [PubMed]
22. Farries, M.C.; Russell, P.St.J.; Fermann, M.E.; Payne, D.N. Second-Harmonic Generation in an Optical Fiber by Self-Written $\chi^{(2)}$ Grating. *Electron. Lett.* **1987**, *23*, 322–324. [CrossRef]
23. Kashyap, R. Phase-Matched Periodic Electric-Field-Induced Second-Harmonic Generation in Optical Fibers. *J. Opt. Soc. Am. B* **1989**, *6*, 313–328. [CrossRef]
24. Mukherjee, N.; Myers, R.A.; Brueck, S.R.J. Dynamics of Second-Harmonic Generation in Fused Silica. *J. Opt. Soc. Am. B* **1994**, *11*, 665–669. [CrossRef]
25. Alley, T.G.; Brueck, S.R.J.; Myers, R.A. Space Charge Dynamics in Thermally Poled Fused Silica. *J. Non-Cryst. Solids* **1998**, *242*, 165–176. [CrossRef]
26. Krieger, U.K.; Lanford, W.A. Field Assisted Transport of Na^+ ions, Ca^{2+} ions and Electrons in Commercial Soda-Lime Glass I: Experimental. *J. Non-Cryst. Solids* **1988**, *102*, 50–61. [CrossRef]
27. Kazansky, P.G.; Smith, A.R.; Russell, P.St.J.; Yang, G.M.; Sessler, G.M. Thermally Poled Silica Glass: Laser Induced Pressure Pulse Probe of Charge Distribution. *Appl. Phys. Lett.* **1996**, *68*, 269–271. [CrossRef]
28. Pureur, D.; Liu, A.C.; Digonnet, M.J.; Kino, G.S. Absolute Measurement of the Second-Order Nonlinearity Profile in Poled Silica. *Opt. Lett.* **1998**, *23*, 588–590. [CrossRef]
29. Kudlinski, A.; Quiquempois, Y.; Martinelli, G. Modeling of the chi(2) Susceptibility Time-Evolution in Thermally Poled Fused Silica. *Opt. Express.* **2005**, *13*, 8015–8024. [CrossRef] [PubMed]
30. Kudlinski, A.; Martinelli, G.; Quiquempois, Y. Time Evolution of Second-Order Nonlinear Profiles Induced within Thermally Poled Silica Samples. *Opt. Lett.* **2005**, *30*, 1039–1041. [CrossRef] [PubMed]
31. Faccio, D.; Pruneri, V.; Kazansky, P.G. Dynamics of the Second-Order Nonlinearity in Thermally Poled Silica Glass. *Appl. Phys. Lett.* **2001**, *79*, 2687–2689. [CrossRef]
32. Fujiwara, T.; Wong, D.; Fleming, S. Large Electrooptic Modulation in a Thermally-Poled Germanosilicate Fiber. *IEEE Photonics Technol. Lett.* **1995**, *7*, 1177–1179. [CrossRef]
33. Myrén, N.; Olsson, H.; Norin, L.; Sjödin, N.; Helander, P.; Svennebrink, J.; Margulis, W. Wide Wedge-Shaped Depletion Region in Thermally Poled Fiber with Alloy Electrodes. *Opt. Express.* **2004**, *12*, 6093–6099. [CrossRef] [PubMed]
34. Kazansky, P.G.; Pruneri, V. Electric-Field Poling of Quasi-Phase-Matched Optical Fibers. *J. Opt. Soc. Am. B* **1997**, *14*, 3170–3179. [CrossRef]
35. Wong, D.; Xu, W.; Fleming, S.; Janos, M.; Lo, K.-M. Frozen-in Electrical Field in Thermally Poled Fibers. *Opt. Fiber Technol.* **1999**, *5*, 235–241. [CrossRef]
36. Blazkiewicz, P.; Xu, W.; Wong, D.; Fleming, S.; Ryan, T. Modification of Thermal Poling Evolution Using Novel Twin-Hole Fibers. *J. Lightwave Technol.* **2001**, *19*, 1149–1154. [CrossRef]
37. Margulis, W.; Tarasenko, O.; Myrén, N. Who Needs a Cathode? Creating a Second-Order Nonlinearity by Charging Glass Fiber with Two Anodes. *Opt. Express* **2009**, *17*, 15534–15540. [CrossRef]
38. An, H.; Fleming, S. Investigating the Effectiveness of Thermally Poling Optical Fibers with Various Internal Electrode Configurations. *Opt. Express* **2012**, *20*, 7436–7444. [CrossRef]
39. Camara, A.; Tarasenko, O.; Margulis, W. Study of Thermally Poled Fibers with a Two-Dimensional Model. *Opt. Express* **2014**, *22*, 17700–17715. [CrossRef] [PubMed]
40. Alley, T.G.; Brueck, S.R.J. Visualization of the Nonlinear Optical Space-Charge Region of Bulk Thermally Poled Fused-Silica Glass. *Opt. Lett.* **1998**, *23*, 1170–1172. [CrossRef] [PubMed]
41. Zhang, J.; Qian, L. Real-time $\chi^{(2)}$ Evolution in Twin-Hole Fiber During Thermal Poling and Repoling. *J. Opt. Soc. Am. B* **2009**, *26*, 1412–1416. [CrossRef]
42. De Lucia, F.; Huang, D.; Corbari, C.; Healy, N.; Sazio, P.J.A. Optical Fiber Poling by Induction. *Opt. Lett.* **2014**, *39*, 6513–6516. [CrossRef] [PubMed]
43. De Lucia, F.; Sazio, P.J.A. Optimized Optical Fiber Poling Configurations. *Opt. Mater. X* **2019**, *1*, 100016. [CrossRef]
44. De Lucia, F.; Huang, D.; Corbari, C.; Healy, N.; Sazio, P.J.A. Optical Fiber Poling by Induction: Analysis by 2D Numerical Modeling. *Opt. Lett.* **2016**, *41*, 1700–1703. [CrossRef]

45. De Lucia, F.; Bannerman, R.; Englebert, N.; Nunez Velazquez, M.M.A.; Leo, F.; Gates, J.; Gorza, S.-P.; Sahu, J.; Sazio, P.J.A. Single is Better than Double: Theoretical and Experimental Comparison between Two Thermal Poling Configurations of Optical Fibers. *Opt. Express* **2019**, *27*, 27761–27776. [CrossRef]
46. De Lucia, F.; Keefer, D.W.; Corbari, C.; Sazio, P.J.A. Thermal Poling of Silica Optical Fibers Using Liquid Electrodes. *Opt. Lett.* **2017**, *42*, 69–72. [CrossRef]

© 2020 by the authors. Licensee MDPI, Basel, Switzerland. This article is an open access article distributed under the terms and conditions of the Creative Commons Attribution (CC BY) license (http://creativecommons.org/licenses/by/4.0/).

Review

Fiber Amplifiers and Fiber Lasers Based on Stimulated Raman Scattering: A Review

Luigi Sirleto * and Maria Antonietta Ferrara

National Research Council (CNR), Institute of Applied Sciences and Intelligent Systems, Via Pietro Castellino 111, 80131 Naples, Italy; antonella.ferrara@na.imm.cnr.it
* Correspondence: luigi.sirleto@cnr.it

Received: 10 January 2020; Accepted: 24 February 2020; Published: 26 February 2020

Abstract: Nowadays, in fiber optic communications the growing demand in terms of transmission capacity has been fulfilling the entire spectral band of the erbium-doped fiber amplifiers (EDFAs). This dramatic increase in bandwidth rules out the use of EDFAs, leaving fiber Raman amplifiers (FRAs) as the key devices for future amplification requirements. On the other hand, in the field of high-power fiber lasers, a very attractive option is provided by fiber Raman lasers (FRLs), due to their high output power, high efficiency and broad gain bandwidth, covering almost the entire near-infrared region. This paper reviews the challenges, achievements and perspectives of both fiber Raman amplifier and fiber Raman laser. They are enabling technologies for implementation of high-capacity optical communication systems and for the realization of high power fiber lasers, respectively.

Keywords: stimulated raman scattering; fiber optics; amplifiers; lasers; optical communication systems

1. Introduction

Optical communication systems require optoelectronic devices, such as sources, detectors and so on, and utilize fiber optics to transmit the light carrying the signals impressed by modulators. Optical fibers are affected by chromatic dispersion, losses, and nonlinearity. Dispersion control is, usually, achieved via fiber geometry and material composition. Losses limit the transmission distance in modern long haul fiber-optic communication systems, so in order to boost a weak signal, optical amplifiers have been developed. The basic idea, behind the fiber amplifier realisation, was to allow to the signal to remain in optical form throughout a link or network. Fiber amplifiers as repeaters offer a number of advantages, including the ability to change system data rates, or to simultaneously transmit multiple rates—all without the need to modify the transmission channel. The most important advantage is that signal power at multiple wavelengths can be simultaneously boosted by a single amplifier—a task that would require a separate electronic repeater for each wavelength. The state-of-the-art of optical amplifiers are erbium-doped fiber amplifiers (EDFAs), which are formed by doping the glass fiber host with erbium ions. In EDFAs, gain by stimulated emission at wavelengths in the vicinity of 1.55 µm is obtained, by optical pumping using light at either 1.48-µm or 0.98-µm wavelengths. EDFAs are lumped amplifier, this means that the signal, monotonically attenuated in the fiber span, is amplified at a point of the EDFA location to recover the original level, before entering the next fiber span. The distance between amplifiers is determined by the span loss, by the limit imposed from the power maximum value allowed in the fiber, without inducing nonlinear effects, and by the minimum acceptable power, avoiding a degradation of the signal-to-noise-ratio. EDFAs are compact and highly efficient devices with high gain and low noise [1].

Optical multiplexing allows to combine multiple optical signals into one to make full use of the immense bandwidth potential of an optical cannel. The basic idea is to divide the huge bandwidth of optical fiber into individual channels of lower bandwidth, so that multiple access with lower-speed electronics is achieved. The large gain bandwidth of EDFAs has enabled the realization of dense wavelength-division multiplexed systems, in which terabit/sec data rates have been demonstrated. EDFAs operate in the C-band (from 1530 to 1565 nm) while extended-band EDFAs can provide gain in the L-band (from 1565 to 1625 nm). However, the capacity of most fibers can be extended by opening up the S band (from 1480 to 1530 nm), and the S+ band (from 1430 to 1480 nm). In addition, we note that earlier transmission systems were deployed in the 1310-nm band, which can stretch between 1280 and 1340 nm, while there could be also a 1400-nm band, which is only useful in new fibers using special drying techniques to reduce the water peak absorption around 1390 nm. Definitely, the available communication range could span from 1270 to 1650 nm, corresponding to about 50 THz bandwidth, but the current strong limitation is that a great part of this range is inaccessible by EDFAs [1–4].

In optical communications, fiber nonlinearities are the basis of a number of devices such as amplifiers, switching and logic elements. The nonlinear effects in optical fiber can be divided in two class. The first one is due to the Kerr-effect, i.e., intensity dependence of refractive index of the medium, which manifests itself in three different effects: Self-Phase Modulation (SPM), Cross-Phase Modulation (CPM) and Four-Wave Mixing (FWM), depending upon the type of input signal. The second one is due to inelastic-scattering phenomenon, in which the optical field transfers part of its energy to the nonlinear medium. Such an inelastic-scattering phenomenon can induce stimulated effects such as Stimulated Brillouin-Scattering (SBS) and Stimulated Raman-Scattering (SRS). We note that each type of stimulated scattering process can be used as a source of gain in the fiber. In both phenomena, at high power level, the intensity of scattered light grows exponentially if the incident power exceeds a certain threshold value. Raman amplification is more useful, because of the relatively large frequency shift and the broader-gain bandwidth. The basic difference between them is that in the case of Brillouin the interaction is between the guided optical wave and low-frequency acoustic phonons, while in the case of Raman the interaction is between the guided wave and high-frequency optical phonons. Another fundamental difference is that SBS in optical fibers occurs only in the backward direction whereas SRS can occur in both directions [2–4].

In long-range high-datarate systems, involving high optical power levels and signals at multiple wavelengths, the consequences of fiber nonlinearities can be: (1) the generation of additional signal bandwidth within a given channel, (2) modifications of the phase and shape of pulses and can cause spectral broadening, which leads to increased dispersion, (3) the generation of light at other wavelengths at the expense of power in the original signal, and (4) crosstalk between signals at different wavelengths and polarizations. The first two arise from self-phase modulation, while the third and fourth effects arise from stimulated Raman or Brillouin scattering or four-wave mixing. These last can be used when generation or amplification of additional wavelengths is desired. We note that in order to obtain an error-free system performance, due to the finite sensitivity of the optical receiver, the signal should have a high-enough level. On the other hand, the nonlinear effects are proportional to the product of the signal power, P, and the transmission distance L. This means that by increasing the signal level, the distance and the transmission bit rate (speed), all the problems—dispersion, noise, and nonlinearity in the fiber are increased, too. It is worth noting that the signal distortions is an issue when the nonlinearity is involved, because it can couple a number of detrimental effects together, such as dispersion, noise, polarization mode dispersion, polarization-dependent loss/gain, etc. [2–4].

SRS is a nonlinear process, observed for the first time in 1962, which lies at the heart of fiber Raman amplifiers and fiber Raman lasers (FRAs and FRLs). Direct fiber optical amplification obtained by SRS, called Raman amplification, was demonstrated in the early 1970s. In Raman amplification, a power transfer from pump (s) to information carrying signals (usually described as probes) can occurs, if there is a sufficient pump power within the fiber. In principle, when optical pump energy is added along with signals in ordinary optical fibers, optical amplification can take place, providing low-noise, flat and wideband signal gain. The more attractive advantage is that FRAs, operating in a signal band outside the EDFA bands, could open new transmission windows in the future [2–4].

High-power fiber lasers have had a significant development, achieving output powers of multiple kilowatts from a single fiber. Due to its inherent material advantages, Ytterbium has been the primary rare-earth-doped gain medium, so fiber lasers are largely confined to its narrow emission wavelength region. It is worth noting that because of its Raman-shifted out, SRS is a workable method for generating coherent radiation at new frequencies. Taking advantages of the technology of high-power lasers, which serve as their input, FRLs can lead to conversion to wavelengths higher than the starting wavelength, using a series of Raman–Stokes shifts. In fiber-lasers, the tight spatial confinement of pump light, which is maintained over relatively large distances, significantly lowers threshold pump powers down to practical levels and enables continuous-wave operation. The main advantage of Raman laser is that essentially any laser wavelength can be achieved from the ultraviolet to the infrared with a suitable choice of the pump wavelength, providing that wavelengths are within the transparency region of the material and sufficiently high nonlinearity and/or optical intensity are reached. For this reason, currently, FRLs are the only wavelength scalable, high-power fiber laser technology that can span the wavelength spectrum [2–4].

Due to the extent of subject, a comprehensive review including mathematical [5–8] and physics aspects of SRS [9–14], architecture of fiber system [15–19] and laser systems would be impossible to realize. Therefore, the aim of our paper is to provide an overview of the field, emphasizing physical effects and working principles of fiber optics amplifier and laser based on SRS. This approach could have the advantage to provide a quick look of the state of art and to allow to scientists, who are new to Raman amplification, to go into the field. In addition, a comprehensive list of references is also provided for readers who wish to pursue any of the topics in more detail.

The paper is organized as follows. In next paragraph, for the sake of completeness an essential theoretical background is reported. In Section 3, Raman amplifiers are described, which are generally divided into two categories, namely, distributed and lumped amplifiers. Of course, the two classes can be combined to form a hybrid amplifier, where the discrete FRA can be replaced by an EDFA [20]. In Section 4, Raman lasers in fiber are introduced, which can be classified into two general categories. In the former, the wavelength is shifted by one Raman–Stokes shift, while, in the latter, called cascaded Raman laser, the wavelength is shifted by multiple Raman–Stokes shifts. In the last section, a brief introduction about Raman soliton laser is reported, too.

2. Theoretical Background

SRS can be obtained by irradiating a sample with two simultaneous light sources: a light wave at frequency ω_L (the pump laser wave) and a light wave at frequency $\omega_S = \omega_L - \omega_v$ (the Stokes Raman wave), where $\hbar\omega_v$ corresponds to a vibrational energy. In SRS phenomenon, when the frequency difference between the pump and the Stokes laser beams matches a given molecular vibrational frequency of the sample under test, there is a transfer of energy from the high power pump beam to the probe beam, which can be co-propagating or counter-propagating. The SRS effect occurs in the form of a gain for the Stokes beam power (stimulated Raman gain, SRG) and a loss for the pump beam power (stimulated Raman loss, SRL), see Figure 1a. It is worth noting that in SRS, because of its coherent nature, the molecular bonds oscillate with a constant phase relation (see Figure 1b) and interfere constructively within a certain macroscopic area (e.g., inside the focus area of laser beam), therefore the SRS signal can be orders of magnitude more sensitive than spontaneous Raman scattering [9–12].

In SRS, high-order Raman sidebands can be produced by a field propagating through a medium optically polarizable. Many Stokes frequencies can be generated at the output when the pump power exceeds the threshold value, thus Stokes frequencies at $\omega_P - \omega_v$, $\omega_P - 2\omega_v$ and of anti-Stokes frequencies at $\omega_P + \omega_v$, $\omega_P + 2\omega_v$ can be observed. If the intensity of the first-Stokes wave is enough high, it can generate a "second-Stokes" beam (see Figure 2). By iterating this process, higher Stokes order can be generated, leading to the so-called "cascaded" SRS. The intensity of the ith Stokes order thus depends on the conversion rate from the $(i-1)$th Stokes order (proportional to $g \cdot I_i \cdot I_{i-1}$) and loss rate to the $(i+1)$th order (proportional to $g \cdot I_i \cdot I_{i+1}$) where g is the Raman gain of material. We note that frequency conversion can be obtained over a wide range of output wavelengths (visible and UV) within a single Raman laser system. For example, it is possible to use standard diode-pumped crystalline solid-state lasers to generate several Stokes order by cascading effect [9–14].

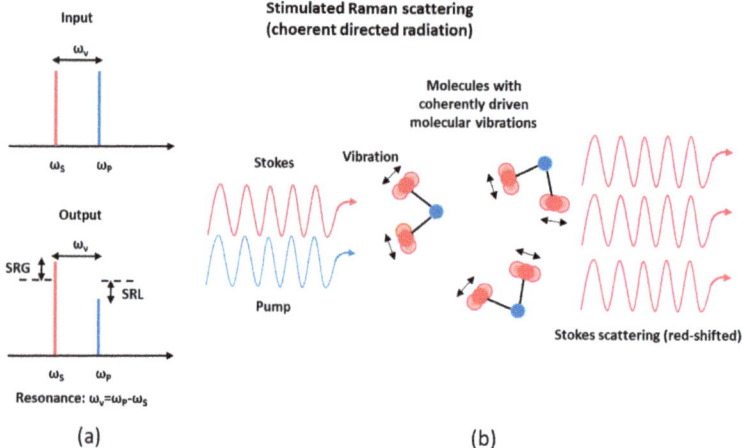

Figure 1. Stimulated Raman-Scattering (SRS) principle. (**a**) Pump–probe modalities associated with the SRS process are pointed out: SRG, stimulated Raman gain; SRL, stimulated Raman loss. (**b**) Stimulated Raman scattering occurs through inelastic scattering of probe photons off from vibrationally excited molecules that interfere coherently.

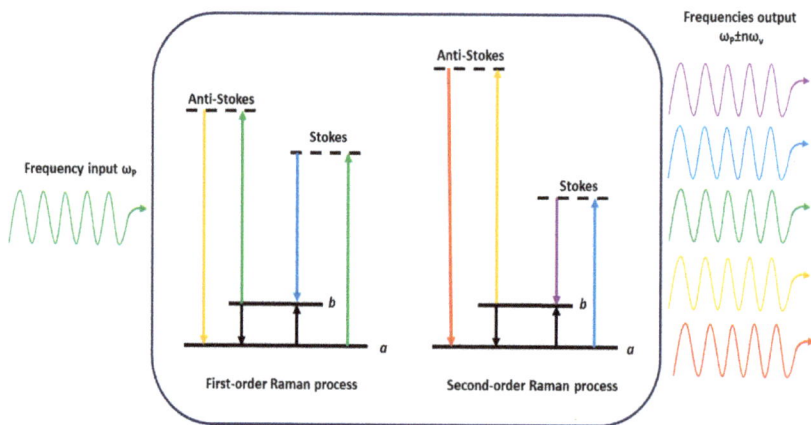

Figure 2. Cascaded Raman scattering principle. An input frequency ω_P (green) can cause the spontaneous emission of frequencies $\omega_P \pm \omega_v$, [Stokes (blue) and anti-Stokes (orange)]. This initial spontaneous event provides the seed photons necessary for subsequent first-order SRS (**left**). The first-order Stokes and anti-Stokes photons then can cause second-order Raman scattering (**right**) to produce new frequencies $\omega_P \pm 2\omega_v$ (red and violet) and so on through propagation, resulting in a broadband ladder of frequencies $\omega_P \pm n\omega_v$.

Optical fibers are an excellent medium for utilizing SRS. There are two key parameters for performances optimization of Raman fiber amplifier and laser. The first, linear losses are fundamentals for enabling long interaction lengths. We note that different types of fibers have different low-loss wavelength windows. For example, the silica fibers, which are used in the majority of optical fiber applications, have extremely low losses (<1dB/km) in the near-IR region, whereas they are very lossy in the mid-IR region. The second, effective area: Because of their small core diameter <10 µm, single-mode fiber has the ability to confine light to small mode areas, which significantly enhances nonlinearity. We note that in silica glass, in spite of intrinsically small values of the third-order nonlinear coefficients, which are smaller by a factor of 100 or more compared to many crystals and liquids, nonlinear effects in optical fibers can be observed at relatively low power levels.

In the past century, fused silica has been the main material used for long and short-haul transmission of optical signals, because of its good optical properties and attractive figure of merit (trade-off between Raman gain and losses). Since in amorphous materials, molecular vibrational frequencies spread out into bands that overlap and create a continuum, in silica fibers, the Raman gain ($g_R(\Omega)$) spectrum extends over a large frequency range (up to 40 Thz). The maximum of the Raman gain spectrum in silica fibers is downshifted from the pump frequency by about 13.2 THz (440 cm^{-1}), corresponding to 100 nm in the telecom window. Therefore, being the signal wavelength usually around 1550 nm, the pump light wavelength has to be about 1450 nm. We note that at high enough powers there can be lasing at all frequencies with sufficient Raman gain [21–27], though the peak gain is at a specific frequency shift. The following fundamental properties of Raman gain are critical for designing Raman fiber amplifiers and lasers:

(1) Raman gain has a spectral shape that depends primarily on the frequency separation between a pump and signal, not on their absolute frequencies. This follows from energy conservation: Their frequency separation must be equal to the frequency of the created optical phonon;

(2) Since Raman gain does not depend on the relative direction of propagation of pump and signal, SRS occurs almost uniformly for all the orientations between the pump and signal propagation direction, as a consequence FRAs can work for both the copropagating, counter-propagating or bidirectional pumps with respect to the signal;

(3) SRS is a fast process, response time of fused Silica is evaluated to be less than 100 fs [28,29]. For most applications, this appears instantaneous, and especially in relation to the application as Raman amplifiers in optical communications systems;

(4) Raman gain is polarization dependent. Raman gain coefficients are usually quoted for parallel linearly polarized pump and Stokes fields. For perpendicular polarizations the gain is more than an order of magnitude smaller than the parallel one. In typical long fibers, which do not maintain linear polarization, the Raman gain will assume some average value which is approximately half of the polarized gain [30,31];

(5) Peak Raman gain in a specific fiber can be different based upon the material composition. For example, in phosphosilicate fiber the peak gain is at 1330 cm^{-1}. This larger Raman shift is attractive for many applications, since fewer Stokes shifts are needed to reach the final desired wavelength.

Concerning materials, two problems are of great interest: (1) Development of glasses with a higher Raman gain coefficient; (2) increase of Raman gain bandwidth [32–36]. The increase of dopant content in Ge- and P-doped silica fibers with a simultaneous reduction of optical losses seems the most straightforward solution to the first problem. The second option is to use multicomponent glasses, including heavy-metal oxides-doped glasses, which also permits the development of glasses with the Raman gain bandwidth of several hundreds of cm^{-1} [37–42]. However, in multicomponent glass approach the current limitation is to obtain low-loss fibers.

In fiber communication systems, two basic issues are related to SRS. First, pump-to-Stokes coupling provides a mechanism for crosstalk from short- to long-wavelength channels, which is more efficiently when the channel frequency spacing is close to the one associated with the maximum Raman gain [43]. In wavelength division multiplexed (WDM) systems, within the 1.53- to 1.56-µm erbium-doped fiber amplifier window, channel spacings on the order of 100 GHz are used, thus due to the Raman gain peak at approximately 500 cm^{-1}, Raman gain is considerably reduced, but is still sufficient to cause appreciable crosstalk, which can lead to system penalties depending on the number of channels [44]. Second, the conversion to Stokes power from the original signal—a mechanism by which signal power can be depleted. A related problem is walkoff [45] occurring between the signal and Stokes pulses, since these will have different group delays. If pulses are of sub-picosecond widths, additional complications arise due to the increased importance of SPM and cross-phase modulation (XPM) [6].

3. Fiber Raman Amplifiers (FRAs)

SRS in a pulsed mode was observed, for the first time, in a liquid core fiber by Ippen in 1970 [46]. The first SRS observation in a single-mode fiber (SMF) was made by Stolen et al. in 1971 [47], whereas the first SRS observation in continuous wave (CW) mode was obtained by Stone [48] in a hollow fused silica fibers filled with benzene and with a pump power value less than 100 mW. All these authors pointed out the fiber geometry allows SRS amplification/generation, due to the confinement of high power pump power density over long distances [49].

The fundamental advantages of FRAs are:

(1) They do not require special dopants. This means that ordinary, transparent, passive optical fibers can be turned into a Raman amplifier;

(2) Raman gain is obtainable in any conventional transmission fiber, which can be used as transmission line and as Raman gain medium, too. Therefore, Raman amplification is compatible with most available transmission systems;

(3) Raman amplification can be provided at any wavelength. Being the Raman gain non-resonant, it is available over the entire transparency region of fiber ranging from approximately 0.3 to 2 micron, if the appropriate pump sources are available.

(4) The broad gain bandwidth, obtained in conventional transmission fibers, is an important property for WDM systems.

(5) Raman gain has a high-speed response. Therefore, in principle, in fiber the entire bit stream can be amplified without any distortion in high bit rate systems.

3.1. Distributed Fiber Raman Amplifiers

In a distributed FRA, pump at the optical frequency ω_P and signal at the frequency $\omega_s = \omega_P - \omega_v$ are jointly launched into the same ordinary optical fiber. Raman optical amplification takes place when their frequency difference $\Omega = \omega_P - \omega_s$ lies within the bandwidth of the Raman-gain spectrum. The distributed FRA guides light at both the signal and pump wavelengths and it is normally single mode to ensure the best overlap of all traveling waves. A distributed FRA unit is a pumping unit at 14xx nm without gain medium and it, usually, consists of pump lasers, optical couplers to combine them, isolators and a few monitor taps, as well as the WDM coupler for signal and pump. Depolarizers are necessary when the pump is polarized, in order to suppress the polarization dependence of Raman amplification.

Figure 3 shows a configuration in which pumping light propagates bi-directionally in distributed FRA. Moreover, the system is provided with band pass filter (BPF) and arrayed waveguide grating (AWG) devices which act as demultiplexing unit in the receiving side.

As mentioned before, in a distributed FRA both forward and backward pumping schemes are possible. In the forward pumping, the signal is amplified near its fiber-input end, where the signal power is relatively high. Thus, this scheme corresponds to a booster amplifier. In the backward pumping, the optical signal is amplified near its fiber-output end before photodetection, where the signal power may be so weak as to be comparable to the noise light power. Since this scheme is used to improve detectivity or to reduce the minimum detectable signal power, it can be considered as a detection preamplifier [50,51].

We note that the process of Raman amplification takes place so rapidly that, unless the intensity noise of the forward pumping light is sufficiently small, the pumping light noise will be transferred to the signal light resulting in increasing transmission bit error rates. Since signals propagating in the opposite direction to the pump will average over the fluctuations, in order to mitigate unwanted crosstalk between the pump and the signal, in many cases only backward pumping is used [52,53]. Typically, high-powered counter propagating Raman pumps are deployed in conjunction with discrete amplifiers, such as EDFAs.

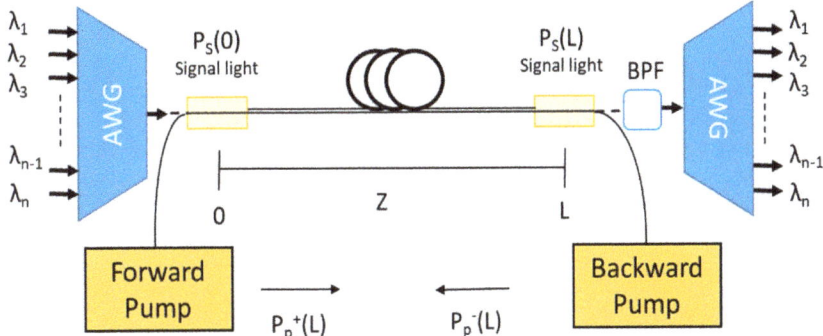

Figure 3. Schematic representation of multiplexing/demultiplexing based distributed optical Raman amplifier.

Distributed FRAs offer several advantages: simplicity, because they provide direct optical amplification of signal light in the transmission fiber; flexibility in the use of signal wavelengths, because the Raman gain peak is dependent on the pump wavelength and not the emission cross section of a dopant; broad gain bandwidth, which allows to employ multiple pumps. However, one of the major benefits, only obtainable through the use of distributed amplification, is that signal gain may be pushed into a transmission span preventing the signal from decaying as much as it otherwise would have if no amplification was provided within the span. As a consequence, the signal-to-noise ratio does not drop as much as it would have in a system based on transmission through a passive fiber followed by a discrete amplifier (see Figure 4). Thus, lower signal powers can be used, nonlinear penalty are reduced and higher loss can be tolerated. This improved noise performance may be used in different ways. One way is to extend the reach between repeaters, another is to extend the total reach of a transmission system, and a third one is to improve the transmission capacity.

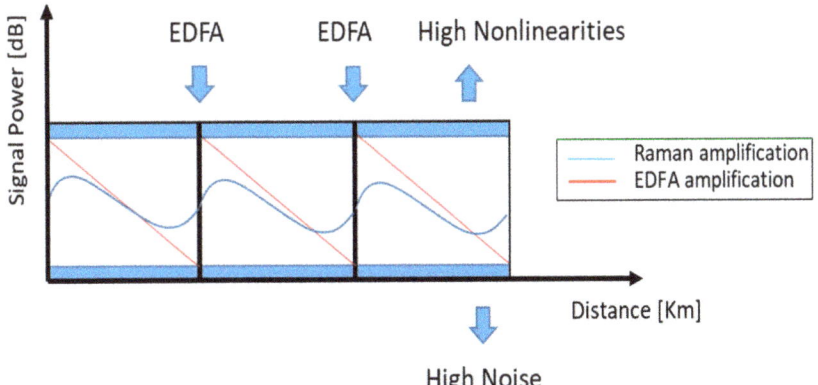

Figure 4. Ideal distributed amplifier: the loss is counterbalanced at every point along the span, leading to an improvement of the signal to noise ratio (SNR) respect to (ideal) discrete amplification.

Using distributed FRAs, the transmission fiber becomes an amplifier, giving an important advantage but, at the same time, the optimization of the fiber design with respect to amplifier performance becomes more difficult due to the need to optimize the fiber for signal transmission, too. Several parameters have to be taken into account to evaluate the performance of a distributed FRAs, such as linear noise accumulation, pulse distortion due to group velocity dispersion, effects due to Rayleigh scattering, nonlinear interactions between pump and signal channels, and pump–signal crosstalk. Considering that in distributed amplifiers the gain is gathered over tens of kilometers and that the Raman effect is a very fast process (response time in the order of femtoseconds), these noise sources become significant.

In Ref. [54] a 25.6 Tb/s experiment was demonstrated over a 240 km repeated span by employing 160 WDM channels with 50 GHz channel spacing in the C+L bands. The authors used backward Raman pumping to increase the received optical signal to noise ratio (OSNR) and simplify the optical repeaters, while an optical equalizer was used in the receiver to minimize channel distortion caused by narrow optical filtering.

3.2. Discrete Fiber Raman Amplifiers

Lumped or discrete FRAs use a fiber medium that is localized before or after transmission to fully or partially compensate for the transmission loss. Unlike a distributed FRA, all of the pump power is confined to the lumped element (see Figure 5). In this particular case, counterpropagating pump power is confined within the unit by the use of isolators surrounding the amplifier. Compared with Figure 3, no pump power enters the transmission line.

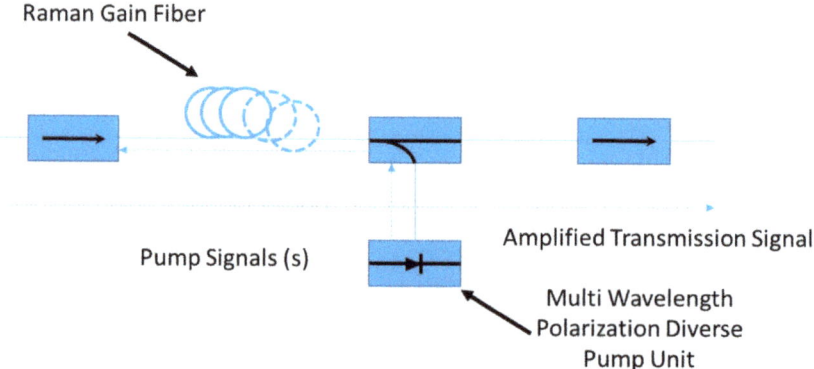

Figure 5. Discrete fiber Raman amplifier (FRA) scheme with pump power confined to the lumped element.

In discrete FRAs, the used fiber is a lot shorter (order tens instead of hundreds of km) compared to distributed FRAs. Since gains accumulate with fiber length, Raman amplification better fits in the distribuited FRAs class. Therefore, discrete FRAs are primarily used to increase the capacity of fiber-optic networks, opening up new wavelength windows, which are inaccessible by EDFAs. In FRAs, the signal is amplified at a wavelegth given by the frequency difference between the pump and the Stokes frequencies, thus by choosing the pump wavelength, gain at any wavelength can be obtained. Moreover, by combining multiple pump wavelengths, the shaping of the gain spectrum can be achieved, whereas by combining several pumps centered at different wavelengths, a flat gain in an ultrawide bandwidth is allowed.

The latter issue, i.e., a flat gain, is highly desirable in WDM systems in order to equally amplify all the channels. For this reason, flat gain of broad-band Raman amplifier, is considered one of the most important goal for the optimization of fiber communications. This task can be obtained by simultaneously using pumps at different wavelegths that generate Raman gain lightly shifted from each other so as to partly overlap each other leading to a composite gain (see Figure 6). The final Raman gain can be very flat over a wide range of wavelegth by accurately designing each pump wavelength and power. The gain-band expansion that can be achieved by multi-wavelength pumping is determined by the magnitude of the Stokes shift. A large Stokes shift is therefore desirable for a wideband FRAs using multi-wavelength pumping [55]. The maximum amplification bandwidth of a FRA is about 100 nm for silica fiber in the 1550-nm band, so the composite gain could cover a nearly 100-nm signal band [56]. They are particularly useful for amplifying (dense) WDM systems.

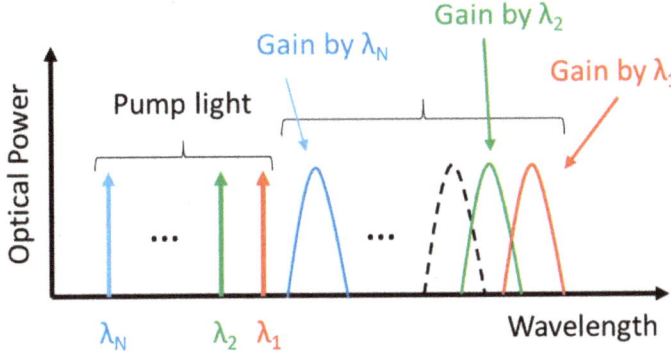

Figure 6. Multi-wavelength pumping technique for wideband Raman amplifiers.

Using a broadband pumping approach, amplifiers with gain bandwidths greater than 100 nm have been demonstrated [57] with average net gain of 2 dB gain for SMF, while for dispersion-shifted fiber and reverse dispersion fiber the net gain was of 6.5 dB. A 12- wavelength-channel WDM LD (high-power laser diode (LD) pumping unit), with a range of wavelength from 1405 to 1510 nm and combined by an Il-MZI-integrated, using a planar lightwave circuit technology, were used to enlarge the gain bandwidth of Raman amplifiers [57] However, due to a strong interaction between pumps, these broadband FRAs typically need of more power at the shortest wavelengths pumps which amplify the longer wavelegth pumps, leading to at an increase of the noise properties.

Dispersion-compensating Raman amplifiers integrate two crucial tasks, dispersion compensation and discrete Raman amplification, into a single component [58–60]. Dispersion compensating fiber (DCF) devices, based on typical dispersion for the propagating mode, are usually in the order of several kilometers and, as a result, show important insertion loss. Hansen et al. demonstrated that Raman gain produced in the DCF was as a simple and attractive method to achieve lossless DCF modules, without affect system performance [58]. Due to a small mode-field diameter, a very low pump powers are required, like those available from semiconductor laser diodes. The DCF shows a net gain higher than loss, resulting in a wider system margin and the possibility to introduce other elements such as optical add-drop multiplexers into the system. In Ref. [61] a polarization-combined four channel WDM 14xx nm laser diode unit was used as pumping light source to demonstrate a broadband Raman amplification in a DCF. A fully compensation of the losses in the DCF were obtained by means of the Raman gain over 50 nm bandwidth within +/− 0.5 dB variation.

4. Fiber Raman Lasers (FRLs)

FRLs are similar to ordinary lasers. They basically involve three key components: A low-effective-area and high-nonlinearity optical fiber for providing Raman gain, mirrors for feedback and components for isolation between the pump laser (usually a rare-earth-doped fiber laser) and the Raman resonator. A first analogy is that, in FRLs, lasing occurs when the Raman-active gain medium is placed inside a cavity, for example, between mirrors reflecting the first Stokes wavelength. A second analogy is that, in FRLs, the threshold power is obtained when Raman amplification during a round trip is as large as to compensate the cavity losses. However, there are also some important differences between FRLs and traditional lasers. A first one is that an amplifier medium based on Raman gain is used rather than on stimulated emission from excited atoms or ions. A second difference is that the required wavelength for pumping Raman laser does not depend on the electronic structure of the medium, so it can be chosen to minimize absorption. In addition, it is worth highlighting the unique property of Raman laser, i.e., they can be operated at several wavelengths simultaneously. Higher-order Stokes wavelengths can be generated inside the active medium at high pump powers and they can be properly dispersed spatially in association with separate mirrors for each Stokes beam [62].

The first continuous-wave Raman laser in fiber was demonstrated in 1976 [63,64]. Pumping at the watts level, a distributed Raman gain, higher than the linear attenuation plus losses at splice/connector points, was induced in a single-mode glass fiber, placed inside a Fabry-Perot cavity. SRS process occurred in fiber resulting in an intense output, due to wavelength-selective feedback for the Stokes light of the Fabry-Perot cavity, which was formed by two partially reflecting mirrors. Higher-order Stokes wavelengths were generated inside the fiber at high pump powers, too. The spatial dispersion of various Stokes wavelengths through an intracavity prism allowed tuning of the laser wavelength.

A single wavelength shift FRL is a fiber resonator at the Stokes wavelength, in which SRS shifts the spectrum of the propagating pump radiation through an optical fiber towards lower frequency Stokes components. Raman lasing is obtainable in conventional single-mode telecom fibers (e.g., SMF-28) as well as other passive fibers, by trapping Stokes components by reflectors and by pumping the laser by a high-power rare-earth-doped fiber laser (commonly ytterbium). In an all-fiber configuration, enough positive feedback for lasing at the first Stokes wavelength is easily achieved. Directionality to the laser emission is provided by a high-reflecting fiber Bragg grating (reflectivity > 99%) on the input side and by a low-reflectivity output coupler (at Stokes wavelength) on the output side. Fiber Bragg gratings (FBGs) are narrow-band reflectors, acting as the mirrors of the laser cavity and inscribed directly into the core of the optical fiber used as the gain medium. However, we note that in Raman laser, when Raman gain of the active medium is very large, the laser can build up from noise to a substantial signal without any feedback. In addition, a single wavelength shift FRL, with a broadband flat Raman gain profile can be obtained using multiple pump wavelengths. Figure 7 shows a schematic diagram of a specific Raman fiber laser with one wavelength shift. In this particular case, the laser uses a 1117 nm ytterbium-doped fiber laser and converts it to 1178 nm.

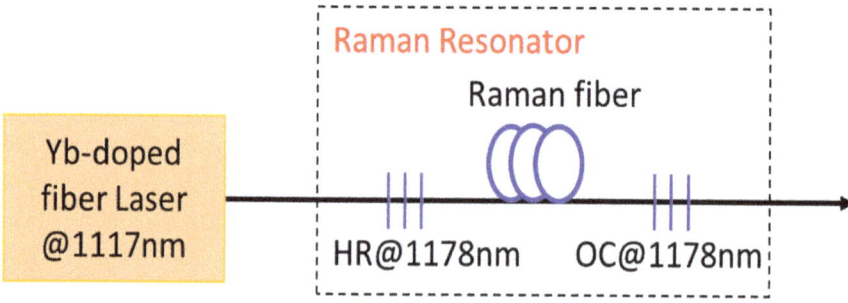

Figure 7. Schematic representation of a Raman fiber laser with one wavelength shift. (HR—high-reflectivity fiber Bragg grating, OC—output coupler—a low-reflectivity fiber Bragg grating).

A multiple wavelength shifts FRL takes advantage of SRS cascading. The pump light gives rise to the "first-order" laser light in a single frequency-shifting step, which remains trapped in the laser resonator. Afterwards, the "first-order" laser light can be pushed to very high power levels becoming itself the pump for the generation of "second-order" laser light, which is shifted by the same vibrational frequency of first order. Using this technique, conversion of the pump light (typically around 1060 nm) to an "arbitrary" desired output wavelength through several discrete steps can be performed by using a single laser resonator. In Figure 8 is schematically reported the wavelength conversion over two or more Stokes shifts obtained by using a cascaded Raman resonator with fiber Bragg gratings incorporated in it, in order to realize nested cavities at each of the intermediate wavelengths. These intermediate wavelengths are chosen close to the peak of the Raman gain of the respective preceding wavelength; finally, a low-reflectivity output coupler terminates the wavelength conversion. The resulting output is at the designed final wavelength while only small fractions are at the intermediate wavelengths.

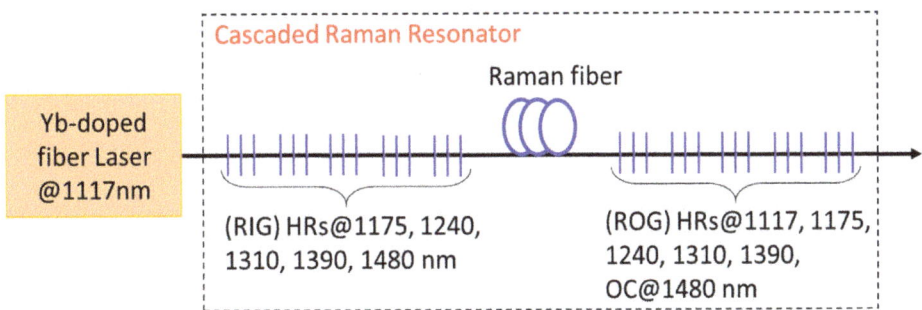

Figure 8. Schematic representation of a Raman fiber laser with multiple wavelength shifts by using a cascaded Raman resonator (RIG—Raman input grating set, ROG—Raman output grating set).

Tunability is an important property for lasers. The first tunable FRL, using a linear cavity configuration, was reported by Jain et al. [65]. A tuning range of 8 nm with an efficiency of 9% were demonstrated. However, the bulk optics (prisms and rotating mirrors), placed inside the laser cavity for tuning, introduced mechanical instabilities as well as coupling losses, which gave rise to high power thresholds and low efficiencies. In order to eliminate losses and to make a compact device, the approach based on bulk optics was replaced by an all-fiber design. In such a design, cavity mirrors were integrated within the fiber by the development of tunable fiber Bragg gratings [66]. In 1988, the first FRL based on fiber Bragg gratings was reported [67].

Since the Raman gain value of germanosilicate fiber is about eight times higher than for silicate fiber, these fibers are extensively used in FRLs. Due to their peak Raman shift of 440–490 cm^{-1}, in an FRL with a 1064 nm pump source, the third and the sixth Stokes orders produce outputs at 1239 nm and 1484 nm, respectively. So, in order to make a FRL at 1484 nm, one would need six cascaded cavities. In Ref. [68], a GeO–doped high-power and widely tunable all-fiber Raman laser was demonstrated. The FRL, using a linear cavity configuration and a purely axial compression of the FBGs, was continuously tuned over 60 nm, from 1075 to 1135 nm. A Stokes output power up to 5 W, for 6.5 W of launched pump power was provided. A high efficiency over the whole tuning range, from 76.1 to 93.1% and laser thresholds varying from 0.78 to 2.59W were obtained, respectively.

It is worth noting that the peak Raman shift of low-loss phosphosilicate (P_2O_5-SiO_2) fiber is 1330 cm^{-1}, so when they are pumped by a 1064 nm pump source, the first Stokes (S1) and the second Stokes (S2) orders occur at 1239 nm and 1484 nm, respectively. This means that to make a FRL at 1484 nm, only two cascaded cavities would be required, thereby greatly increasing the FRL efficiency. In Ref. [69], a CW high-power Raman fiber laser was reported, which was formed by a CW 8.4 W/ 1064 nm Yb-doped double-clad fiber laser as a pump, a 700 m phosphosilicate fiber and cascaded cavities with two pairs of FBG mirrors for the first and the second Stokes orders. A maximum output power of 2.24 W and slope efficiency of 32.8% at 1484 nm were demonstrated. In Ref. [70], Babin et al. demonstrated a simple all-fiber widely tunable phosphosilicate FRL of high efficiency. The laser had more than 50 nm tuning range and generated up to 3.2 W of output power with 72% maximum slope efficiency. The output power was almost constant in the range 1258–1303 nm.

Typically, the operating wavelengths of Raman laser are in the 1–1.5 µm region of the spectrum and have been created using silica glass as the base material with the desired application directed toward telecommunications. However, there are a large number of applications requiring high power laser with wavelengths longer than 2 µm. As a result of the versatility offered by FRLs, these types of sources are attractive for this purpose [71]. Chalcogenide glass fibers have a Raman gain coefficients approximately two orders of magnitude greater than the gain coefficients of silica. For wavelengths longer than 1.8 µm, the development of Raman fiber lasers based on chalcogenide glasses has now become technically achievable [72]. In Ref. [73] a Raman fiber laser, using a nonoxide glass as the gain medium and a 2 µm fiber laser pump source was demonstrated. Due to the low optical losses and high nonlinear index of commercially available glassy As_2Se_3 fiber, a combined output of 1 W at the first and second Stokes wavelengths of 2062 and 2074 nm, were generated respectively.

Single-frequency laser sources have a great number of applications, but are difficult to implement, suffer from poor robustness, poor quality (linewidth and stability) and are expensive. In recent years, there has been much interest in rare-earth doped distributed feedback (DFB) fiber lasers, due to their very low noise and high efficiency [74]. The oscillating wavelength can be accurately defined by a Bragg grating written directly into the fiber and inserting a π phase-shift (PS) at, or close, to the center of the Bragg grating structure. Of course, the operation of π-PS-DFB fiber lasers has been mostly limited to the spectral bands of rare-earth dopants [75].

A Raman gain based DFB fiber laser would have a number of potential advantages. First, the possibility of generating narrow linewidth low-noise oscillation in wavelength bands outside the one of rare-earth doped materials. Second, Raman based fiber laser systems do not suffer from issues associated with high-concentration rare-earth doped fibers, which limit their efficiency due to thermal effects. Using Raman gain in DFB fiber lasers was proposed for uniform FBGs by Perlin and Winful [76] for the first time, while a Raman DFB fiber laser based on a relatively short (<20cm), π-shift fiber Bragg grating were proposed by Hu et al. [77]. Westbrook et al. have realized the first functional DFB using the Raman gain in fiber, but with a very high threshold (40 W) was observed at 1584 nm. The authors showed that watt-level thresholds can be achieved by using a small-core high-numerical aperture (NA) fiber [78]. Shi et al. demonstrated that a polarized pump and an ultralong FBG (30 cm) could yield sub-watt threshold at a lasing wavelength of 1120 nm [79].

In reference [80] highly efficient Raman DFB fiber lasers with CW output power up to 1.6 W was demonstrated, for the first time. The DFB Bragg gratings were written directly into two types of commercially available passive germano-silica fibers. Two lasers of 30 cm length were pumped with CW power at 1068 nm up to 15 W. The threshold power was ~2 W for a Raman-DFB laser written in standard low-NA fiber, and only ~1 W for a laser written in a high-NA fiber, both of which oscillate in a narrow linewidth at ~1117 nm and ~1109 nm, respectively. The slope efficiencies were ~74% and ~93% with respect to absorbed pump power in the low-NA fiber and high-NA fiber respectively. Recently a Raman-DFB with lower thresholds at 1120 nm and 1178 nm was demonstrated [81]. Despite these developments, the performance in terms of slope efficiency has remained poor (1–7% for single-ended output) compared to the expected simulated efficiencies (40–50%).

By exploiting the multiple scattering of photons in a disordered gain medium, a random laser can be obtained allowing a coherent light source without a traditional cavity [82]. For different scientific and medical applications, ultrafast Raman fiber lasers are an interesting option. An efficient way to obtain a high power ultrafast Raman fiber laser is pulse pumping, but it requires a real-time synchronization between the pump pulses and laser cavity. In order to overcome this limitation, recently, random fiber lasers with distributed Raman gain and Rayleigh feedback in standard telecommunication optical fibers were demonstrated [83,84]. Their advantages are a simple and flexible design, quasi-CW operation, narrow spectrum generation, high beam quality and pump energy conversion efficiencies comparable to the ones of conventional cavity lasers [85]. In recent years, many works have been reported on the mechanism [84,85], the improvement of performance including power scaling [86,87], wavelength tuning [88,89], and Q-switching [90,91]. Nowadays, random fiber is a unique laser technology with various advantages in the fields of fiber sensing and optical communications [92,93].

A promising approach for the further development of devices based on SRS is represented by Photonic crystal fiber (PCF). The most important advantage of this technology is the reduction of the fiber length and the power levels required. The first preliminary results in photonic crystal fiber for Raman amplification and for an all optical Raman modulator was presented by Yusoff et al. [94]. An analysis of the Raman properties in triangular photonic crystal fiber was reported by Fuochi et al. [95], which investigated the influence of presence of the germanium doped core on the Raman properties. A model of PCF amplifiers was developed by Bottaccini et al. [96], which demonstrated that germanium doped PCF can enhance the gain for the same pump power. Finally, SRS in ethanol core PCF was demonstrated by Yiou et al. [97]. The approach of PCF with liquid could open new perspectives for SRS nonlinear properties with applications in optical sensing.

5. Raman Soliton Laser

Nonlinear effects are usually favored when pulsed operation are used, since high peak powers can be achieved with modest average powers. During their propagation inside optical fibers, short optical pulses (less than 10 ps) are affected by combination of chromatic dispersion and by the nonlinear effects [2–4], giving rise to a variety of phenomena, such as supercontinuum generation and soliton formation.

In the simplest situation, a single input pulse at the carrier frequency ω_0 excites a single mode of the fiber. Each spectral component of the input field propagates as a plane wave and acquires a slightly different phase shift because of the frequency dependence of the propagation constant $\beta\omega$. For this reason, it is useful to expand $\beta\omega$ in a Taylor series around the carrier frequency ω_0, and, depending on the pulse bandwidth, the second-order dispersion term (group velocity dispersion (GVD)) can be considered or third- or higher-order dispersion terms can be included. The GVD parameter (called β_2) can be positive or negative with values in the range of 0.1–20 ps^2/km, depending on how close the pulse wavelength is to the zero-dispersion wavelength of the fiber. When the dispersive term can be neglected, spectral changes induced by SPM are a direct consequence of the time dependence of nonlinear phase shift. The time dependence of $\delta\omega$ is referred to as frequency chirping. The chirp induced by SPM increases in magnitude with the propagated distance. In other words, new frequency components are generated continuously as the pulse propagates down the fiber. The situation changes drastically when both the GVD and SPM are equally important and must be considered simultaneously inside the fiber. In the case of normal dispersion ($\beta_2 > 0$), both the pulse shape and spectrum change as the pulse propagates through the fiber and the combined effects of GVD and SPM can be used for pulse compression.

An interesting situation occurs for anomalous dispersion ($\beta_2 < 0$), where the generation of optical solitons can be obtained. Solitons are special types of optical wavepackets formed by the balance between nonlinearity (positive SPM) and dispersion (and/or diffraction). This could be surprising, since GVD affects the pulse in the time domain while the SPM effect is in the frequency domain. However, a small time-dependent phase shift added to a Fourier transform-limited pulse does not change the spectrum to first-order. If this phase shift is cancelled by GVD in the same fiber, the pulse does not change its shape or its spectrum as it propagates. In the context of optical fibers, the use of solitons for optical communications was, for the first time, suggested in 1973 [98] and observed in an experiment in 1980 [99]. Temporal solitons have recently been observed in optical fiber waveguides [98–100] in laser resonators [101], and in dielectric microcavities [102]. In each of these cases nonlinear compensation of GVD is provided by the Kerr effect.

Optical communication systems are often limited by fiber dispersion that broadens the pulse and by fiber losses. The use of fundamental solitons as an information bit solves the prolem of dispersion, because in a fiber channel, nonlinear phase modulation can compensate for linear group dispersion leading to pulses that propagate without changing temporal shape and spectrum. This mutual compensation of dispersion and nonlinear effects takes place continuously with the distance in the case of "classical" solitons and periodically with the so-called dispersion map length in the case of dispersion-managed solitons. Due to the fiber losses, soliton width begins to increase because of a decrease in the peak power during propagation inside the fiber, so amplification is required in order to recover original width and peak power. A proposed scheme makes use of SRS [103–106]. Figure 9 showes the basic idea. Solitons are launched into a fiber link consisting of many segments of length L. At the end of each segment, pump lasers inject CW light, upshifted in frequency from the soliton carrier frequency by about 13 THz, in both directions through wavelength dependent directional couplers. Since the Raman gain is distributed over the entire fiber length, the soliton can be adiabatically amplified.

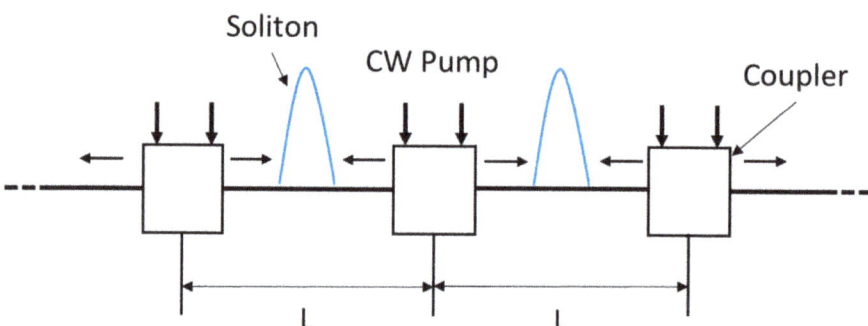

Figure 9. Schematic illustration of a soliton communication system. Solitons are amplified through SRS by injecting continuous wave (CW) pump radiation periodically.

During their propagation in optical fiber, soliton pulses are affected by SRS. The Intrapulse Stimulated Raman Scattering is a phenomenon that appears for short pulses with a relatively wide spectrum, in which Raman gain can amplify the low-frequency components of a pulse by transferring energy from the high-frequency components of the same pulse. As a result, the pulse spectrum shifts continuously toward the red side as the pulse propagates through the fiber [107,108]. This shift, called Raman-induced frequency shift (RIFS), was experimentally observed in 1986 [109], its Raman origin was also pointed out soon after [110], while a more general theory was developed later [7]. With the advent of microstructured fibers, much larger values of the RIFS (>50THz) were observed [111]. The RIFS is useful to generate Raman solitons whose carrier wavelength can be tuned by changing fiber length or input peak power. The effects of intrapulse Raman scattering can be dramatic in the context of solitons, where they lead to new phenomena such as decay and self-frequency shift of solitons.

Besides the Kerr nonlinearity, a secondary effect associated with soliton propagation is caused by Raman interaction. Equations describing SRS can be solved by inverse scattering methods and are found to have solutions. The so called Raman soliton was experimentally observed for the first time in 1983 [112]. When the wavelength of the pump pulse is close to or inside the anomalous dispersion region of an optical fiber, the Raman pulse should experience the soliton effects, i.e., under suitable conditions, almost all of the pump-pulse energy can be transferred to a Raman pulse that propagates undistorted as a fundamental soliton. We note that the soliton pulse is a bright soliton, while the Stokes pulse is a dark soliton and it has been demonstrated that quantum fluctuaction can induce them [113,114]. Raman solitons have also been generated by using a conventional fiber with the zero-dispersion wavelength near 1.3 µm led to 100-fs Raman pulses near 1.4 µm [115].

An interesting application of the soliton effects has led to the development of Raman soliton laser [116–121]. Such lasers provide their output in the form of solitons of widths 100 fs, but at a wavelength corresponding to the first-order Stokes wavelength., which can be tuned over a considerable range (10 nm). A ring-cavity configuration is commonly employed (see Figure 10).

A dichroic beamsplitter, highly reflective at the pump wavelength and partially reflective at the Stokes wavelength, is used to couple pump pulses into the ring cavity and to provide the laser output. In a 1987 experimental demonstration of the Raman soliton laser [122], 10 ps pulses from a mode-locked color-center laser operating near 1.48 µm were used to pump the Raman laser synchronously. Even though Raman soliton lasers are capable of generating femtosecond solitons useful for many applications, they suffer from a noise problem that limits their usefulness [123]. The performance of Raman soliton lasers can be significantly improved if the Raman-induced frequency shift can somehow be suppressed [124].

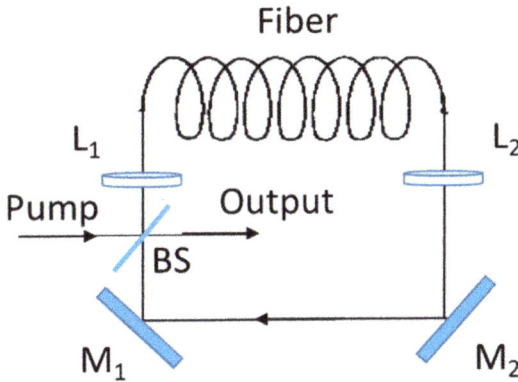

Figure 10. Schematic representation of the ring-cavity geometry implemented for Raman soliton lasers. BS is a dichroic beamsplitter, M_1 and M_2 are mirrors of 100% reflectivity, L_1 and L_2 are microscope objective lenses.

6. Conclusions

In this paper the state of the art of fiber Raman amplifier and fiber Raman laser is described. Today, Raman amplifiers are widely commercialized devices. Almost every long-haul (typically defined 300 to 800 km) or ultralong-haul (typically defined above 800 km) fiber-optic transmission system uses Raman amplification. On the other hand, nowadays, commercially available fiber-based Raman lasers can deliver output powers in the range of a few tens of Watts in continuous-wave operation. In addition, due to their tunability, compactness, and capability for multi-wavelength operation, FRLs have great commercial potential in a variety of applications.

Author Contributions: L.S. and M.A.F.; writing—original draft preparation, review and editing. All authors have read and agreed to the published version of the manuscript.

Funding: This research received no external funding.

Conflicts of Interest: The authors declare no conflict of interest.

References

1. Desurvire, E. *Erbium-Doped Fiber Amplifiers: Principles and Applications*; John Wiley & Sons: New York, NY, USA, 1994.
2. Agrawal, G. *Nonlinear Fiber Optics*, 5th ed.; Academic Press: Cambridge, MA, USA, 2012; ISBN 9780123970237.
3. Agrawal, G.P. Nonlinear fiber optics: Its history and recent progress. *J. Opt. Soc. Am. B* **2011**, *28*, A1–A10. [CrossRef]
4. Headley, C.; Agrawal, G.P. *Raman Amplification in Fiber Optical Communication Systems*; Elsevier: Amsterdam, The Netherlands; Academic Press: Cambridge, MA, USA, 2004.
5. Blow, K.J.; Wood, D. Theoretical description of transient stimulated Raman scattering in optical fibers. *IEEE J. Quantum Electron.* **1989**, *25*, 2665–2673. [CrossRef]

6. Headley, C.; Agrawal, G.P. Unified description of ultrafast stimulated Raman scattering in optical fibers. *J. Opt. Soc. Am. B* **1996**, *13*, 2170–2177. [CrossRef]
7. Santhanam, J.; Agrawal, G.P. Raman-induced spectral shifts in optical fibers: General theory based on the moment method. *Opt. Commun.* **2003**, *222*, 413–420. [CrossRef]
8. Lin, Q.; Agrawal, G.P. Vector theory of stimulated Raman scattering and its application to fiber-based Raman amplifiers. *J. Opt. Soc. Am. B* **2003**, *20*, 1616–1631. [CrossRef]
9. Armstrong, J.A.; Bloembergen, N.; Ducuing, J.; Pershan, P.S. Interactions between light waves in a nonlinear dielectric. *Phys. Rev.* **1962**, *127*, 1918–1939. [CrossRef]
10. Bloembergen, N.; Shen, Y.R. Multimode effects in stimulated Raman emission. *Phys. Rev. Lett.* **1964**, *13*, 720–724. [CrossRef]
11. Shen, Y.R.; Bloembergen, N. Theory of stimulated Brillouin and Raman scattering. *Phys. Rev.* **1965**, *137*, A1787–A1805. [CrossRef]
12. Bloembergen, N. The stimulated Raman effect. *Am. J. Phys.* **1967**, *35*, 989–1023. [CrossRef]
13. Sirleto, L.; Ferrara, M.A.; Nikitin, T.; Novikov, S.; Khriachtchev, L. Giant Raman gain in silicon nanocrystals. *Nat. Commun.* **2012**, *3*, 1–6. [CrossRef]
14. Sirleto, L.; Vergara, A.; Ferrara, M.A. Advances in stimulated Raman scattering in nanostructures. *Adv. Opt. Photonics* **2017**, *9*, 169–217. [CrossRef]
15. Islam, M.N. Raman amplifiers for telecommunications. *IEEE J. Sel. Top. Quantum Electron.* **2002**, *8*, 548–559. [CrossRef]
16. Islam, M.N. *Raman Amplifiers for Telecommunications 1*; Springer: New York, NY, USA, 2003.
17. Bromage, J. Raman amplification for fiber communications systems. *J. Lightwave Technol.* **2004**, *22*, 79–93. [CrossRef]
18. Namiki, S.; Seo, K.; Tsukiji, N.; Shikii, S. Challenges of Raman amplification. *Proc. IEEE* **2006**, *94*, 1024–1035. [CrossRef]
19. Pelouch, W.S. Raman Amplification: An Enabling Technology for Long-Haul Coherent Transmission Systems. *J. Lightwave Technol.* **2016**, *34*, 1. [CrossRef]
20. Carena, A.; Curri, V.; Poggiolini, P. On the optimization of hybrid Raman/erbiumdoped fiber amplifiers. *IEEE Photonics Technol. Lett.* **2001**, *13*, 11702. [CrossRef]
21. Walrafen, G.E.; Krishnan, P.N. Model analysis of the Raman spectrum from fused silica optical fibers. *Appl. Opt.* **1982**, *21*, 350–359. [CrossRef] [PubMed]
22. Stolen, R.H.; Lee, C.; Jain, R.K. Development of the stimulated Raman spectrum in single-mode silica fibers. *J. Opt. Soc. Am.* **1984**, *1*, 652. [CrossRef]
23. Liu, K.X.; Garmire, E. Understanding the formation of the SRS spectrum in fused silica fibers. *IEEE J. Quantum Electron.* **1991**, *27*, 1022–1030. [CrossRef]
24. Dougherty, D.J.; Kartner, F.X.; Haus, H.A.; Ippen, E.P. Measurement of the Raman gain spectrum of optical fibers. *Opt. Lett.* **1995**, *20*, 31. [CrossRef]
25. Dianov, E. Advances in Raman Fibers. *J. Lightwave Technol.* **2002**, *20*, 1457. [CrossRef]
26. Oguama, F.A.; Garcia, H.; Johnson, A.M. Simultaneous measurement of the Raman gain coefficient and the nonlinear refractive index of optical fibers: Theory and experiment. *J. Opt. Soc. Am. B* **2005**, *22*, 426. [CrossRef]
27. Mandelbaum, I.; Bolshtyansky, M.; Heinz, T.F.; Hight Walker, A.R. Method for measuring the Raman gain tensor in optical fibers. *J. Opt. Soc. Am. B* **2006**, *23*, 621. [CrossRef]
28. Stolen, R.H.; Gordon, J.P.; Tomlinson, W.J.; Haus, H.A. Raman response function of silica-core fibers. *J. Opt. Soc. Am. B* **1989**, *6*, 1159–1166. [CrossRef]
29. Rottwitt, K.; Povlsen, J.H. Analyzing the Fundamental Properties of Raman Amplification in Optical Fibers. *J. Lightwave Technol.* **2005**, *23*, 3597–3605. [CrossRef]
30. Stolen, R.H. Polarization effects in fiber Raman and Brillouin lasers. *IEEE J. Quantum Electron.* **1979**, *15*, 1157–1160. [CrossRef]
31. Linand, Q.; Agrawal, G. Statistics of polarization-dependent gain in fiber based Raman amplifiers. *Opt. Lett.* **2003**, *28*, 227–229.
32. Ferrara, M.A.; Sirleto, L.; Nicotra, G.; Spinella, C.; Rendina, I. Enhanced gain coefficient in Raman amplifier based on silicon nanocomposites. *Photonics Nanostruct. Fundam. Appl.* **2011**, *9*, 1–7. [CrossRef]

33. Sirleto, L.; Ferrara, M.A.; Rendina, I.; Nicotra, G.; Spinella, C. Observation of stimulated Raman scattering in silicon nanocompisties. *Appl. Phys. Lett.* **2009**, *94*, 221106. [CrossRef]
34. Sirleto, L.; Ferrara, M.A.; Rendina, I.; Basu, S.N.; Warga, J.; Li, R.; Dal Negro, L. Enhanced stimulated Raman scattering in silicon nanocrystals embedded in silicon-rich nitride/silicon superlattice structures. *Appl. Phys. Lett.* **2008**, *93*, 251104. [CrossRef]
35. Ferrara, M.A.; Donato, M.G.; Sirleto, L.; Messina, G.; Santangelo, S.; Rendina, I. Study of Strain and Wetting Phenomena in Porous Silicon by Raman scattering. *J. Raman Spectrosc.* **2008**, *39*, 199–204. [CrossRef]
36. Sirleto, L.; Ferrara, M.A.; Rendina, I.; Jalali, B. Broadening and tuning of Spontaneous Raman Emission in Porous Silicon at 1.5 micron. *Appl. Phys. Lett.* **2006**, *88*, 211105. [CrossRef]
37. Sirleto, L.; Ferrara, M.A.; Vergara, A. Toward an ideal nanomaterial for on-chip Raman laser. *J. Nonlinear Opt. Phys. Mater.* **2017**, *26*. [CrossRef]
38. Sirleto, L.; Aronne, A.; Gioffrè, M.; Fanelli, E.; Righini, G.C.; Pernice, P.; Vergara, A. Compositional and thermal treatment effects on Raman gain and bandwidth in nanostructured silica based glasses. *Opt. Mater.* **2013**, *36*, 408–413. [CrossRef]
39. Lipovskii, A.A.; Tagantsev, D.K.; Apakova, I.E.; Markova, T.S.; Yanush, O.V.; Donato, M.G.; Sirleto, L.; Messina, G.; Righini, G.C. Mid-Range Structure of Niobium–Sodium–Phosphate Electro-Optic Glasses. *J. Phys. Chem. B* **2013**, *117*, 1444–1450. [CrossRef] [PubMed]
40. Pernice, P.; Sirleto, L.; Vergara, A.; Aronne, A.; Gagliardi, M.; Fanelli, E.; Righini, G.C. Large Raman Gain in a Stable Nanocomposite Based on Niobiosilicate Glass. *J. Phys. Chem. C* **2011**, *115*, 17314–17319. [CrossRef]
41. Donato, M.G.; Gagliardi, M.; Sirleto, L.; Messina, G.; Lipovskii, A.A.; Tagantsev, D.K.; Righini, G.C. Raman optical amplification properties of sodium–niobium–phosphate glass. *Appl. Phys. Lett.* **2010**, *97*, 231111. [CrossRef]
42. Sirleto, L.; Donato, M.G.; Messina, G.; Santangelo, S.; Lipovskii, A.A.; Tagantsev, D.K.; Pelli, S.; Righini, G.C. Raman gain in Niobium-Phosphate Glasses. *Appl. Phys. Lett.* **2009**, *94*, 031105. [CrossRef]
43. Chraplyvy, A.R. Optical Power Limits in Multi-Channel Wavelength Division Multiplexed Systems due to Stimulated Raman Scattering. *Electron. Lett.* **1984**, *20*, 58–59. [CrossRef]
44. Forghieri, F.; Tkach, R.W.; Chraplyvy, A.R. Effect of Modulation Statistics on Raman Crosstalk in WDM Systems. *IEEE Photonics Technol. Lett.* **1995**, *7*, 101–103. [CrossRef]
45. Stolen, R.H.; Johnson, A.M. The Effect of Pulse Walkoff on Stimulated Raman Scattering in Fibers. *IEEE J. Quantum Electron.* **1986**, *22*, 2154–2160. [CrossRef]
46. Ippen, E.P. Low-power quasi-CW Raman oscillator. *Appl. Phys. Lett.* **1970**, *16*, 303–305. [CrossRef]
47. Stolen, R.H.; Ippen, E.P. Raman gain in glass optical waveguides. *Appl. Phys. Lett.* **1973**, *22*, 276–281. [CrossRef]
48. Stone, J. CW Raman fiber amplifier. *Appl. Phys. Lett.* **1975**, *26*, 163. [CrossRef]
49. Lin, C.; Stolen, R.H. Backward Raman amplification and pulse steepening in silica fibers. *Appl. Phys. Lett.* **1976**, *29*, 428–431. [CrossRef]
50. Aoki, Y. Properties of Fiber Raman Amplifiers and Their Applicability to Digital Optical Communication Systems. *J. Lightwave Technol.* **1988**, *6*, 1225. [CrossRef]
51. Mandelbaum, I.; Bolshtyansky, M. Raman amplifier model in single mode optical fiber. *IEEE Photonics Technol. Lett.* **2003**, *15*, 17046. [CrossRef]
52. Siddiqui, A.S.; Vienne, G.G. The effect of pump and signal laser fluctuations on the output signal from Raman and Brillouin optical fiber amplifiers. *J. Opt. Commun.* **1992**, *13*, 33–36. [CrossRef]
53. Chinn, S.R. Analysis of counter-pumped small-signal fiber Raman amplifiers. *Electron. Lett.* **1997**, *33*, 607–608. [CrossRef]
54. Gnauck, A.H.; Charlet, G.; Tran, P.; Winzer, P.J.; Doerr, C.R.; Centanni, J.C.; Burrows, E.C.; Kawanishi, T.; Sakamoto, T.; Higuma, K. 25.6-Tb/s WDM Transmission of Polarization Multiplexed RZ-DQPSK Signals. *J. Lightwave Technol.* **2008**, *26*, 79–84. [CrossRef]
55. Kobtsev, S.M.; Pustovskikh, A.A. Improvement of Raman Amplifier Gain Flatness by Broadband Pumping Sources. *Laser Phys.* **2004**, *14*, 1488–1491.
56. Kidorf, H.; Rottwitt, K.; Nissov, M.; Ma, M.; Rabarijaona, E. Pump interactions in a 100-nm bandwidth Raman amplifier. *IEEE Photonics Technol. Lett.* **1999**, *11*, 530–532. [CrossRef]
57. Emori, Y.; Tanaka, K.; Namiki, S. 100 nm bandwidth flat-gain Raman amplifiers pumped and gain equalised by 12-wavelength-channel laser diode unit. *Electron. Lett.* **1999**, *35*, 1355–1356. [CrossRef]

58. Hansen, P.B.; Jacobovitz-Veselka, G.; Gruner-Nielsen, L.; Stentz, A.J. Raman amplification for loss compensation in dispersion compensating fiber modules. *Electron. Lett.* **1998**, *34*, 1136–1137. [CrossRef]
59. Masuda, H.; Kawai, S.; Suzuki, K.-I. Optical SNR enhanced amplification in long-distance recirculating-loop WDM transmission experiment using 1580nm band hybrid amplifier. *Electron. Lett.* **1999**, *35*, 411–412. [CrossRef]
60. Lewis, S.A.E.; Chrenikov, S.V.; Taylor, J.R. Broadband high-gain dispersion compensating Raman amplifier. *Electron. Lett.* **2000**, *36*, 1355–1356. [CrossRef]
61. Emori, Y.; Akasaka, Y.; Namiki, S. Broadband lossless DCF using Raman amplification pumped by multichannel WDM laser diodes. *Electron. Lett.* **1998**, *34*, 21456. [CrossRef]
62. Supradeepa, V.R.; Feng, Y.; Nicholson, J.W. Raman fiber lasers. *J. Opt.* **2017**, *19*, 023001. [CrossRef]
63. Stolen, R.H.; Ippen, E.P.; Tynes, A.R. Raman oscillation in glass optical waveguide. *Appl. Phys. Lett.* **1972**, *20*, 62–64. [CrossRef]
64. Hill, K.O.; Kawasaki, B.S.; Johnson, D.C. Low-threshold cw Raman laser. *Appl. Phys. Lett.* **1976**, *29*, 181–183. [CrossRef]
65. Jain, R.K.; Lin, C.; Stolen, R.H.; Pleibel, W.; Kaiser, P. A high efficiency tunable CW Raman oscillator. *Appl. Phys. Lett.* **1977**, *30*, 162–164. [CrossRef]
66. Bélanger, E.; Déry, B.; Bernier, M.; Bérubé, J.P.; Vallée, R. Long term stable device for tuning fiber Bragg gratings. *Appl. Opt.* **2007**, *46*, 3089–3095. [CrossRef] [PubMed]
67. Kean, P.N.; Sinclair, B.D.; Smith, K.; Sibbett, W.; Rowe, C.J.; Reid, D.C.J. Experimental evaluation of a fiber Raman oscillator having fiber grating reflectors. *J. Mod. Opt.* **1988**, *35*, 397–406. [CrossRef]
68. Bélanger, E.; Bernier, M.; Faucher, D.; Côté, D.; Vallée, R. High-Power and Widely Tunable All-Fiber Raman Laser. *J. Lightwave Technol.* **2008**, *26*, 1696. [CrossRef]
69. Prabhu, M.; Kim, N.S.; Jianren, L.; Ueda, K. Output Characteristics of High-Power Continuous Wave Raman Laser at 1484 nm Using Phosphosilicate Fiber. *Opt. Rev.* **2000**, *7*, 455–461. [CrossRef]
70. Babin, S.A.; Churkin, D.V.; Kablukov, S.I.; Rybakov, M.A.; Vlasov, A.A. All-fiber widely tunable Raman fiber laser with controlled output spectrum. *Opt. Express* **2007**, *15*, 8438–8443. [CrossRef]
71. Dianov, E.M.; Prokhorov, A.M. Medium-power CW Raman fiber lasers. *IEEE J. Sel. Top. Quantum Electron.* **2000**, *6*, 1022–1028. [CrossRef]
72. Thielen, P.A.; Shaw, L.B.; Pureza, P.C.; Nguyen, V.Q.; Sanghera, J.S.; Aggarwal, I.D. Small-core As–Se fiber for Raman amplification. *Opt. Lett.* **2003**, *28*, 1406–1408. [CrossRef]
73. Jackson, S.D.; Anzueto-Sánchez, G. Chalcogenide glass Raman fiber laser. *Appl. Phys. Lett.* **2006**, *88*, 221106. [CrossRef]
74. Asseh, A.; Storøy, H.; Kringlebotn, J.T.; Margulis, W.; Sahlgren, B.; Sandgren, S.; Stubbe, R.; Edwall, G. 10 cm Yb^{3+} DFB fiber laser with permanent phase shifted grating. *Electron. Lett.* **1995**, *31*, 969–970. [CrossRef]
75. Kringlebotn, J.T.; Archambault, J.L.; Reekie, L.; Payne, D.N. Er^{3+}: Yb^{3+}-codoped fiber distributed-feedback laser. *Opt. Lett.* **1994**, *19*, 2101–2103. [CrossRef] [PubMed]
76. Perlin, V.E.; Winful, H.G. Distributed feedback fiber Raman laser. *IEEE J. Quantum Electron.* **2001**, *37*, 38–47. [CrossRef]
77. Hu, Y.; Broderick, N.G.R. Improved design of a DFB Raman fiber laser. *Opt. Commun.* **2009**, *282*, 3356–3359. [CrossRef]
78. Westbrook, P.S.; Abedin, K.S.; Nicholson, J.W.; Kremp, T.; Porque, J. Raman fiber distributed feedback lasers. *Opt. Lett.* **2011**, *36*, 2895–2897. [CrossRef] [PubMed]
79. Shi, J.; Alam, S.-U.; Ibsen, M. Sub-watt threshold, kilohertz-linewidth Raman distributed-feedback fiber laser. *Opt. Lett.* **2012**, *37*, 1544–1546. [CrossRef]
80. Shi, J.; Alam, S.; Ibsen, M. Highly efficient Raman distributed feedback fiber lasers. *Opt. Express* **2012**, *20*, 5082. [CrossRef]
81. Loranger, S.; Karpov, V.; Schinn, G.W.; Kashyap, R. Single frequency low threshold linearly polarized DFB Raman fiber lasers. *Opt. Lett.* **2017**, *42*, 3864–3867. [CrossRef]
82. Wiersma, D.S. The physics and applications of random lasers. *Nat. Phys.* **2008**, *4*, 359–367. [CrossRef]
83. Turitsyn, S.K.; Babin, S.A.; El-Taher, A.E.; Harper, P.; Churkin, D.V.; Kablukov, S.I.; Ania-Castañón, J.D.; Karalekas, V.; Podivilov, E.V. Random distributed feedback fiber laser. *Nat. Photonics* **2010**, *4*, 231–235. [CrossRef]

84. Turitsyn, S.K.; Babin, S.A.; Churkin, D.V.; Vatnik, I.D.; Nikulin, M.; Podivilov, E.V. Random distributed feedback fiber laser. *Phys. Rep.* **2014**, *542*, 133–193. [CrossRef]
85. Churkin, D.V.; Kolokolov, V.; Podivilov, E.V.; Vatnik, I.D.; Nikulin, M.A.; Vergeles, S.S.; Terekhov, I.S.; Lebedev, V.V.; Falkovich, G.; Babin, S.A.; et al. Wave kinetics of random fiber lasers. *Nat. Commun.* **2015**, *2*, 6214. [CrossRef] [PubMed]
86. Du, X.; Zhang, H.; Wang, X.; Zhou, P.; Liu, Z. Short cavity-length random fiber laser with record power and ultrahigh efficiency. *Opt. Lett.* **2016**, *41*, 571–574. [CrossRef] [PubMed]
87. Wang, Z.; Wu, H.; Fan, M.; Zhang, L.; Rao, Y.; Zhang, W.; Jia, X. High power random fiber laser with short cavity length: Theoretical and experimental investigations. *IEEE J. Sel. Top. Quantum Electron.* **2015**, *21*, 0900506. [CrossRef]
88. Babin, S.A.; El-Taher, A.E.; Harper, P.; Podivilov, E.V.; Turitsyn, S.K. Tunable random fiber laser. *Phys. Rev. A* **2011**, *84*, 21805. [CrossRef]
89. Zhang, L.; Jiang, H.; Yang, X.; Pan, W.; Feng, Y. Ultra-wide wavelength tuning of a cascaded Raman random fiber laser. *Opt. Lett.* **2016**, *41*, 215–218. [CrossRef]
90. Kuznetsov, A.G.; Podivilov, E.V.; Babin, S.A. Actively Q-switched Raman fiber laser. *Laser Phys. Lett.* **2015**, *12*, 35102. [CrossRef]
91. Xu, J.; Ye, J.; Xiao, H.; Leng, J.; Wu, J.; Zhang, H.; Zhou, P. Narrow-linewidth Q-switched random distributed feedback fiber laser. *Opt. Express* **2016**, *24*, 19203–19210. [CrossRef]
92. Wang, Z.N.; Rao, Y.J.; Wu, H.; Li, P.Y.; Jiang, Y.; Jia, X.H.; Zhang, W.L. Long-distance fiber-optic point-sensing systems based on random fiber lasers. *Opt. Express* **2012**, *20*, 17695–17700. [CrossRef]
93. Churkin, D.V.; Sugavanam, S.; Vatnik, I.D.; Wang, Z.; Podivilov, E.V.; Babin, S.A.; Rao, Y.; Turitsyn, S.K. Recent advances in fundamentals and applications of random fiber lasers. *Adv. Opt. Photonics* **2015**, *7*, 516–569. [CrossRef]
94. Yusoff, Z.; Lee, J.H.; Belardi, W.; Monro, T.M.; The, P.C.; Richardson, D.J. Raman effects in a highly nonlinear holey fiber: Amplification and modulation. *Opt. Lett.* **2002**, *27*, 424–426. [CrossRef]
95. Fuochi, M.; Poli, F.; Selleri, S.; Cucinotta, A.; Vincetti, L. Study of Raman amplification properties in triangular photonic crystal fibers. *J. Lightwave Technol.* **2003**, *21*, 2247–2254. [CrossRef]
96. Bottacini, M.; Poli, F.; Cucinotta, A.; Selleri, S. Modeling of photonic crystal fiber Raman amplifiers. *J. Lightwave Technol.* **2004**, *22*, 1707–1713. [CrossRef]
97. Yiou, S.; Delaye, P.; Rouvie, A.; Chinaud, J.; Frey, R.; Roosen, G.; Viale, P.; Février, S.; Roy, P.; Auguste, J.L.; et al. Stimulated Raman scattering in an ethanol core microstructured optical fiber. *Opt. Express* **2005**, *13*, 4786–4791. [CrossRef] [PubMed]
98. Hasegawa, A.; Tappert, F. Transmission of stationary nonlinear optical pulses in dispersive dielectric fibers. I. Anomalous dispersion. *Appl. Phys. Lett.* **1973**, *23*, 142–144. [CrossRef]
99. Mollenauer, L.F.; Stolen, R.H.; Gordon, J.P. Experimental Observation of Picosecond Pulse Narrowing and Solitons in Optical Fibers. *Phys. Rev. Lett.* **1980**, *45*, 1095–1098. [CrossRef]
100. Haus, H.A.; Wong, W.S. Solitons in optical communications. *Rev. Mod. Phys.* **1996**, *68*, 423–444. [CrossRef]
101. Haus, H.A. Mode-locking of lasers. *IEEE J. Sel. Top. Quantum Electron.* **2000**, *6*, 1173–1185. [CrossRef]
102. Herr, T. Temporal solitons in optical microresonators. *Nat. Photonics* **2014**, *8*, 145–152. [CrossRef]
103. Hasegawa, A. Amplification and reshaping of optical solitons in a glass fiber—IV: Use of the stimulated Raman process. *Opt. Lett.* **1983**, *8*, 650–652. [CrossRef]
104. Molleanauer, L.F.; Stolen, R.H.; Islam, M.N. Experimental demonstration of soliton propagation in long fibers: Loss compensated by Raman gain. *Opt. Lett.* **1985**, *10*, 229–231. [CrossRef]
105. Molleanauer, L.F.; Gordon, J.P.; Islam, M.N. Soliton propagation in long fibers with periodically compensated loss. *IEEE J. Quantum Electron.* **1986**, *22*, 157–173. [CrossRef]
106. Molleanauer, L.F.; Smith, K. Demonstration of soliton transmission over more than 4000 km in fiber with loss periodically compensated by Raman gain. *Opt. Lett.* **1988**, *13*, 657–677. [CrossRef] [PubMed]
107. Karpov, M.; Guo, H.; Kordts, A.; Brasch, V.; Pfeiffer, M.H.P.; Zervas, M.; Geiselmann, M.; Kippenberg, T.J. Raman self-frequency shift of dissipative Kerr solitons in an optical microresonator. *Phys. Rev. Lett.* **2016**, *116*, 103902. [CrossRef] [PubMed]
108. Yi, X.; Yang, Q.-F.; Yang, K.Y.; Vahala, K. Theory and measurement of the soliton self-frequency shift and efficiency in optical microcavities. *Opt. Lett.* **2016**, *41*, 3419–3422. [CrossRef] [PubMed]

109. Mitschke, F.M.; Mollenauer, L.F. Discovery of the soliton self-frequency shift. *Opt. Lett.* **1986**, *11*, 659–661. [CrossRef]
110. Gordon, J.P. Theory of the soliton self-frequency shift. *Opt. Lett.* **1986**, *11*, 662–664. [CrossRef] [PubMed]
111. Liu, X.; Xu, C.; Knox, W.H.; Chandalia, J.K.; Eggleton, B.J.; Kosinski, S.G.; Windeler, R.S. Soliton self-frequency shift in a short tapered air–silica microstructure fiber. *Opt. Lett.* **2001**, *26*, 358–360. [CrossRef] [PubMed]
112. Druhl, K.; Wenzel, R.G.; Carlsten, J.L. Observation of Solitons in Stimulated Raman Scatterin. *Phys. Rev. Lett.* **1983**, *51*, 1171. [CrossRef]
113. Englund, J.C.; Bowden, C.M. Spontaneous generation of Raman solitons from quantum noise. *Phys. Rev. Lett.* **1986**, *57*, 2661–2663. [CrossRef]
114. MacPherson, D.C.; Carlstein, J.L.; Druhl, J.K. Quantum fluctuation and correlations in the stimulated Raman scattering spectrum. *Phys. Rev. A* **1989**, *39*, 3487–3497. [CrossRef]
115. Gouveia-Neto, A.S.; Gomes, A.S.L.; Taylor, J.R. High-efficiency single-pass solitonlike compression of Raman radiation in an optical fiber around 1.4 µm. *Opt. Lett.* **1987**, *12*, 1035–1037. [CrossRef] [PubMed]
116. Islam, M.N.; Mollenauer, L.F.; Stolen, R.H.; Simpson, J.R.; Shang, H.T. Amplifier/compressor fiber Raman lasers. *Opt. Lett.* **1987**, *12*, 814–816. [CrossRef] [PubMed]
117. Kafka, J.D.; Baer, T. Fiber Raman soliton laser pumped by a Nd: YAG laser. *Opt. Lett.* **1987**, *12*, 181–183. [CrossRef] [PubMed]
118. Gouveia-Neto, A.S.; Gomes, A.S.L.; Taylor, J.R. A femtosecond soliton Raman ring laser. *Electron. Lett.* **1987**, *23*, 537–538. [CrossRef]
119. Gouveia-Neto, A.S.; Gomes, A.S.L.; Taylor, J.R. Soliton Raman fiber-ring oscillators. *Opt. Quantum Electron.* **1988**, *20*, 165–174. [CrossRef]
120. Haus, H.A.; Nakazawa, M. Theory of the fiber Raman soliton laser. *J. Opt. Soc. Am. B* **1987**, *4*, 652–660. [CrossRef]
121. Gouveia-Neto, A.S.; Gomes, A.S.L.; Ainslie, B.J.; Craig, S.P.; Taylor, J.R. Cascade Raman soliton fiber ring laser. *Opt. Lett.* **1987**, *12*, 927. [CrossRef]
122. Da Silva, V.L.; Gomes, A.S.L.; Taylor, J.R. Hogh order cascade Raman generation and soliton-like pulse formation in a 1.06 µm pumped fiber laser. *Opt. Commun.* **1988**, *66*, 231–234. [CrossRef]
123. Keller, U.; Li, K.D.; Rodwell, M.J.W.; Bloom, D.M. Noise characterization of femtosecond fiber Raman soliton lasers. *IEEE J. Quantum Electron.* **1989**, *25*, 280–288. [CrossRef]
124. Ding, M.; Kikuchi, K. Improvement of the fiber raman soliton laser for femtosecond optical pulse generation. *Fiber Integr. Opt.* **1994**, *13*, 337. [CrossRef]

© 2020 by the authors. Licensee MDPI, Basel, Switzerland. This article is an open access article distributed under the terms and conditions of the Creative Commons Attribution (CC BY) license (http://creativecommons.org/licenses/by/4.0/).

Review

Integrated Raman Laser: A Review of the Last Two Decades

Maria Antonietta Ferrara and Luigi Sirleto *

National Research Council (CNR), Institute of Applied Sciences and Intelligent Systems,
Via Pietro Castellino 111, 80131 Naples, Italy; antonella.ferrara@na.imm.cnr.it
* Correspondence: luigi.sirleto@cnr.it

Received: 26 February 2020; Accepted: 22 March 2020; Published: 23 March 2020

Abstract: Important accomplishments concerning an integrated laser source based on stimulated Raman scattering (SRS) have been achieved in the last two decades in the fields of photonics, microphotonics and nanophotonics. In 2005, the first integrated silicon laser based upon SRS was realized in the nonlinear waveguide. This breakthrough promoted an intense research activity addressed to the realization of integrated Raman sources in photonics microstructures, like microcavities and photonics crystals. In 2012, a giant Raman gain in silicon nanocrystals was measured for the first time. Starting from this impressive result, some promising devices have recently been realized combining nanocrystals and microphotonics structures. Of course, the development of integrated Raman sources has been influenced by the trend of photonics towards the nano-world, which started from the nonlinear waveguide, going through microphotonics structures, and finally coming to nanophotonics. Therefore, in this review, the challenges, achievements and perspectives of an integrated laser source based on SRS in the last two decades are reviewed, side by side with the trend towards nanophotonics. The reported results point out promising perspectives for integrated micro- and/or nano-Raman lasers.

Keywords: nonlinear optics; stimulated Raman scattering; lasers; microphotonics; nanophotonics; nonlinear waveguide; optical microcavity; photonics crystals; nanocrystals

1. Introduction

Nonlinear optical devices enable the control of light by light. Since photons do not interact directly, interaction is possible only by taking advantage of a suitable nonlinear optical material. This is the key element in any nonlinear process, governing the type of nonlinear phenomena supported, the efficiency, size speed, and power characteristics [1–3].

In nonlinear optical devices, third order nonlinear effects play a fundamental role. They are due to an induced material polarization, which is proportional to the third power of the electric field, and they can be divided in two class [1–4]. The first one is related to the Kerr-effect, i.e., the intensity dependence of the refractive index of the medium, occurring in three effects: Self-Phase Modulation (SPM), Cross-Phase Modulation (CPM) and Four-Wave Mixing (FWM). The second one is due to inelastic-scattering phenomenon, which can induce a stimulated effect such as Stimulated Raman-Scattering (SRS). SRS, observed in 1962 for the first time [5–7], is given by the interaction between the guided wave and high-frequency optical phonons. SRS depends on the pump intensity and on a gain coefficient, which is proportional to the spontaneous Raman scattering cross section, and inversely proportional to the linewidth of the corresponding Raman line. Commonly, SRS is observed in two forms. The first, Raman generation, describes the Stokes-beam growth in the material from spontaneously scattered Raman-shifted radiation. The second, Raman amplification, is obtained when the energy from an intense pump beam is transferred to a weaker signal beam, which can be copropagating or counterpropagating [8–16].

Integrated nonlinear optics devices have been investigated since the 1970s. The attractive features of nonlinear waveguides are: 1) light intensity can be confined within an area comparable to the wavelength of light; 2) the diffractionless propagation in one or two dimensions results in interaction lengths over a distance (about a few cm) much longer than the one obtained with a bulk material. Since the nonlinear interaction efficiency depends nonlinearly upon the interacting beam intensities (power/area), and it is also proportional (either linearly or quadratically) to the interaction distance, the waveguide geometries offer the best prospects for optimizing the efficiency of nonlinear devices. Various types of all-optical functionalities, which can be significant for all-optical telecommunication networks, can be implemented by nonlinear integrated optical devices, based on different kinds of optical nonlinearities. They include signal regeneration, wavelength conversion, optical switching, routing optical demultiplexing and optical delay/buffering. At present the issue to be overcome is the materials' performance. In the last decades, there has been a significant progress in this area with semiconductors and organic materials. However, there is still a key trade-off among the propagation losses, the nonlinearity and the nonlinearity response time, which does not allow the realization of high-speed, nonlinear devices performing at low optical power. The propagation losses limit the effective device length, and combined with nonlinearity, define the required device power. Due to the diffraction limit, a further limitation is that waveguide components do not allow confining light to the microscale or nanoscale dimension [17,18].

Microphotonics explore light behavior on the microscale and its interaction with micro-objects. The aim of microphotonics development is to go beyond the limit of photonics, offering a reliable platform for dense integration. The key challenges for microphotonics are a reduction in the size of integrated optical devices, and an improvement of performances with respect to nonlinear waveguide devices. During the last decades, the fast growth of micro-scales fabrication techniques has enabled the successful demonstration of various types of microphotonics devices, for example ring resonators and photonic crystals (PhCs) [19]. In a microphotonics device, photons are trapped in small volumes close to the diffraction limit for sufficiently long times [20], so that these photons strongly interact with the host material, creating enhanced nonlinear [21], quantum [22] and optomechanical [23] effects. We note that in this microphotonics device, although the physical phenomena observed are similar to the ones reported in a resonator etalon, the performance of micro structures involved is boosted by orders of magnitude. As a consequence, many physical phenomena have been observed with high compactness and integration, such as the Purcell effect [24], strong coupling between quantum dot, and cavity modes [25]. In nonlinear optical applications, microphotonics devices exhibit two interesting and useful aspects: the micrometers dimension and the increasing of the local field, combining a small modal volume with high optical quality-factors (Q) [26–28]. As an important consequence, a significant reduction of the power threshold of nonlinear optical effects is obtained.

Nanophotonics is a fascinating field, investigating the light behavior on the nanometer scale and its interaction with nanometer-scale objects [29,30]. We will have a big demand in the near future for devices, which should allow us to control light with light in a very thin nanoscale layer, or in a single nanoparticle of nonlinear material. In principle, in order to control a signal light in a nonlinear optical device, the intensity or phase of light has to be changed by a control signal, thus changing the optical characteristics of the medium. Of course, the stronger the nonlinearity of the material, the shorter the required interaction length L. We note that in nanoscale devices, the nonlinear effects cannot be enhanced using photon confinement effects, thus they only depend on the nonlinearity of the medium itself [31]. Therefore, a development of nanostructured materials with large nonlinearities, and satisfying also various technological and economical requirements [32], is mandatory. This is both an applicative issue, for an efficient device realization and design, and a fundamental issue, since the interplay between light and nanostructures is not yet understood.

Local enhancement and strong resonance of electromagnetic (EM) radiation in metallic nanoparticles and films continues to attract significant attention [33]. In these structures, a collective motion of conduction electrons (the surface plasmon polariton) becomes resonantly excited by visible light.

Bright, visible light emission in "bulk-sized" silicon coupled with plasmon nanocavities from non-thermalized carrier recombination was demonstrated by Cho, et al. [34]. However, although plasmonics allows a significant size reduction of optical components, the main drawback are still optical losses. On the other hand, owing to the big interest for the monolithic integration of photonic technology and semiconductor electronics, the EM radiation enhancement obtained from semiconducting and insulating materials is considerable [35].

In this review, the impressive progress, concerning an integrated laser source based on SRS and obtained in the last two decades, is described. In Section 2, for the sake of completeness, an essential theoretical background about SRS and a short introduction about the Raman laser in bulk are reported. In Section 3, the breakthrough of the first silicon laser is described. The possibility to obtain micrometer dimensions and the reduction of the power threshold to the microwatt pushed towards microphotonics structures. So, in Section 4 the most successful Raman lasers in microcavities and photonic crystals are reported. Finally, in Section 5, we discuss about SRS in nanostructures. We note that there are significant implications from both a fundamental and applicative point of view. From the fundamental one, a number of investigations, both experimental and theoretical, have been proposed in literature, but the "question is still open", while from the applicative one, in order to realize micro-/nano-sources with improved performance, some encouraging perspectives have been pointed out [36–39]. Finally, recently proposed devices, combining silicon nanocrystals and microstructures, are reported, too.

2. Theoretical Background of SRS and Introduction to Raman Laser

The SRS effect happens when a transfer of energy from a high power pump beam to a probe or Stokes beam (copropagating or counterpropagating) takes place. Precisely, this energy exchange occurs only if the frequency difference between the pump (at frequency ω_P) and the Stokes laser beams (at frequency ω_S) matches a molecular vibrational frequency of the considered material (i.e., $\omega_S = \omega_P - \omega_v$, where $\hbar\omega_v$ corresponds to a vibrational energy); the SRS effect leads to a gain of the Stokes beam power (stimulated Raman gain, SRG) and to a loss of the pump beam power (stimulated Raman loss, SRL) [8–16], as showed in Figure 1a.

SRS belongs to a class of nonlinear, optical processes called quasi-resonant. Although none of the two fields is in resonance with the vibrations in the lattice of the medium (optical phonons), their difference equals the transition frequency. The origin of SRS can be understood in terms of a two-step process: Frst, the pump produces frequency sidebands (Stokes and anti-Stokes) due to molecular vibrations; second, the Stokes wave, beating with the pump wave, produces a modulation of the total intensity, which coherently excites the molecular vibrations. These two steps strengthen each other, so the pump effect induces to a stronger Stokes wave, which in turn brings to stronger molecular vibrations. Because of its coherent nature, in SRS, the molecular bonds oscillate with a constant phase relation (see Figure 1b), interfering constructively inside the focus area of the laser beam, and therefore the SRS signal is orders of magnitude stronger than spontaneous Raman scattering. In SRS, the detection of a macromolecule with thousands of identical vibrational modes that interfere coherently is feasible. (see Figure 1b).

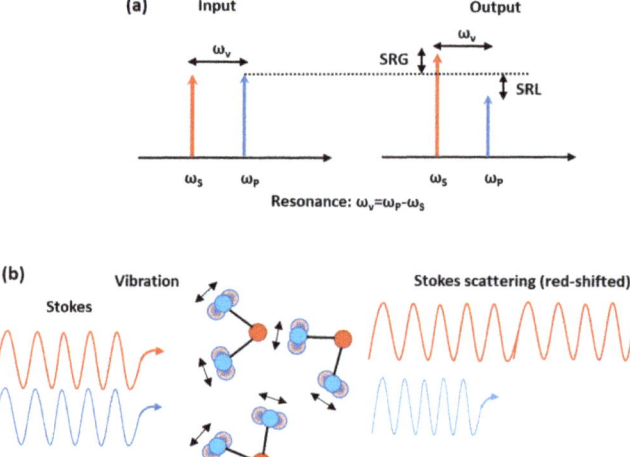

Figure 1. (**a**) SRS modalities: SRG, stimulated Raman gain; SRL, stimulated Raman loss; (**b**) inelastic scattering of probe photons obtained from vibrationally-excited molecules interfering coherently.

In SRS, high-order Raman sidebands can be produced by a field propagating through a medium optically polarizable. Many Stokes frequencies can be generated at the output when the pump power exceeds the threshold value, thus the Stokes frequencies at $\omega_P - \omega_v$, $\omega_P - 2\omega_v$ and those of anti-Stokes frequencies at $\omega_P + \omega_v$, $\omega_P + 2\omega_v$ can be observed. If the intensity of the first-Stokes wave is high enough, it can generate a "second-Stokes" beam. By iterating this process, a higher Stokes order can be generated, leading to the so-called "cascaded" SRS. The intensity of the ith Stokes order thus depends on the conversion rate from the $(i-1)^{th}$ Stokes order (proportional to $g \cdot I_i \cdot I_{i-1}$) and the loss rate to the $(i+1)^{th}$ order (proportional to $g \cdot I_i \cdot I_{i+1}$), where g is the Raman gain of material [8–16]. The generation of a coherent radiation in a wide interval of wavelengths, from the ultraviolet to the infrared can be obtained within a single Raman laser system, by combining conventional diode-pumped, solid-state lasers and different Stokes orders of SRS provided by a cascading effect [40–43].

In bulk semiconductors, lasing by SRS was first discovered in GaP [44]. Raman lasers (RLs) are similar to ordinary lasers. A first analogy is that lasing in RLs occurs when the Raman-active gain medium is placed inside a cavity to achieve the laser threshold. A second analogy is that the threshold power in RLs is obtained when the Raman amplification during a round trip is as large as to compensate the cavity losses. However, there are also some important differences between RLs and traditional lasers. A first one is that an amplifier medium based on Raman gain is used rather than on stimulated emission from excited atoms or ions. A second difference is that the required wavelength for pumping the Raman laser does not depend on the electronic structure of the medium, so it can be chosen to minimize absorption.

A Raman-laser-based approach takes advantage of the huge developments achieved in the engineering of diode-pumped, solid-state lasers, and the existing number of excellent Raman laser crystals. Moreover, this approach offers many attractive features. The main advantage is that in principle, any Raman laser wavelength can be obtained by a proper choice of the pump wavelength, when both wavelengths are within the transparency region of the material, and an adequately high nonlinearity and/or optical intensity are provided. The tunable output wavelength is a very important feature of crystalline Raman laser sources, since it can lead to several applications for Raman-laser-based sources working in the UV range, such as in remote sensing or the detection of explosives and biological agents, biomedical, including drug discovery, mass spectrometry and fluorescence imaging. Regarding the visible range, applications are in areas such as ophthalmology, biomedicine, dermatology, display and remote sensing [43]. Moreover, the Raman-laser-based approach shows excellent prospective for miniaturization, high beam quality, relative easy design based on commercially available components (and hence low cost), possibility to tailor output properties, potential to scale to multi-watt powers, and possibility to implement both CW and pulsed mode.

Numerous design options for lasers employing SRS, which can be used to tailor Raman laser characteristics and performance, have been developed. The basic configuration is the Raman generator, where a pump beam is focused (or telescoped) into a Raman crystal in order to generate a Stokes beam. This configuration can be used when the Raman gain is very large, the laser intensity can increase from noise to a significant signal without feedback. Typically used with short pulse (picosecond (ps)) lasers, this process is capable of generating a large number of Stokes and anti-Stokes lines, and yields efficiencies as high as 95%.

For pump pulses longer than the transit time through the active medium, the Raman crystal is placed inside a cavity resonating the Stokes field. Basically, there are two main configurations to implement RL: (i) external-resonator RL, where the Raman crystal is placed inside a cavity, resonating the Stokes beam (Figure 2a); (ii) the intracavity RL (Figure 2b), where both a Raman medium and the laser medium are combined inside a single cavity, so the fundamental and Stokes fields are both resonating inside the cavity [41]. Incorporating the Raman material inside the laser cavity, the high intracavity powers can be used to allow the conversion of low power lasers. Moreover, fiber optics amplification has been obtained by means of the Raman effect [45]. In this case, being that optical fiber used as the Raman gain medium, both pump and Stokes beams are launched into it. (Figure 2c). However, considering the currently-available communication range of about 50 THz, the existing silica fiber amplifiers are limited by their narrow usable bandwidth for Raman amplification (5 THz, approx. 150 cm^{-1}).

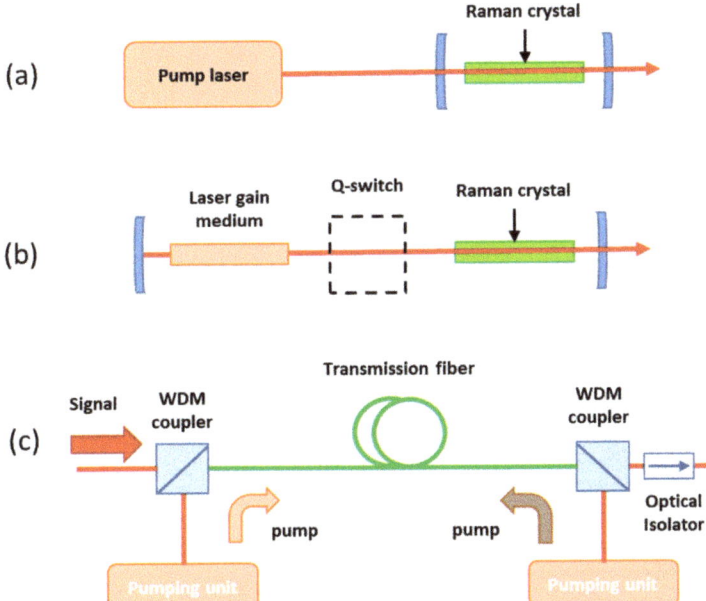

Figure 2. Typical configurations of the Raman laser: (**a**) external-resonator Raman laser; (**b**) intracavity Raman laser; and (**c**) fiber Raman amplifiers.

Cascading is an interesting approach for the multiple-wavelength and wavelength-selectable Raman lasers. In order to select the output wavelength from a single device, the cascading effect, in combination with second-harmonic generation and or sum-frequency generation (SHG/SFG), provides a high degree of flexibility. SRS and SFG/SHG can be carried out either intracavity or extracavity, therefore a range of configurations can be developed to provide the selection of wavelengths.

As general rule, all laser gain bulk materials have a tradeoff between gain and bandwidth, such that linewidth may be increased to the detriment of peak gain. This trade-off is a significant limitation towards micro-/nano-sources realizations with large emission spectra, for which a nonlinear material with wide, flat and high Raman gain in the range of interest [46,47] should be individuated. We consider, for example, glasses and silicon, which have been two leading materials for application in fiber and integrated photonics, respectively. Silicon has a high Raman gain and small bandwidth, whereas silica has a large bandwidth, but a small Raman gain; in Figure 3 we reported as an example the Raman spectra of the two extreme cases of crystalline silicon (c-Si) and silica. In order to answer the telecommunications demands, the investigation of new materials with both large Raman gain coefficients and spectral bandwidth is required. [48–58].

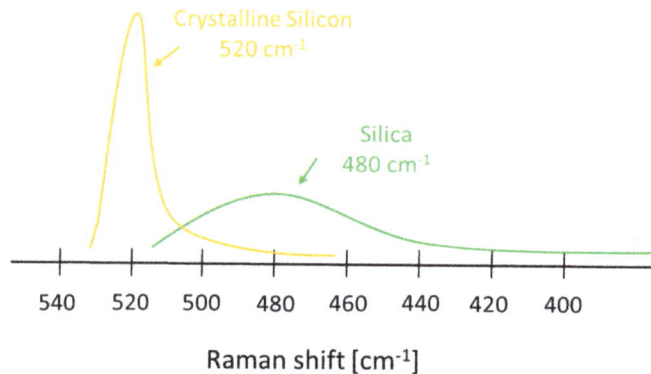

Figure 3. Comparison of Raman bandwidth and the efficiency of silicon and silica.

3. The First Silicon Laser

Silicon (Si) is considered the principal material for the microelectronics industry, but this is not yet true in photonics, due to its indirect bandgap. However, due to its wide bandwidth, high speed and low power dissipation, optical interconnect is emerging as an encouraging approach for on- and off-chip communications; therefore, it is viewed as the natural replacement of the electronic component in data transmission. For these reasons, silicon photonics is considered a technology on which to invest, because silicon-based optical components could be realized, taking advantage of existing complementary metal–oxide–semiconductor (CMOS) silicon fabrication techniques. The final goal is to integrate both electronic and optical functions on the same chip. With this aim, a silicon-based light source [59], a silicon waveguide, a silicon optical modulator [60], and a silicon-based photodetector [61] should be developed. While passive silicon devices were proposed and developed in the 1990s [62], the design of active devices seem to be much more challenging, due to silicon hostile physical properties, such as an indirect bandgap that does not allow efficient optical transitions, and the almost lack of Pockel's effect caused by a centrosymmetric crystal structure [63]. Therefore, crystalline silicon shows relatively poor linear optical properties, thus light emission is precluded. In the last years, several approaches were proposed to realize Si-based light sources and amplifiers; they can be classified in three main categories:

- Using spatial confinement of the electron in order to overcome the indirect band structure;
- Using optically active dopants obtained by the introduction of rare earth impurities;

- Take advantage of Raman scattering in order to achieve optical gain.

Taking advantage by nonlinear effects shown in silicon, nonlinear silicon photonics [64,65] has attracted a lot of interest to obtain light amplification, light generation and wavelength conversion [66–70]. Being high quality silicon-on-insulator (SOI) wafers currently commercially available, the silicon waveguide in SOI can be easily produced by a standard CMOS fabrication technique.

In SOI waveguides, the optical field can be confined to an area that is about 100 times smaller than the modal area of a standard single-mode, optical fiber, additionally due to the difference between crystalline and amorphous materials [71], the Raman gain cross-section in c-Si is five orders of magnitude larger than in silica. Thus, the silicon waveguide could be used to fabricate integrable and efficient amplifiers. The waveguide approach is schematically reported in Figure 4. We note that this Raman silicon laser was limited to centimeter-sized cavities with thresholds higher than 20 milliwatts [72,73].

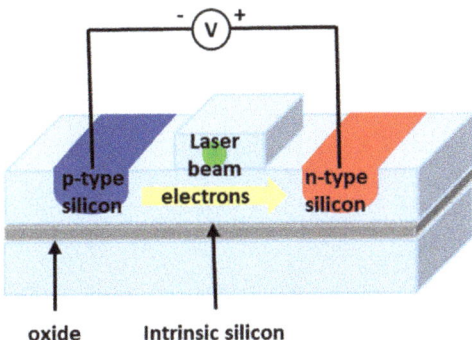

Figure 4. Typical configurations of silicon-on-insulator (SOI) waveguide Raman laser.

The Raman approach allows one to avoid rare earth doping, such as erbium, making it fully compatible with silicon microelectronics manufacturing. On the other hand, the main limitation of the Raman approach is that electrical excitation is needed, requiring an off-chip pump. Another limitation is related to the two-photon-absorption (TPA), which gives rise to pump depletion and the generation of free carriers. TPA, through the free carrier plasma effect, leads to a broadband absorption spectrum. While the pump depletion is negligible since the TPA coefficient, β, is rather small (~0.5 cm/GW) [74], absorption by TPA-generated free carriers is a broadband competing process with respect to the Raman gain, becoming a limiting factor in all-optical switching in III-V semiconductor waveguides, which is added to the linear optical scattering loss due to the waveguide sidewall roughness [75]. Moreover, because of the nonlinear optical loss associated with TPA-induced free carrier absorption (FCA) [76–78] in silicon waveguides, both the pump and signal beam decrease, making SRS not possible.

In order to overcome free carrier absorption, the first example of a silicon amplifier was based on a pulsed pump beam with a pulse duration shorter than the lifetime of the free carriers. The realized waveguide was 3 cm long, and its output signal was looped back into the entry by using an 8 m long single mode optical fiber; thus, an optical cavity was formed to achieve lasing [79].

The first observation of net optical gain in a low loss silicon waveguide in SOI, based on stimulated Raman scattering with a pulsed pump laser at 1.545 µm, was reported from the Intel Corporation [79]. A silicon rib waveguide was designed and fabricated on the (100) surface of a SOI substrate using standard photolithographic patterning and reactive ion etching techniques, and its effective core area was evaluated to be ~1.5 µm^2. With the aim to increase the interaction length, but at the same time have a micrometer-sized device, the waveguide was realized in an S-shaped curve with a bend radius of 400 µm and a total length of 4.8 cm. The measured linear loss was of 0.22 ± 0.05 dB/cm for both TE and TM modes; nevertheless, due to the strong pump beam, additional nonlinear optical loss in the silicon waveguide was observed and ascribed to TPA, since the two-photon energy of the pump beam is greater than the energy band gap of silicon. By using pulsed excitation, a decrease of the TPA-induced FCA allows obtain a Raman gain greater than the nonlinear optical loss. The reported net gain for such structure was of 2 dB, with a peak pump power of 470 mW and a pulse width of ~17 ns [80].

In 2005, after one year from the previous result, the same group demonstrated the first all-silicon-pulsed Raman laser [72]. The laser cavity was made by placing a mirror on one side of the 4.8 cm long, S-shaped silicon rib waveguide. The mirror was broadband, and had a high reflectivity (~90%) for both the pump and the Stokes wavelengths, of 1.536 µm and 1.67 µm, respectively. The other side of the waveguide was used to couple the pulsed pump beam into the cavity, and had a reflectivity of ~30% for both pump and Raman wavelengths. This solution allows us to weakly overcome the narrow-band (105 GHz) of stimulated Raman gain in Si that makes it inappropriate for use in wavelength division multiplexing (WDM) applications, unless expensive multi-pump schemes should be implemented. Indeed, by changing the cavity length it is possible to obtain different wavelengths out of a single pump wavelength within a very narrow interval. To overcome the TPA-induced FCA associated with Raman scattering, and thus to definitely reduce the accumulated carrier density inside the silicon waveguide, the authors proposed the use of a reversely biased p-i-n diode. They realized a p-i-n diode structure along the rib waveguide, and demonstrated that when a reverse bias is applied to it, the TPA-generated electron–hole pairs can be carried away from the silicon waveguide between the p- and n-doped regions by the applied electric field. Therefore, by changing the reverse bias voltage, the carrier transit time or effective carrier lifetime can be modified. The laser output power was measured at the uncoated side of the silicon waveguide cavity with a reverse bias of 25 V; in this condition, the obtained laser threshold was at ~0.4 mW, and the slope efficiency (single side output) was 9.4%.

Afterwards, always the same group reported a continuous-wave (CW), all-silicon Raman laser [81]. The previous presented optical cavity was suitably modified to obtain CW lasing. In detail, both sides of the low-loss, S-shaped silicon waveguide were coated with multilayer dielectric films to reduce cavity loss. The front facet coating was a dichroic with a reflectivity of ~71% for the Stokes wavelength (1686 nm) and ~24% for the pump wavelength (1550 nm). Regarding the back side, it was designed to have a broadband high-reflectivity coating (~90%) for both pump and Raman wavelengths. With these values of reflectivity, a low finesse cavity at the pump wavelength was obtained, allowing us to get the pump power cavity enhancement effect to lower the lasing threshold. If the pump laser matches the resonance of the cavity, the power inside the waveguide cavity is improved, leading to a decrease of the power threshold. Here an electric field generated by a reverse biased p-i-n diode drives away the free carrier generated by TPA in the waveguide channel. These free carriers are then collected by two electrodes, leading to a stable CW laser emission. The reported lasing thresholds were of ~180 mW with a 25-V bias and ~280 mW with a 5-V bias. This difference in the threshold depending on the bias voltage is due to lower nonlinear loss linked to the reduction of the effective carrier lifetime, as expected. The slope efficiency (single side output) above threshold was ~4.3%, with a reverse bias of 25 V and 2% with a 5-V reverse bias. The p-i-n diode has been used also with a forward bias to inject free carriers in the silicon waveguide to absorb light and switch off the laser emission [82].

Tyszka–Zawadzka et al. have presented a semi-analytical model of Raman generation in the SOI rib waveguide with a DBR/F–P resonator [83]. The authors have proposed an approximate, semi-analytical expression relating the Raman output power to the pump power and system parameters, taking into account the Raman amplification, as well as the linearly-distributed losses in the laser cavity, the linear effects related to the FCA, and the nonlinear absorption associated with the TPA-induced free carrier absorption. A slight influence of TPA on the threshold pump power and output power of the SOI laser was highlighted by their numerical results obtained for the threshold laser operation and the above threshold laser operation. Another sophisticated and integrated multiphysics algorithm procedure was developed to accurately predict wave evolution and power transfer (pump, Stokes) propagating in the SOI microcavity resonator by taking into account the thermal and stress influences on the SRS and on all other linear and nonlinear physical effects involved in the Raman lasing mechanism [84].

4. Raman Laser in Microphotonics

In microstructures, the SRS enhancement, which is attributed to the photon confinement effect, can be evaluated by an equivalent gain (g_{micro}), given by $g_{micro} = f \times g_{bulk}$, where g_{bulk} is the gain of bulk material making up the microstructures, and f is the optical field enhancement related to the presence of microstructures. We note that the bandwidth of the Raman gain does not change, and therefore, using microstructures, the fundamental trade-off between the gain and bandwidth of bulk materials cannot be overcome.

4.1. Raman Laser in Microcavities

Regarding SRS, it was observed that, within a droplet microcavity, the Raman gain improvement is inversely proportional both to the linewidth of the Raman process and to the square of the radius of the spherical cavity [85].

Spillane et al. studied SRS in spherical droplets and silica microspheres, with diameters of the order of tens of micrometers, and optically coupled by the use of a tapered optical fiber [35]. The authors measured the threshold power, whereas the coupling air gap between the taper and the microsphere was changed, and they found that in such way a nonlinear Raman source with a pump threshold approximately 1000 times lower than reported before, and a pump-signal conversion higher than 35%, can be obtained [46,86].

A glycerol–water droplet on a superhydrophobic surface coated with silica nanoparticles was used, as stationary microdroplets to achieve Raman lasing, by Sennaroglu, et al. [87]. By exciting a 12.4-µm-diameter droplet with a 532 nm Nd:YAG laser, both cavity-enhanced Raman scattering and Raman lasing centered at 632.3 nm were observed. Moreover, a typical on/off behavior of Raman lasing in microdroplets was reported and ascribed to the thermally-induced fluctuations during lasing. An improvement of the lasing effect can be obtained by increasing the rate of convective cooling, making this system useful as a light source for short-haul communications systems [87].

A complete study of Raman oscillation in fiber-taper-coupled microspheres and microtoroids on-a-chip, from both theoretical and experimental points of view, was carried out by Kippenberg et al. [88]. With respect to Raman oscillation in microspheres, microtoroids show both power efficiency and spectral benefits, as well as having advantages with respect to their chip-based fabrication. Additionally, in the microtoroid-based device, single-mode oscillations are allowed due to a high reduction in the complexity of the mode spectrum, and for comparable Q-factors, lower threshold pump powers can be achieved due to a controllable and reduced mode volume. In particular, the threshold power reduction will depend upon the "aspect ratio" of the toroid given by D/d, where $D = 50$ µm is the outer diameter, and d is the minor one. The case of $d = 50$ µm corresponds to a sphere. A schematic example of the microtoroid-based device is reported in Figure 5.

In 2011, the influence on the Raman signal in a single microsphere of several parameters, such as the pump wavelength, size and refractive index of the microsphere, and the numerical aperture of the microscopic objective lens, was analyzed [89]. Results showed that, due to the increased field of the photonic nanojet emerging from the single microsphere, an enhancement ratio of silicon wafer and cadmium ditelluride Raman peaks, approximately of two orders of magnitude, can be obtained by suitable selection of the experimental parameters [89].

Taking advantage of the third-order nonlinearity $\chi(3)$ in a microresonator, the four-wave mixing (FWM) process can be used to generate a broadband frequency comb using parametric oscillation. Using a silicon microresonator, the first low-noise coherent mid-IR frequency comb source was obtained [90,91]. As a result of the interaction between the stimulated Raman effect and FWM, phase locking of the generated comb can be observed, leading to strong comb lines separated by the Raman shift in silicon [90,91].

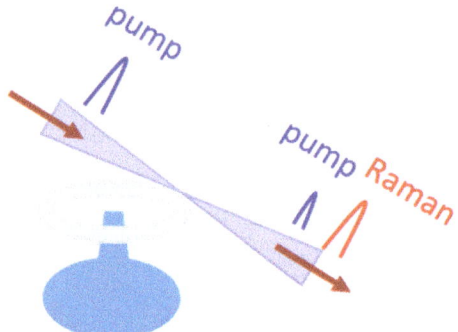

Figure 5. Schematic example of a taper-toroid coupling system.

Alternatively, diamond was suggested as a reasonable material for compact, on-chip Raman lasers over a wide spectrum [92]. A CW low-threshold Raman laser, based on waveguide-integrated diamond racetrack microresonators embedded in silica on a silicon chip, was demonstrated in [93]. Pumping at telecom wavelengths, a tunable Stokes output over a ~100 nm bandwidth around 2 μm, with output power > 250 μW, was reported.

In 2017, a tunable Raman laser in the hollow bottle-like microresonator (BLMR) with a high-Q factor of 2.2×10^8 was demonstrated [94]. Continuous output frequency tuning was obtained by controlling the pump laser frequency or power through the thermal effect; the tuning range was of 1.2 nm, corresponding to a frequency range of 132 GHz, with a minimum tuning step of about 85 MHz. Additionally, a large range frequency tuning of the Raman laser, with the tuning range of 132 GHz and a resolution of about 85 MHz, was also demonstrated by mechanically stretching the resonator. These approaches could open the way to future optical applications of WGM microresonators [95,96].

4.2. Raman Laser in Photonics Crystals

Photonic crystals were introduced for the first time in 2005 to enhance stimulated Raman amplification and lasing in monolithic silicon [97]. Indeed, due to their increased light–matter interactions, these structures seem to be good candidates to permit the realization of ultracompact silicon Raman light amplifiers and lasers.

An L5 photonic bandgap nanocavity in two-dimensional photonic-crystal slabs was realized, and the power threshold estimated was about 130 μW, considering parameters g_s=70 cm/GW, $\xi = 1$ (modal overlap), $\lambda_{s,p}$ = 1550 nm, Q_{pump} = 1550 and $Q_{Stokes} = 4.2 \times 10^4$ ($Q_{pump,Stokes}$ are the cavity quality factors). The threshold can be further reduced to tens of microwatts by designing ultrasmall, photonic, bandgap nanocavities with higher-Q factors [98,99]. Indeed, a drastic enhancement of the Raman gain beyond that predicted theoretically can be achieved by designing nanocavities considering that the strength of light–matter interactions is proportional to the ratio of the quality factor to the cavity volume [99].

Regarding the photonic crystal waveguide (PCWG), a number of experiments have been performed in the so-called W1 waveguides obtained by removing a single line of holes in the ΓK direction of an otherwise perfect photonic crystal [100,101]. A theoretical study of the SRS enhancement in a slow-light silicon PCWG through a four wave-mixing formalism from the computed modes of the line-defect waveguide was carried out by McMillan et al. [100]. In detail, by comparing the group velocities of Stokes and pump signals, an enhancement of the stimulated Raman gain up to approximately 10^4 times compared to bulk Si was demonstrated. Afterwards, the same group demonstrated an enhancement of the spontaneous Raman scattering coefficient higher than six times, due to the slow light in silicon PCWG [101].

However, these structures are resonant only at the Stokes frequency, thus SRS occurs under pulsed excitation, while in a continuous regime the Raman-scattering efficiency is not high enough with respect to the efficiency of other competitive nonlinear effects, such as TPA and FCA.

In order to increase the spontaneous and stimulated Raman-scattering efficiency, structures resonating at both the pump and Stokes wavelengths should be implemented. In 2010, Checoury et al. investigated spontaneous and stimulated Raman scattering in very narrow W0.66 PCWG (width of the W0.66 = 2/3 of the W1 standard width) [102]. The realized waveguides were oriented along the (100) crystallographic direction of silicon, and allow both a 60% decrease in the Raman volume with respect to W1 waveguides oriented along the (110) direction, and a decrease in both the pump and Stokes modes group velocities. SRS was observed at room temperature for a continuous incident power as low as 20 mW, in good agreement with the simulations performed considering FCA negligible.

A Raman laser was made in 2013 by a photonic crystal high-quality-factor nanocavity with a dimension less than 10 µm [103]. The photonic crystal had a triangular lattice structure composed by circular air holes in a suspended silicon membrane. The nanocavity was realized by a linedefect waveguide that shows two types of propagation modes within the photonic bandgap, an odd-waveguide mode and an even-waveguide mode, allowing us to confine pump light and Stokes–Raman-scattered light, respectively. This structure gives rise to a CW Raman silicon laser with a low lasing threshold of 1 µW. Moreover, this Raman laser does not require p–i–n diodes, because the nanocavity produces net Raman gain in the low-excitation range before TPA-induced FCA becomes dominant, allowing this low lasing threshold [103].

5. Raman Laser in Nanophotonics

During the past few decades, a significant number of nanomaterials have shown to have notable optical properties, motivating the fabrication and design of nanoscale photonic devices [104]. These nanoscale devices are important in order to obtain a smooth integration with electronic devices, for example, transistors in sub-100 nm length scales. In this context, one of the most promising materials for light emission applications in microphotonics is based on silicon nanocrystals (Si-nc). The main idea is to induce quantum confinement effects by limiting carriers into very small silicon nanoclusters (1–4 nm in size), leading to change in the physical properties of bulk silicon. In detail, the reduction of dimensionality affects not only the linear optical properties, such as emission efficiency and band gap, but also the nonlinear optical properties, which are usually enhanced [93,105]. Recently, third-order, nonlinear, optical properties of Si-nc have been investigated, and a significant variation of the nonlinear refractive-index (n_2) values has been shown up to two orders of magnitude in SiO_2 films containing Si-nc and/or Si nanoclusters, with respect to crystalline silicon. Therefore, the comparison between experimental and theoretical results is still an issue [106]. From a theoretical point of view, it has been demonstrated [107], that in order to explain the origin of large values of third-order, nonlinear, optical properties, which are affected by their structural parameters (crystallinity, density, size and distribution), both the defect states and the quantized electronic states should be taken into account.

Concerning spontaneous Raman scattering, a strong enhancement ($\sim 10^3$) obtained from individual silicon nanowires and nanocones, as compared with bulk Si, was reported by Cao et al. [108]. The increase in Raman-scattering intensity with decreasing diameter was explained by structural resonances in the local field.

Concerning SRS in silicon nanocrystals (Si-nc), a giant Raman gain was measured at the wavelength of interest for telecommunications. The film of silicon nanocrystals (Si-nc), embedded in a silica matrix and obtained by the molecular beam deposition method, had an increasing Si concentration along the longer dimension. An impressive enhancement of the Raman gain in Si-nc up to four orders of magnitude, compared with bulk silicon, as a function of Si concentration, was observed [56]. We note that the volume of Si-nc does not exceed 10% of the volume of the sample, which makes the difference between the gain properties of Si-nc and bulk silicon more significant. From the applicative point of view, these experimental results open the way towards Raman–laser-based nanostructured material.

Since the SRS effect in Si-nc about 10^4 times larger with respect to silicon, a Raman laser with typical dimensions of a few micrometers could be obtained.

Therefore, all the advantages of combining optical and electronic functions on a single chip [56] could be experienced.

Although a general theory on the relation between nanostructuring and Raman gain is not established, we expect that when the particle dimensions are of a few nanometers, the phonon confinement effect plays a significant role; therefore, SRS enhancement can be attributed to the quantum confinement effect. Therefore, the gain of nanomaterials should be different from bulk, and related to the intrinsic properties of materials, and the essential trade-off between gain and bandwidth should be overcome, too. We guess that in Si-nanocrystals (Si-nc), the physical insight is similar to the one reported by Peng in [37]. When the mean free path of an electron is larger with respect to a phonon, the electron can collide with the phonon many times, and a strong phonon amplification can be obtained. From the point of view of energy transfer, first the energy of the laser field is absorbed by the electrons, and then it is transferred to the phonons by the electron–phonon interaction. If a resonance condition is obtained, for example, due to the interface levels, the movement of electrons is strongly confined, and due to the confinement effect, we expect that all light-generated electrons have to be involved in the electron–phonon interaction, resulting in significant amplification of phonons. Definitely, the efficiency of the electron–phonon interaction in a nanocrystal is much higher than in a bulk crystal [38].

In reference [109], Agarwal et al. reported a strong SRS and very high Raman gain in optical cavities made of Si nanowire of various diameters in the visible region [109]. The authors evaluated by electromagnetic calculations an enhancement of the Raman gain coefficient of Si nanowire by a factor greater than 10^6 at 532 nm excitation with respect to the gain value at the 1.55 µm wavelength reported in literature [56], even though the losses are 10^8 higher at 532 nm. They ascribed this behavior to the higher electric field intensity and lower mode volume of the electromagnetic modes inside the Si nanowire, with respect to bulk Si at the pump and the Stokes wavelength, as well as the spatial overlap of these modes inside the cavity. Moreover, they measured the SRS threshold as low as 30 kW/cm^2. These results could allow the realization in the next future of a monolithically-integrable, nanoscale low-powered Si Raman laser.

Recently, Rukhlenko and Kalavally presented the first theoretical study of CW Raman amplification in silicon–nanocrystal waveguides with enhanced mode confinement [110]. In detail, they studied Raman amplification in silicon nanocrystal waveguides, and observed that it strongly depends on the composition and geometric parameters of the waveguide. Fixing the geometric parameters of the waveguide, the maximal Stokes intensity peaks can be obtained for a given optimal density of silicon nanocrystals. Moreover, the optimal length and peak Stokes intensity depends on the height and width of the waveguide's cross-section and input conditions. However, the amplifier again requires a moderate power off-chip laser pump, and also the waveguide length is yet not small enough in view of dense integration.

Interestingly, slotted PCWGs performance are highly dependent on the width and refractive index of the slot. This dependence was studied and formalized by Datta's group [111], by calculating the respective dispersion diagrams using the three-dimensional full-vector Plane Wave Expansion method. They also characterized the nonlinear performance of these slotted PCWGs by using a low-index, highly nonlinear material in the slot. In particular, a silicon nanocrystal-embedded PCWG was designed to enhance the SRS gain, and this enhancement shows a net gain of the order of 11 dB in a ~7 µm long slotted PCWG.

Thereafter, the same group, exploiting their theoretical study, as well as the giant Raman gain of silicon nanocrystal material [56], proposed an all-silicon, micron-scale, Raman amplifier based on SiNC/SiO$_2$ embedded in slotted PCWG [112]. In detail, the structure consisted in a suspended silicon slab, which is pierced by holes of air to form the photonic crystal. The holes were organized in a triangular lattice (lattice constant a = 0.426 µm), and their radius, the thickness of the slab and the width of the SiNC/SiO$_2$ slot, have been taken as 0.3a, 0.7a and 0.2a, respectively. A volumetric proportion of approximately 1:10 of Si to the SiO$_2$ leads to a refractive index of 1.98 of the SiNC/SiO$_2$. In order to obtain both a low-cost optical pumping and the possibility of on-chip integration, a Light Emitting Diode (LED) with 6 mW pump power has been considered.

The authors demonstrated that an amplifier made with a 4 µm long slotted PCWG yields to an overall gain of ~3.22 dB at a bit rate of 400 Gbps, whereas with a 10 µm long slotted PCWG the gain is increased to ~7.93 dB at 200 Gbps. Finally, they suggest to exploit the strong electroluminescence from SiNC/SiO$_2$ in order to integrate the pump source on the same platform, slowing in the next future to realize the micron-scale silicon Raman laser without any external pump source.

In reference [113], Pradhan et al. reported the design of an integrable all-silicon Raman laser of a foot print of 7µm, based on a slotted photonic crystal nanocavity, which takes advantage the giant Raman gain coefficient of a silicon nanocrystal [56]. The device exhibits a lasing efficiency of 18% at a wavelength of 1552 nm, with an optical threshold power of the order of 0.5 µW. The effect of imperfections introduced during fabrication on the performance of the device was evaluated, too. Considering random variation in radii and in-plane positions of the air holes, the device performance was tolerant up to a 6% random variation of the structural parameters. In addition, the submicrowatt threshold of the device as a function of Q-factors and modal volume was evaluated, and it was demonstrated that it is tolerant for a significant deviation (over 30%) of these parameters from their optimized values.

More recently, a heterostructure nanocavity has been used to implement a Raman laser [114]. A line defect in a photonic crystal, made with a triangular lattice of circular air holes with a lattice constant a of 410 nm, defines the nanocavity, while the heterostructure is formed by increasing the lattice constant a by 5 and 10 nm in the x-direction in the areas closer to the center (Figure 6a, light and dark green areas, respectively) [115]. This change of the lattice parameter allows two propagation band frequencies that are the Stokes and pump modes, respectively (see Figure 6b), thus leading to the formation of two nanocavity modes with high Q. Figure 6c reports the in-plane Raman scattering in a microscopic model for the x-direction of the cavity chosen parallel to the [100] direction of Si. The polarization directions of the pump light and the Raman scattered light are orthogonal in the x–y plane in this geometry. The Si Raman laser was the realized on a modified (100) SOI wafer, in which the top Si layer has been changed by rotating the crystal orientation in the plane by 45° with respect to the crystal direction of the substrate. This device was characterized, and room temperature CW laser oscillation with a very impressive sub-microwatt threshold (0.53 µW), and a maximum energy efficiency of 5.6%, is reported [114].

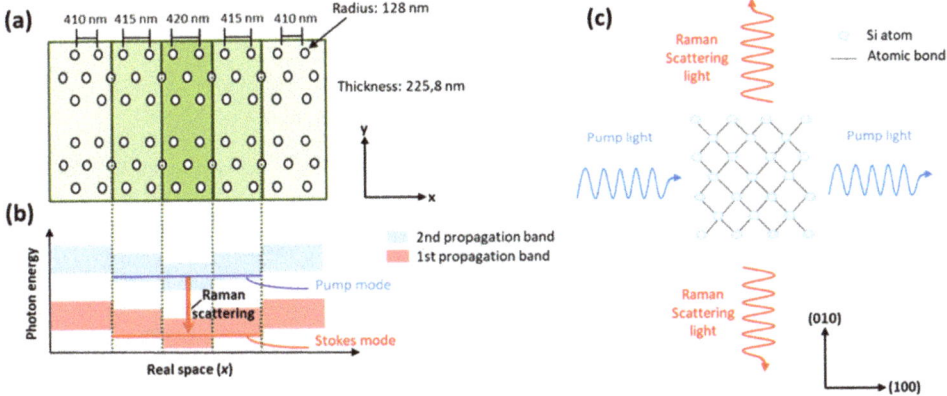

Figure 6. (a) Schematic of a heterosctructure nanocavity. (b) Band diagram of the nanocavity. (c) Schematic of the in-plane Raman scattering for the cavity's x-direction being parallel to the (100) direction of crystalline Si.

6. Conclusions

In this review, the most significant results, concerning laser sources based on SRS obtained in the last two decades, are reviewed. After the description of first silicon laser, which was based on the nonlinear waveguide, we focus on microcavity and photonics crystals, which are able to enhance the nonlinear interaction between light and matter, allowing us to obtain promising integrated Raman active devices. Finally, some recent interesting investigations, combining the potentiality of silicon nanocrystals and microphotonics structures, have been reported, too.

Here, we try to highlight, not only the development of integrated Raman source, but also the transition between photonics devices realised in waveguide, and in microphotonics structures based on photon confinement effects. We try to elucidate the main difference related to their working principle and the advantages for their applications. From a theoretical point of view, we note that a general theory on the relation between nanostructuring and Raman gain is not completely established [116–130], while from an applicative point of view, reported results open new perspectives for the realization of more efficient Raman lasers with ultra-small sizes, which would increase the synergy between nanoelectronic and nanophotonic devices.

We note that while the transition between integrated waveguide and microphotonics structures has been facilitated by the development of micro-scales' fabrication techniques, the one between micro and nano is still an issue. Right now, silicon nanocrystals are the more promising options. However, we note that SRS in nano is in its infancy, many investigations are not mature, and in some cases, only pioneering works or preliminary results on devices have been reported in the literature. Therefore the big challenge of the future is a reduction in the size of integrated optical devices towards nano dimensions, while maintaining a high level of performance.

We really hope that these encouraging perspectives can stimulate further theoretical and experimental works required to finally achieve the crucial milestone of a monolithically integrable, nanoscale, low-powered Si Raman laser, which could be integrated with other nanoscale electronic and optical components, leading to the development of next generation of nanosystems.

Author Contributions: L.S. and M.A.F. writing original draft preparation, review and editing. All authors have read and agreed to the published version of the manuscript.

Funding: This research received no external funding.

Conflicts of Interest: The authors declare no conflict of interest.

References

1. Boyd, R.W. *Nonlinear Optics*; Academic Press: Cambridge, MA, USA, 2003.
2. Shen, Y.R. *The Principles of Nonlinear Optics*; Wiley: Hoboken, NJ, USA, 2003.
3. Sasian, J.M. Design of null lens correctors for the testing of astronomical optics. *Opt. Eng.* **1988**, *27*, 121051. [CrossRef]
4. Agrawal, G.P.; Boyd, R.W. *Contemporary Nonlinear Optics*; Academic Press: Cambridge, MA, USA, 1992.
5. Armstrong, J.A.; Bloembergen, N.; Ducuing, J.; Pershan, P.S. Interactions between Light Waves in a Nonlinear Dielectric. *Phys. Rev.* **1962**, *127*, 1918–1939. [CrossRef]
6. McClung, F.J.; Hellwarth, R.W. Giant optical pulsations from ruby. *J. Appl. Phys.* **1962**, *33*, 828–829. [CrossRef]
7. Eckhardt, G.; Hellwarth, R.W.; McClung, F.J.; Schwarz, S.E.; Weiner, D.; Woodbury, E.J. Stimulated Raman scattering from organicliquids. *Phys. Rev. Lett.* **1962**, *9*, 455–457. [CrossRef]
8. Eckhardt, G.; Bortfeld, D.P.; Geller, M. Stimulated emission of stokes and anti-stokes raman lines from diamond, calcite, and α-sulfur single crystals. *Appl. Phys. Lett.* **1963**, *3*, 137–138. [CrossRef]
9. Hellwarth, R.W. Theory of Stimulated Raman Scattering. *Phys. Rev.* **1963**, *130*, 1850–1852. [CrossRef]
10. Bloembergen, N.; Shen, Y.R. Coupling Between Vibrations and Light Waves in Raman Laser Media. *Phys. Rev. Lett.* **1964**, *12*, 504–507. [CrossRef]
11. Bloembergen, N.; Shen, Y.R. Multimode Effects in Stimulated Raman Emission. *Phys. Rev. Lett.* **1964**, *13*, 720–724. [CrossRef]

12. Shen, Y.R.; Bloembergen, N. Theory of Stimulated Brillouin and Raman Scattering. *Phys. Rev.* **1965**, *137*, A1787–A1805. [CrossRef]
13. Bloembergen, N. The Stimulated Raman Effect. *Am. J. Phys.* **1967**, *35*, 989. [CrossRef]
14. Wang, C.-S. Theory of Stimulated Raman Scattering. *Phys. Rev.* **1969**, *182*, 482–494. [CrossRef]
15. Raymer, M.G.; Mostowski, J. Stimulated Raman scattering: Unified treatment of spontaneous initiation and spatial propagation. *Phys. Rev. A* **1981**, *24*, 1980–1993. [CrossRef]
16. Raymer, M.G.; Walmsley, I.A.; Mostowski, J.; Sobolewska, B. Quantum theory of spatial and temporal coherence properties of stimulated Raman scattering. *Phys. Rev. A* **1985**, *32*, 332–344. [CrossRef] [PubMed]
17. Stegeman, G.I.; Seaton, C.T. Nonlinear integrated optics. *J. Appl. Phys.* **1985**, *58*, R57–R78. [CrossRef]
18. Guo, Y.; Kao, C.K.; Li, H.E.; Chiang, K.S. *Nonlinear Photonics: Nonlinearities in Optics, Optoelectronics and Fiber Communications*; Springer-Verlag: Heidelberg, Germany, 2002.
19. Vahala, K. Optical microcavities. *Nature* **2003**, *424*, 839–846. [CrossRef]
20. Shields, A.J. Semiconductor quantum light sources. *Nat. Photon.* **2007**, *1*, 215–223. [CrossRef]
21. Gibbs, H.M. *Optical Bistability: Controlling Light with Light*; Academic Press: Orlando, FL, USA, 1985.
22. Yokoyama, H.; Ujihara, K. *Spontaneous Emission and Laser Oscillation in Microcavities*; CRC Press: Boca Raton, FL, USA, 1995.
23. Kippenberg, T.J.; Vahala, K.J. Cavity Opto-Mechanics. *Opt. Express* **2007**, *15*, 17172. [CrossRef]
24. Gerard, J.-M.; Sermage, B.; Gayral, B.; Legrand, B.; Costard, E.; Thierry-Mieg, V. Enhanced Spontaneous Emission by Quantum Boxes in a Monolithic Optical Microcavity. *Phys. Rev. Lett.* **1998**, *81*, 1110–1113. [CrossRef]
25. Reithmaier, J.P.; Sęk, G.; Loffler, A.; Hofmann, C.; Kühn, S.; Reitzenstein, S.; Keldysh, L.V.; Kulakovskii, V.D.; Reinecke, T.L.; Forchel, A. Strong coupling in a single quantum dot–semiconductor microcavity system. *Nature* **2004**, *432*, 197–200. [CrossRef]
26. Xu, Q.; Dong, P.; Lipson, M. Breaking the delay bandwidth product in a photonic structure. *Nature Phys.* **2007**, *3*, 406–410. [CrossRef]
27. Little, B.; Haus, H.; Foresi, J.; Kimerling, L.; Ippen, E.; Ripin, D. Wavelength switching and routing using absorption and resonance. *IEEE Photon-Technol. Lett.* **1998**, *10*, 816–818. [CrossRef]
28. Notomi, M.; Tanabe, T.; Shinya, A.; Kuramochi, E.; Taniyama, H.; Mitsugi, S.; Morita, M. Nonlinear and adiabatic control of high-Q photonic crystal nanocavities. *Opt. Express* **2007**, *15*, 17458–17481. [CrossRef] [PubMed]
29. Prasad, P.N. *Nanophotonics*; Wiley: Hoboken, NJ, USA, 2004.
30. Haus, J.W. *Fundamentals and Applications of Nanophotonics*; Woodhead: Sawston, UK, 2016.
31. Zheludev, N.I. Nonlinear optics on the nano scale. *Contemp. Phys.* **2010**, *43*, 365–377. [CrossRef]
32. Banfi, G.; DeGiorgio, V.; Ricard, D. Nonlinear optical properties of semiconductor nanocrystals. *Adv. Phys.* **1998**, *47*, 447–510. [CrossRef]
33. Kawata, S.; Inouye, Y.; Verma, P. Plasmonics for near-field nano-imaging and superlensing. *Nat. Photon.* **2009**, *3*, 388–394. [CrossRef]
34. Cho, C.-H.; Aspetti, C.O.; Park, J.; Agarwal, R. Silicon coupled with plasmon nanocavities generates bright visible hot luminescence. *Nat. Photon.* **2013**, *7*, 285–289. [CrossRef] [PubMed]
35. Gaponenko, S.V. Effects of photon density of states on Raman scattering in mesoscopic structures. *Phys. Rev. B* **2002**, *65*, 140303. [CrossRef]
36. Maeda, A.; Matsumoto, S.; Kishida, H.; Takenobu, T.; Iwasa, Y.; Shiraishi, M.; Ata, M.; Okamoto, H. Large Optical Nonlinearity of Semiconducting Single-Walled Carbon Nanotubes under Resonant Excitations. *Phys. Rev. Lett.* **2005**, *94*, 047404. [CrossRef]
37. Peng, F. Laser-induced phonon amplification in a single-walled nano tube. *Europhys. Lett.* **2006**, *73*, 116–120. [CrossRef]
38. Zhang, B.; Shimazaki, K.; Shiokawa, T.; Suzuki, M.; Ishibashi, K.; Saito, R. Stimulated Raman scattering from individual single-wall carbon nanotubes. *Appl. Phys. Lett.* **2006**, *88*, 241101. [CrossRef]
39. Wu, J.; Gupta, A.K.; Gutierrez, H.R.; Eklund, P.C. Cavity-Enhanced Stimulated Raman Scattering from Short GaP Nanowires. *Nano Lett.* **2009**, *9*, 3252–3257. [CrossRef] [PubMed]
40. Pask, H. The design and operation of solid-state Raman lasers. *Prog. Quantum Electron.* **2003**, *27*, 3–56. [CrossRef]

41. Černý, P. Solid state lasers with Raman frequency conversion. *Prog. Quantum Electron.* **2004**, *28*, 113–143. [CrossRef]
42. Piper, J.A.; Pask, H. Crystalline Raman Lasers. *IEEE J. Sel. Top. Quantum Electron.* **2007**, *13*, 692–704. [CrossRef]
43. Pask, H.; Dekker, P.; Mildren, R.P.; Spence, D.; Piper, J.A. Wavelength-versatile visible and UV sources based on crystalline Raman lasers. *Prog. Quantum Electron.* **2008**, *32*, 121–158. [CrossRef]
44. Nishizawa, J.-I.; Suto, K. Semiconductor Raman laser. *J. Appl. Phys.* **1980**, *51*, 2429. [CrossRef]
45. Stolen, R.H. Fundamentals of Raman Amplification in Fibers. In *Springer Series in Optical Sciences*; Springer Science and Business Media LLC: Berlin, Germany, 2007; Volume 90, pp. 35–39.
46. Sirleto, L.; Vergara, A.; Ferrara, M.A. Advances in stimulated Raman scattering in nanostructures. *Adv. Opt. Photon.* **2017**, *9*, 169. [CrossRef]
47. Sirleto, L.; Ferrara, M.A.; Vergara, A. Toward an ideal nanomaterial for on-chip Raman laser. *J. Nonlinear Opt. Phys. Mater.* **2017**, *26*, 1750039. [CrossRef]
48. Sirleto, L.; Ferrara, M.A.; Rendina, I.; Jalali, B. Broadening and tuning of spontaneous Raman emission in porous silicon at 1.5 µm. *Appl. Phys. Lett.* **2006**, *88*, 211105. [CrossRef]
49. Sirleto, L.; Ferrara, M.A.; Rendina, I.; Basu, S.N.; Warga, J.; Li, R.; Negro, L.D. Enhanced stimulated Raman scattering in silicon nanocrystals embedded in silicon-rich nitride/silicon superlattice structures. *Appl. Phys. Lett.* **2008**, *93*, 251104. [CrossRef]
50. Ferrara, M.A.; Donato, M.G.; Sirleto, L.; Messina, G.; Santangelo, S.; Rendina, I. Study of strain and wetting phenomena in porous silicon by Raman scattering. *J. Raman Spectrosc.* **2008**, *39*, 199–204. [CrossRef]
51. Sirleto, L.; Ferrara, M.A.; Nicotra, G.; Spinella, C.; Rendina, I. Observation of stimulated Raman scattering in silicon nanocomposites. *Appl. Phys. Lett.* **2009**, *94*, 221106. [CrossRef]
52. Sirleto, L.; Donato, M.G.; Messina, G.; Santangelo, S.; Lipovskii, A.; Tagantsev, D.; Pelli, S.; Righini, G.C. Raman gain in niobium-phosphate glasses. *Appl. Phys. Lett.* **2009**, *94*, 31105. [CrossRef]
53. Donato, M.G.; Gagliardi, M.; Sirleto, L.; Messina, G.; Lipovskii, A.; Tagantsev, D.; Righini, G.C. Raman optical amplification properties of sodium–niobium–phosphate glasses. *Appl. Phys. Lett.* **2010**, *97*, 231111. [CrossRef]
54. Ferrara, M.A.; Sirleto, L.; Nicotra, G.; Spinella, C.; Rendina, I. Enhanced gain coefficient in Raman amplifier based on silicon nanocomposites. *Photon. Nanostructures—Fundam. Appl.* **2011**, *9*, 1–7. [CrossRef]
55. Pernice, P.; Sirleto, L.; Vergara, A.; Aronne, A.; Gagliardi, M.; Fanelli, E.; Righini, G.C. Large Raman Gain in a Stable Nanocomposite Based on Niobiosilicate Glass. *J. Phys. Chem. C* **2011**, *115*, 17314–17319. [CrossRef]
56. Sirleto, L.; Ferrara, M.A.; Nikitin, T.; Novikov, S.; Khriachtchev, L. Giant Raman gain in silicon nanocrystals. *Nat. Commun.* **2012**, *3*, 1220. [CrossRef]
57. Sirleto, L.; Aronne, A.; Gioffré, M.; Fanelli, E.; Righini, G.C.; Pernice, P.; Vergara, A. Compositional and thermal treatment effects on Raman gain and bandwidth in nanostructured silica based glasses. *Opt. Mater.* **2013**, *36*, 408–413. [CrossRef]
58. Lipovskii, A.; Tagantsev, D.; Apakova, I.E.; Markova, T.S.; Yanush, O.V.; Donato, M.; Sirleto, L.; Messina, G.; Righini, G.C. Mid-Range Structure of Niobium–Sodium–Phosphate Electro-Optic Glasses. *J. Phys. Chem. B* **2013**, *117*, 1444–1450. [CrossRef]
59. Liang, D.; Bowers, J.E. Recent progress in lasers on silicon. *Nat. Photon.* **2010**, *4*, 511–517. [CrossRef]
60. Reed, G.T.; Mashanovich, G.; Gardes, F.Y.; Thomson, D.J. Silicon optical modulators. *Nat. Photon.* **2010**, *4*, 518–526. [CrossRef]
61. Michel, J.; Liu, J.; Kimerling, L.C. High-performance Ge-on-Si photodetectors. *Nat. Photon.* **2010**, *4*, 527–534. [CrossRef]
62. Jalali, B.; Yegnanarayanan, S.; Yoshimoto, T.; Coppinger, F.; Yoon, T.; Rendina, I. Advances in silicon-on-insulator optoelectronics. *IEEE J. Sel. Top. Quantum Electron.* **1998**, *4*, 938–947. [CrossRef]
63. Jalali, B.; Raghunathan, V.; Dimitropoulos, D.; Boyraz, O. Raman-based silicon photonics. *IEEE J. Sel. Top. Quantum Electron.* **2006**, *12*, 412–421. [CrossRef]
64. Leuthold, J.; Koos, C.; Freude, W. Nonlinear silicon photonics. *Nat. Photon.* **2010**, *4*, 535–544. [CrossRef]
65. Jalali, B. Silicon photonics: Nonlinear optics in the mid-infrared. *Nat. Photonics* **2010**, *4*, 506–508. [CrossRef]
66. Xu, Q.; Almeida, V.; Lipson, M. Micrometer-scale all-optical wavelength converter on silicon. *Opt. Lett.* **2005**, *30*, 2733–2735. [CrossRef]

67. Sharping, J.E.; Lee, K.F.; Foster, M.A.; Turner, A.C.; Schmidt, B.S.; Lipson, M.; Gaeta, A.L.; Kumar, P. Generation of correlated photons in nanoscale silicon waveguides. *Opt. Express* **2006**, *14*, 12388–12393. [CrossRef]
68. Rukhlenko, I.D.; Premaratne, M.; Agrawal, G.P. Nonlinear Silicon Photonics: Analytical Tools. *IEEE J. Sel. Top. Quantum Electron.* **2009**, *16*, 200–215. [CrossRef]
69. Foster, M.A.; Turner, A.C.; Lipson, M.; Gaeta, A.L. Nonlinear optics in photonic nanowires. *Opt. Express* **2008**, *16*, 1300–1320. [CrossRef]
70. Liu, X.; Osgood, R.M.; Vlasov, Y.; Green, W.M.J. Mid-infrared optical parametric amplifier using silicon nanophotonic waveguides. *Nat. Photon.* **2010**, *4*, 557–560. [CrossRef]
71. Daldosso, N.; Pavesi, L. Low-dimensional silicon as a photonic material. In *Nanosilicon*; Elsevier: Amsterdam, The Netherlands, 2008; pp. 314–334.
72. Rong, H.; Liu, A.; Jones, R.; Cohen, O.; Hak, D.; Nicolaescu, R.; Fang, A.; Paniccia, M. An all-silicon Raman laser. *Nature* **2005**, *433*, 292–294. [CrossRef] [PubMed]
73. Rong, H.; Xu, S.; Kuo, Y.-H.; Sih, V.; Cohen, O.; Raday, O.; Paniccia, M. Low-threshold continuous-wave Raman silicon laser. *Nat. Photon.* **2007**, *1*, 232–237. [CrossRef]
74. Claps, R.; Dimitropoulos, D.; Raghunathan, V.; Han, Y.; Jalali, B. Observation of stimulated Raman amplification in silicon waveguides. *Opt. Express* **2003**, *11*, 1731–1739. [CrossRef] [PubMed]
75. Delong, K.W.; Stegeman, G.I. Two-photon absorption as a limitation to all-optical waveguide switching in semiconductors. *Appl. Phys. Lett.* **1990**, *57*, 2063–2064. [CrossRef]
76. Liang, T.; Tsang, H.K. Role of free carriers from two-photon absorption in Raman amplification in silicon-on-insulator waveguides. *Appl. Phys. Lett.* **2004**, *84*, 2745–2747. [CrossRef]
77. Rong, H.; Liu, A.; Nicolaescu, R.; Paniccia, M.; Cohen, O.; Hak, D. Raman gain and nonlinear optical absorption measurements in a low-loss silicon waveguide. *Appl. Phys. Lett.* **2004**, *85*, 2196. [CrossRef]
78. DeLong, R.; Raghunathan, V.; Dimitropoulos, D.; Jalali, B. Role of nonlinear absorption on Raman amplification in Silicon waveguides. *Opt. Expr.* **2004**, *12*, 2774–2780.
79. Boyraz, O.; Jalali, B. Demonstration of a silicon Raman laser. *Opt. Express* **2004**, *12*, 5269–5273. [CrossRef]
80. Liu, A.; Rong, H.; Paniccia, M.; Cohen, O.; Hak, D. Net optical gain in a low loss silicon-on-insulator waveguide by stimulated Raman scattering. *Opt. Express* **2004**, *12*, 4261–4268. [CrossRef]
81. Rong, H.; Jones, R.; Liu, A.; Cohen, O.; Hak, D.; Fang, A.; Paniccia, M. A continuous-wave Raman silicon laser. *Nature* **2005**, *433*, 725–728. [CrossRef] [PubMed]
82. Boyraz, O.; Jalali, B. Demonstration of directly modulated silicon Raman laser. *Opt. Express* **2005**, *13*, 796–800. [CrossRef] [PubMed]
83. Tyszka-Zawadzka, A.; Szczepanski, P.; Mossakowska-Wyszyńska, A.; Karpierz, M.; Bugaj, M. Semi-analytical model of Raman generation in silicon-on-insulator rib waveguide with DBR/F-P resonator. *Opto-Electronics Rev.* **2013**, *21*, 382–389. [CrossRef]
84. De Leonardis, F.; Troia, B.; Campanella, C.E.; Passaro, V.M.N. Thermal and stress influence on performance of SOI racetrack resonator Raman lasers. *J. Opt.* **2014**, *16*, 85501. [CrossRef]
85. Lin, H.-B.; Campillo, A. Microcavity enhanced Raman gain. *Opt. Commun.* **1997**, *133*, 287–292. [CrossRef]
86. Spillane, S.M.; Kippenberg, T.J.; Vahala, K.J. Ultralow-threshold Raman laser using a spherical dielectric microcavity. *Nature.* **2002**, *415*, 621–623. [CrossRef]
87. Sennaroglu, A.; Kiraz, A.; Dündar, M.A.; Kurt, A.; Demirel, A.L. Raman lasing near 630 nm from stationary glycerol-water microdroplets on a superhydrophobic surface. *Opt. Lett.* **2007**, *32*, 2197–2199. [CrossRef]
88. Kippenberg, T.J.; Spillane, S.M.; Min, B.; Vahala, K.J. Theoretical and experimental study of stimulated and cascaded Raman scattering in ultra-high-Q optical microcavities. *IEEE J. Sel. Top. Quantum Electron* **2004**, *10*, 1219–1228. [CrossRef]
89. Dantham, V.R.; Bisht, P.; Namboodiri, C.K.R. Enhancement of Raman scattering by two orders of magnitude using photonic nanojet of a microsphere. *J. Appl. Phys.* **2011**, *109*, 103103. [CrossRef]
90. Foster, M.A.; Turner, A.C.; Sharping, J.E.; Schmidt, B.S.; Lipson, M.; Gaeta, A.L. Broad-band optical parametric gain on a silicon photonic chip. *Nature* **2006**, *441*, 960–963. [CrossRef]
91. Griffith, A.G.; Yu, M.; Okawachi, Y.; Cardenas, J.; Mohanty, A.; Gaeta, A.L.; Lipson, M. Coherent mid-infrared frequency combs in silicon-microresonators in the presence of Raman effects. *Opt. Express* **2016**, *24*, 13044. [CrossRef] [PubMed]

92. Latawiec, P.; Venkataraman, V.; Burek, M.J.; Hausmann, B.J.M.; Bulu, I.; Lončar, M. On-chip diamond Raman laser. *Optica* **2015**, *2*, 924–928. [CrossRef]
93. Daldosso, N.; Pavesi, L. Nanosilicon photonics. *Laser Photon. Rev.* **2009**, *3*, 508–534. [CrossRef]
94. Chen, Y.; Zhou, Z.-H.; Zou, C.-L.; Shen, Z.; Guo, G.-C.; Dong, C.-H. Tunable Raman laser in a hollow bottle-like microresonator. *Opt. Express* **2017**, *25*, 16879. [CrossRef] [PubMed]
95. Li, B.; Clements, W.R.; Yu, X.-C.; Shi, K.; Gong, Q.; Xiao, Y.-F. Single nanoparticle detection using split-mode microcavity Raman lasers. *Proc. Natl. Acad. Sci. USA* **2014**, *111*, 14657–14662. [CrossRef]
96. Ozdemir, S.; Zhu, J.; Yang, X.; Peng, B.; Yilmaz, H.; He, L.; Monifi, F.; Huang, S.H.; Long, G.-L.; Yang, L. Highly sensitive detection of nanoparticles with a self-referenced and self-heterodyned whispering-gallery Raman microlaser. *Proc. Natl. Acad. Sci. USA* **2014**, *111*, E3836–E3844. [CrossRef]
97. Yang, X.; Wong, C.W. Design of photonic band gap nanocavities for stimulated Raman amplification and lasing in monolithic silicon. *Opt. Express* **2005**, *13*, 4723–4730. [CrossRef]
98. Yang, X.; Wong, C.W. Coupled-mode theory for stimulated Raman scattering in high-Q/Vm photonic band gap defect cavity lasers. *Opt. Express* **2007**, *15*, 4763–4780. [CrossRef]
99. Yang, X.; Wong, C.W. Stimulated Raman amplification and lasing in silicon photonic band gap nanocavities. *Sensors Actuators A: Phys.* **2007**, *133*, 278–282. [CrossRef]
100. McMillan, J.F.; Yang, X.; Panoiu, N.C.; Osgood, R.M.; Wong, C.W. Enhanced stimulated Raman scattering in slow-light photonic crystal waveguides. *Opt. Lett.* **2006**, *31*, 1235. [CrossRef]
101. McMillan, J.F.; Yu, M.; Kwong, D.-L.; Wong, C.W. Observation of spontaneous Raman scattering in silicon slow-light photonic crystal waveguides. *Appl. Phys. Lett.* **2008**, *93*, 251105. [CrossRef]
102. Checoury, X.; Han, Z.; Boucaud, P. Stimulated Raman scattering in silicon photonic crystal waveguides under continuous excitation. *Phys. Rev. B* **2010**, *82*, 041308. [CrossRef]
103. Takahashi, Y.; Inui, Y.; Chihara, M.; Asano, T.; Terawaki, R.; Noda, S. A micrometre-scale Raman silicon laser with a microwatt threshold. *Nature* **2013**, *498*, 470–474. [CrossRef] [PubMed]
104. Suresh, S.; Arivuoli, D. Nanomaterials for nonlinear optical applications: A review. *Rev. Adv. Mater. Sci.* **2012**, *30*, 243–253.
105. Martinez, A.; Blasco, J.; Sanchis, P.; Galán, J.V.; García-Rupérez, J.; Jordana, E.; Gautier, P.; Lebour, Y.; Hernández, S.; Spano, R.; et al. Ultrafast All-Optical Switching in a Silicon-Nanocrystal-Based Silicon Slot Waveguide at Telecom Wavelengths. *Nano Lett.* **2010**, *10*, 1506–1511. [CrossRef] [PubMed]
106. Yıldırım, H.; Bulutay, C. Enhancement of optical switching parameter and third-order optical nonlinearities in embedded Si nanocrystals: A theoretical assessment. *Opt. Commun.* **2008**, *281*, 4118–4120. [CrossRef]
107. Ito, M.; Imakita, K.; Fujii, M.; Hayashi, S. Nonlinear optical properties of silicon nanoclusters/nanocrystals doped SiO[sub 2] films: Annealing temperature dependence. *J. Appl. Phys.* **2010**, *108*, 63512. [CrossRef]
108. Cao, L.; Nabet, B.; Spanier, J.E. Enhanced Raman Scattering from Individual Semiconductor Nanocones and Nanowires. *Phys. Rev. Lett.* **2006**, *96*, 1–4. [CrossRef]
109. Agarwal, D.; Ren, M.-L.; Berger, J.S.; Yoo, J.; Pan, A.; Agarwal, R. Nanocavity-Enhanced Giant Stimulated Raman Scattering in Si Nanowires in the Visible Light Region. *Nano Lett.* **2019**, *19*, 1204–1209. [CrossRef]
110. Rukhlenko, I.D.; Kalavally, V. Raman Amplification in Silicon-Nanocrystal Waveguides. *J. Light. Technol.* **2013**, *32*, 130–134. [CrossRef]
111. Datta, T.; Sen, M. Characterization of slotted photonic crystal waveguide and its application in nonlinear optics. *Superlattices Microstruct.* **2017**, *109*, 107–116. [CrossRef]
112. Datta, T.; Sen, M. LED pumped micron-scale all-silicon Raman amplifier. *Superlattices Microstruct.* **2017**, *110*, 273–280. [CrossRef]
113. Pradhan, A.K.; Sen, M. An integrable all-silicon slotted photonic crystal Raman laser. *J. Appl. Phys.* **2019**, *126*, 233103. [CrossRef]
114. Yamauchi, Y.; Okano, M.; Shishido, H.; Noda, S.; Takahashi, Y. Implementing a Raman silicon nanocavity laser for integrated optical circuits by using a (100) SOI wafer with a 45-degree-rotated top silicon layer. *OSA Contin.* **2019**, *2*, 2098–2112. [CrossRef]
115. Song, B.-S.; Noda, S.; Asano, T.; Akahane, Y. Ultra-high-Q photonic double-heterostructure nanocavity. *Nat. Mater.* **2005**, *4*, 207–210. [CrossRef]
116. Ruan, Z.; Fan, S. Superscattering of Light from Subwavelength Nanostructures. *Phys. Rev. Lett.* **2010**, *105*, 013901. [CrossRef]

117. Shan, L.; Pauliat, G.; Vienne, G.; Tong, L.; Lebrun, S. Design of nanofibres for efficient stimulated Raman scattering in the evanescent field. *J. Eur. Opt. Soc. Rapid Publ.* **2013**, *8*, 13030. [CrossRef]
118. Zatryb, G.; Wilson, P.R.J.; Wojcik, J.; Misiewicz, J.; Mascher, P.; Podhorodecki, A. Raman scattering from confined acoustic phonons of silicon nanocrystals in silicon oxide matrix. *Phys. Rev. B* **2015**, *91*, 235444. [CrossRef]
119. Hillenbrand, R.; Taubner, T.; Keilmann, F. Phonon-enhanced light–matter interaction at the nanometer scale. *Nature* **2002**, *418*, 159–162. [CrossRef]
120. Hillenbrand, R.; Keilmann, F. Complex Optical Constants on a Subwavelength Scale. *Phys. Rev. Lett.* **2000**, *85*, 3029–3032. [CrossRef]
121. Weeber, J.; Dereux, A.; Girard, C.; Francs, G.C.D.; Krenn, J.; Goudonnet, J.P. Optical addressing at the subwavelength scale. *Phys. Rev. E* **2000**, *62*, 7381–7388. [CrossRef] [PubMed]
122. Prince, R.; Frontiera, R.R.; Potma, E.O. Stimulated Raman Scattering: From Bulk to Nano. *Chem. Rev.* **2016**, *117*, 5070–5094. [CrossRef] [PubMed]
123. Zhang, Y.; Zhen, Y.-R.; Neumann, O.; Day, J.K.; Nordlander, P.; Halas, N.J. Coherent anti-Stokes Raman scattering with single-molecule sensitivity using a plasmonic Fano resonance. *Nat. Commun.* **2014**, *5*, 4424. [CrossRef] [PubMed]
124. Yampolsky, S.; Fishman, D.A.; Dey, S.; Hulkko, E.; Banik, M.; Potma, E.O.; Apkarian, V.A. Seeing a single molecule vibrate through time-resolved coherent anti-Stokes Raman scattering. *Nat. Photon.* **2014**, *8*, 650–656. [CrossRef]
125. Wickramasinghe, H.; Chaigneau, M.; Yasukuni, R.; Picardi, G.; Ossikovski, R. Billion-Fold Increase in Tip-Enhanced Raman Signal. *ACS Nano* **2014**, *8*, 3421–3426. [CrossRef]
126. Tonndorf, P.; Schmidt, R.; Böttger, P.; Zhang, X.; Börner, J.; Liebig, A.; Albrecht, M.; Kloc, C.; Gordan, O.; Zahn, D.R.; et al. Photoluminescence emission and Raman response of monolayer MoS_2, $MoSe_2$, and WSe_2. *Opt. Express* **2013**, *21*, 4908–4916. [CrossRef]
127. Nie, S. Probing Single Molecules and Single Nanoparticles by Surface-Enhanced Raman Scattering. *Sci.* **1997**, *275*, 1102–1106. [CrossRef]
128. Schuller, J.A.; Taubner, T.; Brongersma, M.L. Optical antenna thermal emitters. *Nat. Photon.* **2009**, *3*, 658–661. [CrossRef]
129. Miroshnichenko, A.E. Non-Rayleigh limit of the Lorenz-Mie solution and suppression of scattering by spheres of negative refractive index. *Phys. Rev. A* **2009**, *80*, 013808. [CrossRef]
130. Tribelsky, M.; Luk'Yanchuk, B. Anomalous Light Scattering by Small Particles. *Phys. Rev. Lett.* **2006**, *97*, 263902. [CrossRef]

© 2020 by the authors. Licensee MDPI, Basel, Switzerland. This article is an open access article distributed under the terms and conditions of the Creative Commons Attribution (CC BY) license (http://creativecommons.org/licenses/by/4.0/).

Review

Optical Frequency Combs in Quadratically Nonlinear Resonators

Iolanda Ricciardi [1,2], **Simona Mosca** [1], **Maria Parisi** [1], **François Leo** [3], **Tobias Hansson** [4], **Miro Erkintalo** [5,6], **Pasquale Maddaloni** [1,2], **Paolo De Natale** [7], **Stefan Wabnitz** [1,8,9] and **Maurizio De Rosa** [1,2,*]

1. CNR-INO, Istituto Nazionale di Ottica, Via Campi Flegrei 34, I-80078 Pozzuoli (NA), Italy; iolanda.ricciardi@ino.cnr.it (I.R.); simona.mosca@ino.cnr.it (S.M.); maria.parisi@ino.cnr.it (M.P.); pasquale.maddaloni@ino.cnr.it (P.M.); stefan.wabnitz@uniroma1.it (S.W.)
2. INFN, Istituto Nazionale di Fisica Nucleare, Sez. di Napoli, Complesso Universitario di M.S. Angelo, Via Cintia, 80126 Napoli, Italy
3. OPERA-photonics, Université libre de Bruxelles, 50 Avenue F. D. Roosevelt, CP 194/5, B-1050 Bruxelles, Belgium; francleo@ulb.ac.be
4. Department of Physics, Chemistry and Biology, Linköping University, SE-581 83 Linköping, Sweden; tobias.hansson@liu.se
5. The Dodd-Walls Centre for Photonic and Quantum Technologies, Auckland 1142, New Zealand; m.erkintalo@auckland.ac.nz
6. Physics Department, The University of Auckland, Auckland 1142, New Zealand
7. CNR-INO, Istituto Nazionale di Ottica, Largo E. Fermi 6, I-50125 Firenze, Italy; paolo.denatale@ino.cnr.it
8. Dipartimento di Ingegneria dell'Informazione, Elettronica e Telecomunicazioni, Sapienza Università di Roma- Via Eudossiana 18, I-00184 Roma, Italy
9. Department of Physics, Novosibirsk State University, 1 Pirogova Street, Novosibirsk 630090, Russia
* Correspondence: maurizio.derosa@ino.cnr.it

Received: 17 January 2020; Accepted: 19 February 2020; Published: 24 February 2020

Abstract: Optical frequency combs are one of the most remarkable inventions in recent decades. Originally conceived as the spectral counterpart of the train of short pulses emitted by mode-locked lasers, frequency combs have also been subsequently generated in continuously pumped microresonators, through third-order parametric processes. Quite recently, direct generation of optical frequency combs has been demonstrated in continuous-wave laser-pumped optical resonators with a second-order nonlinear medium inside. Here, we present a concise introduction to such quadratic combs and the physical mechanism that underlies their formation. We mainly review our recent experimental and theoretical work on formation and dynamics of quadratic frequency combs. We experimentally demonstrated comb generation in two configurations: a cavity for second harmonic generation, where combs are generated both around the pump frequency and its second harmonic and a degenerate optical parametric oscillator, where combs are generated around the pump frequency and its subharmonic. The experiments have been supported by a thorough theoretical analysis, aimed at modelling the dynamics of quadratic combs, both in frequency and time domains, providing useful insights into the physics of this new class of optical frequency comb synthesizers. Quadratic combs establish a new class of efficient frequency comb synthesizers, with unique features, which could enable straightforward access to new spectral regions and stimulate novel applications.

Keywords: optical frequency combs; quadratic nonlinearity; second harmonic generation; optical parametric oscillator; modulation instability

1. Introduction

Twenty years ago, optical frequency combs (OFCs) were established as powerful tools for accurate measurements of optical frequencies and timekeeping [1,2], a result of a long-standing effort, which was recognized with the Nobel Prize in Physics in 2005 [3,4]. The discrete ensemble of equally spaced laser frequencies that distinguish OFCs from other light sources is the spectral counterpart of the regular train of short pulses emitted by mode-locked lasers, which were initially used for comb generation. OFCs have become a critical component in many scientific and technological applications [5], from highly accurate optical frequency measurements for fundamental tests of physics [6–9] to exoplanet exploration [10–12] from air pollution detection [13–18] to telecommunication systems [19–21], while a growing interest has arisen in the quantum properties of OFCs [22–26].

Thereafter, comb emission was also demonstrated in continuous-wave (cw) laser-pumped resonators through cascaded third-order $\chi^{(3)}$ parametric processes [27]. In such Kerr resonators, a first pair of sidebands is generated around the pump frequency by cavity modulation instability or degenerate four-wave mixing (FWM); subsequently, cascaded four-wave mixing processes lead to the formation, around the pump frequency, of a uniform frequency comb, where self- and cross-phase modulation act to compensate for the unequal cavity mode spacing induced by the group velocity dispersion (GVD) [28,29]. Because of the relatively low strength of third-order nonlinearity, generation of Kerr combs requires small interaction volumes and high-Q resonators. For these reasons, small size resonators are particularly suited to reach broadband comb generation with quite moderate pump power [30]. Moreover, when the mode size is comparable with the light wavelength, a careful design of the resonator geometry can effectively modify the GVD of the resonator, leading to a broader spectral emission.

While $\chi^{(2)}$ three-wave mixing processes have been widely used for spectral conversion of femtosecond laser combs since their inception [31–36], only in recent years it was demonstrated that quadratic $\chi^{(2)}$ processes can lead to direct generation of optical frequency combs in cw-pumped quadratic nonlinear resonators. Actually, in 1999 Diddams at al. generated an OFC in a second-order nonlinear system, by actively inducing intracavity phase modulation inside a cw-pumped nearly degenerate optical parametric oscillator (OPO) [37], following a long development of phase modulation in lithium niobate for comb generation [38]. According to this scheme, besides the nonlinear crystal for parametric amplification, a phase modulator was placed inside the OPO cavity and driven at a modulation frequency equal to the cavity free spectral range. The modulator thus generated a family of phase-coupled sidebands, around the nearly degenerate signal and idler waves, which coincided with the resonator mode frequencies. Unlike other works presented in the following, where combs arise through purely $\chi^{(2)}$ optical processes, in that work combs were initially seeded by the sidebands generated in the intracavity modulator. Optical parametric amplification further increased the number of resonant sidebands, eventually leading to an 18-nm wide comb of equally spaced, mode-locked lines around the degenerate OPO frequency, only limited by the dispersive shift of the cavity modes, where mode-locking is imposed by phase modulation.

More recently, an optical frequency comb was produced by adding a second nonlinear crystal in a nondegenerate OPO [39]. The authors observed comb formation around the signal wavelength when the second crystal was phase mismatched for second harmonic generation (SHG) of the signal wave. Subsequent investigations of the same system reported experimental evidence of a comb around the second harmonic of the signal wave, whereas the comb around the signal was simultaneously transferred to the idler spectral range by parametric amplification [40]. In this case, the phase mismatched crystal behaves like a Kerr medium, producing a phase shift of the fundamental wave, which is proportional to the field intensity [41–43]. This phase shift can be explained as the consequence of cascaded quadratic processes which occur in the crystal when SHG is not phase matched. Indeed, when the fundamental pump wave, at frequency $\omega/2\pi$ enters a nonlinear crystal, a second harmonic field is generated, $\omega + \omega \to 2\omega$. If the process is not phase matched, the second harmonic (SH) field travels at a different phase velocity and, after half a coherence length, down-converts back to the fundamental frequency, $2\omega - \omega \to \omega$, with a different phase from that of the unconverted pump field.

As we will see later, a different cascaded three-wave-mixing process is decisive for the onset of frequency combs in phase-matched intracavity SHG—namely, internally pumped optical parametric oscillation [44,45]. In fact, degenerate optical parametric oscillation and SHG are mutually inverse processes, which satisfy the same phase matching condition, $\Delta k = 2k_1 - k_2 = 0$, between wave vectors $k_1 = k(\omega)$ of the fundamental field and $k_2 = k(2\omega)$ of the second harmonic field, respectively. Therefore, a properly phase-matched crystal placed inside an optical resonator can work either for SHG or parametric oscillation, depending on whether it is pumped at the fundamental or second harmonic frequency, respectively. However, the harmonic field generated in the first case can act as a pump for a nondegenerate cascaded OPO, and a pair of parametric fields start to oscillate with frequencies symmetrically placed around the fundamental pump. Although internally pumped OPO was observed and investigated for a long time, before the importance of OFCs was established [44–47], the observation of frequency combs in quadratic nonlinear media was postponed to recent years.

Here, we present a concise introduction to the physical mechanism that underlies quadratic comb formation, as well as an extended theoretical framework that has been developed so far. We particularly focus on our recent activity in this field, discussing our experimental and theoretical work on direct generation of quadratic combs. As a whole, it represents a systematic and coherent, although not exhaustive, approach to this new field. After the work of Ref. [39], Ricciardi et al. experimentally demonstrated direct frequency comb generation in an optical resonator with a single nonlinear crystal inside, originally conceived for cavity-enhanced SHG. OFCs were observed in the case of both phase-matched and phase-mismatched SHG. Moreover, the authors presented a simple theoretical model, which explained comb generation as the result of cascaded $\chi^{(2)}$:$\chi^{(2)}$ processes [48,49]. A more general theoretical description of comb generation in SHG cavity was successively developed by Leo et al., who modeled the dynamics of the cavity field in the time domain [50–52], and described comb formation in the framework of a modulation instability (MI), i.e., the growth of sidebands around a carrier frequency by amplification of small modulations on the carrier wave [53]. A similar theoretical description was adopted to describe the dynamics of quadratic combs observed in a degenerate OPO [54]. Finally, the most general approach, based on a single-envelope equation, has been also developed in order to study multi-octave, quadratic comb formation [55].

Quadratically nonlinear resonators thus emerge as the basis of an entirely new class of highly efficient synthesizers of OFCs, with unique features, such as the simultaneous generation of frequency combs in spectral regions far from the pump frequency and the role of phase matching in mitigating the effect of dispersion. Compared to Kerr combs, quadratic combs exploit the intrinsically higher efficiency of second-order nonlinearity, reducing the requirement in terms of pump power. Quadratic combs are still at an early stage but they are attracting the interest of an increasing number of research groups. More recent works are briefly reviewed in Section 6, where we conclude by giving an overview of promising developments of quadratic combs in terms of material platforms for chip-scale devices, steady low-noise dynamical regimes, and their potential interest for quantum optics.

2. Intracavity Second Harmonic Generation

The first system that we investigated for the generation of quadratic OFCs was a cw-pumped, cavity enhanced SHG system. The system, shown in Figure 1a, was based on a 15-mm-long periodically poled LiNbO$_3$ crystal, placed inside a traveling-wave optical resonator (free spectral range FSR = 493 MHz, quality factor $Q = 10^8$), resonating at the fundamental laser frequency ω_0. Mirror reflectivities were chosen in order to facilitate the onset of an internally pumped OPO [48]. The crystal was pumped by a narrow-line, 1064-nm-wavelength Nd:YAG laser, amplified by a Yb-doped fiber amplifier. Frequency locking of a cavity resonance to the laser was achieved by the Pound–Drever–Hall technique [56].

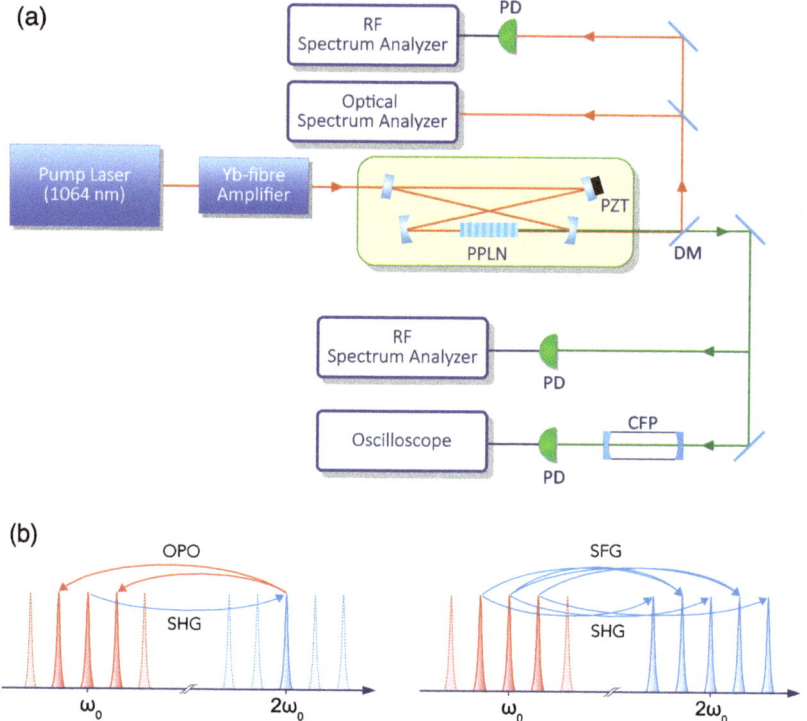

Figure 1. Singly resonant cavity second harmonic generation (SHG). (**a**) Experimental setup: periodically poled lithium niobate crystal (PPLN), piezoelectric actuator (PZT), photodiode (PD), dichroic mirror (DM). The output beams are detected and processed by radio-frequency (RF) analyzers, while optical spectral analysis is performed by an optical spectrum analyzer in the infrared range and a confocal Fabry-Pèrot interferometer (CFP) in the visible range. (**b**) Schematic representation of the first steps leading to the formation of a dual optical frequency comb in cavity-enhanced second-harmonic generation: (left) second-harmonic generation with cascaded nondegenerate optical parametric oscillator (OPO) gives rise to two subharmonic sidebands, which in turn (right) lead to successive, multiple second-harmonic and sum-frequency generations. Adapted with permission from [48]. Copyrighted by the American Physical Society.

The phase-matching condition for SHG was achieved by properly adjusting the crystal temperature. Under this condition, we observed a first regime of pure harmonic generation, where the harmonic power increased with the input pump power. As shown in Figure 2a, when the input power exceeded the threshold for internally pumped OPO, the second harmonic power ceased to grow, and two parametric waves started to oscillate at frequencies $\omega_0 \pm \Delta\omega$, symmetrically placed around the fundamental frequency (FF). As the power was further increased, additional sidebands appeared, displaced by multiples of $\Delta\omega$, leading to a multiple-FSR-spaced frequency comb, as sketched in Figure 2b. Finally, when the input power exceeded 5 W, secondary combs appeared around each of the primary comb lines, shown in Figure 2c. These secondary combs were spaced by 1 cavity FSR, as confirmed by the intermodal beat notes detected by fast photodetectors, both in the IR and in the visible spectral regions and processed by a radio frequency (RF) spectrum analyzer.

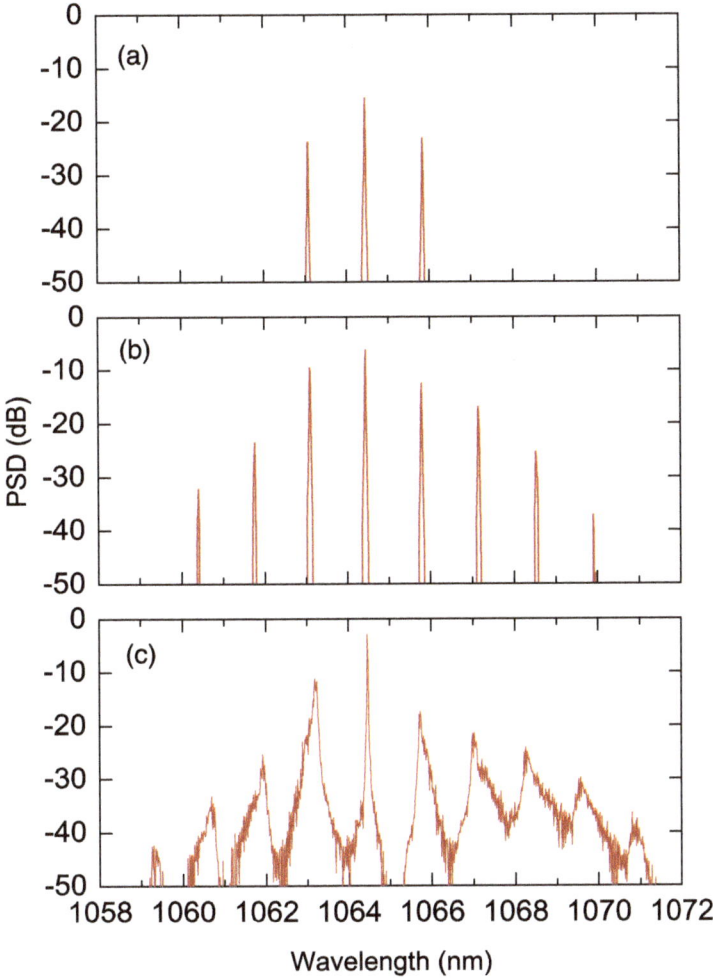

Figure 2. Optical spectral power around the fundamental mode for (**a**) 170 mW, (**b**) 2 W, and (**c**) 9 W of input powers. Adapted with permission from [48]. Copyrighted by the American Physical Society.

Subsequently, wave vector mismatch Δk was changed to finite values by varying the crystal temperature. Figure 3 shows infrared spectra observed for different values of the mismatch vector. For a positive mismatch, $\Delta k > 0$, the spectra (a)–(d) show widely separated sidebands, similar to the spectra observed at $\Delta k = 0$ (see Figure 2b). The spacing between sidebands, as well as the pump power threshold for cascaded optical parametric oscillation, rapidly increases with the mismatch. For $\Delta k < 0$, the spectra (e)–(h) consist of closely spaced (1 FSR) comb lines, and the spectral bandwidth increases with the magnitude of the mismatch. Larger negative phase mismatches are precluded by the limited accessible temperature range. Figure 3i,j show the beat notes corresponding to the comb in Figure 2c and the comb in Figure 3g, respectively. The broad feature of the beat note (i) reveals a strong intermodal phase noise and, as a consequence, a low degree of coherence between the comb teeth. This feature is consistent with a scenario where comb modes are weakly coupled with each other, as they originate independently from each other. On the contrary, the beat note (j) is extremely narrow, being limited by the detection resolution bandwidth and indicates a low intermodal phase noise and thus a strong phase coupling between all the comb teeth.

It is worth noting that the nonlinear resonator exhibits a noticeable thermal effect, mainly due to light absorption in the nonlinear crystal, which generates heat and leads to an increase of the cavity optical path, via thermal expansion and thermo-optic effect [57]. The photothermal effect introduces an additional nonlinear dynamical mechanism, with a temporal scale determined by the thermal diffusion time over the typical optical beam size [58]. In our case, the photothermal effect was helpful in thermally locking a cavity resonance to the laser frequency [59] when, especially at higher power, the PDH locking scheme was less effective. However, a better comprehension of the effect of thermal dynamics on comb formation requires further investigations.

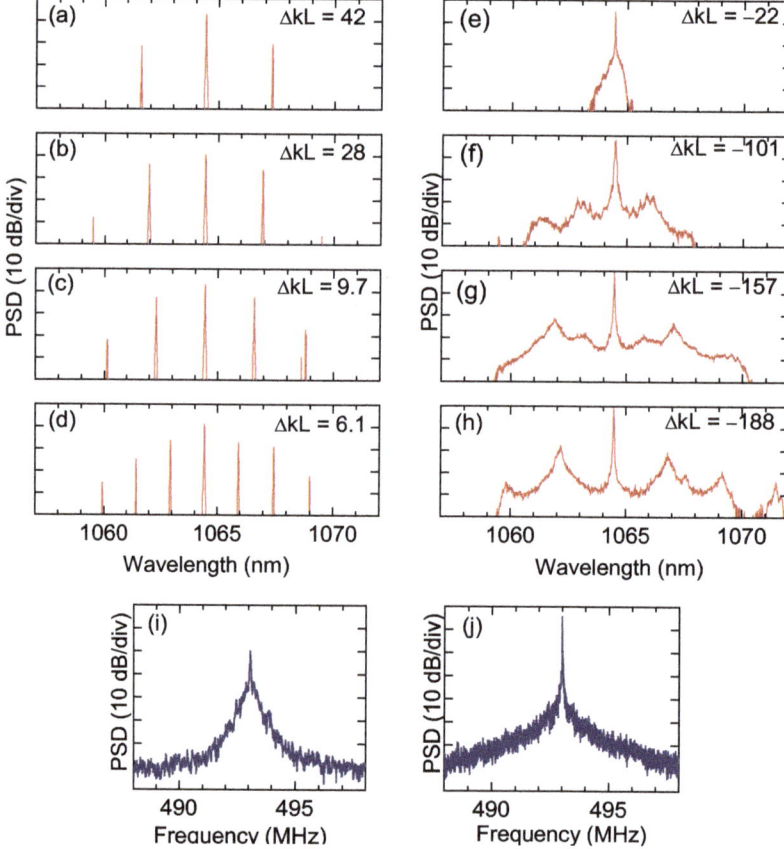

Figure 3. Optical spectra for phase-mismatched singly resonant cavity SHG. (**a**–**d**) Positive phase mismatch; (**e**–**h**) negative phase mismatch. Intermodal beat notes corresponding to (**i**) comb spectrum of Figure 2c, (**j**) comb spectrum in panel (**g**).

As anticipated in the introduction, the onset of internally pumped OPO marks the beginning of a cascade of second-order nonlinear processes, which eventually produces a comb of equally spaced frequencies. As depicted in Figure 1b, once generated, each parametric mode can generate new field modes through second harmonic, $(\omega + \Delta\omega) + (\omega + \Delta\omega) \to 2\omega + 2\Delta\omega$ and sum frequency with the fundamental wave $\omega + (\omega + \Delta\omega) \to 2\omega + \Delta\omega$, processes, respectively. All these processes have been considered for the derivation of a simple system of coupled mode equations for the three intracavity subharmonic electric field amplitudes, the fundamental A_0, at ω_0, and the parametric intracavity fields A_μ and $A_{\bar\mu}$, at $\omega_\mu = \omega_0 + \Delta\omega$ and $\omega_{\bar\mu} = \omega_0 - \Delta\omega$, respectively, which read [48]

$$\dot{A}_0 = -(\gamma_0 + i\Delta_0)A_0 - 2g\,\eta_{00\mu\bar\mu}A_0^* A_\mu A_{\bar\mu} - g(\eta_{0000}|A_0|^2 + 2\eta_{0\mu0\mu}|A_\mu|^2 + 2\eta_{0\bar\mu0\bar\mu}|A_{\bar\mu}|^2)A_0 + F_{\text{in}} \quad (1)$$

$$\dot{A}_\mu = -(\gamma_\mu + i\Delta_\mu)A_\mu - g\,\eta_{\mu\bar\mu00}A_0^2 A_{\bar\mu}^* - g(2\eta_{\mu00\mu}|A_0|^2 + \eta_{\mu\mu\mu\mu}|A_\mu|^2 + 2\eta_{\mu\bar\mu\mu\bar\mu}|A_{\bar\mu}|^2)A_\mu \quad (2)$$

$$\dot{A}_{\bar\mu} = -(\gamma_{\bar\mu} + i\Delta_{\bar\mu})A_{\bar\mu} - g\,\eta_{\bar\mu\mu00}A_0^2 A_\mu^* - g(2\eta_{\bar\mu00\bar\mu}|A_0|^2 + 2\eta_{\bar\mu\mu\bar\mu\mu}|A_\mu|^2 + \eta_{\bar\mu\bar\mu\bar\mu\bar\mu}|A_{\bar\mu}|^2)A_{\bar\mu}. \quad (3)$$

Here, $F_{\text{in}} = \sqrt{2\gamma_0/t_R}\,A_{\text{in}}$ is the cavity coupled amplitude of the constant input driving field A_{in}, at frequency ω_0; the γ's are the cavity decay constants; the Δ's are the cavity detunings of the respective modes; $g = (\kappa L)^2/2t_R$ is a gain factor depending on the crystal length L (hereafter we consider the cavity length equal to the crystal length); t_R is the cavity round trip time; $\kappa = \sqrt{8\omega_0}\chi^{(2)}_{\text{eff}}/\sqrt{c^3 n_1^2 n_2 \epsilon_0}$ is the second-order coupling strength. The latter is normalized so that the square modulus of the field amplitudes is measured in watts, with $\chi^{(2)}_{\text{eff}}$ the effective second-order susceptibility, c the speed of light, $n_{1,2}$ the refractive indices, and ϵ_0 the vacuum permittivity. The integer mode number μ denotes the μth cavity mode, starting from the central mode at ω_0, and overline stands for negative (lower frequencies). The η's are complex nonlinear coupling constants, depending on the wave-vector mismatches associated with a pair of cascaded second-order processes,

$$\eta_{\mu\sigma\rho\nu} = \frac{2}{L^2}\int_0^L\int_0^z \exp\left[-i(\xi_{\mu\sigma}z - \xi_{\rho\nu}z')\right] dz'\,dz \quad (4)$$

where $\xi_{jk} = k_{\omega_j} + k_{\omega_k} - k_{\omega_j + \omega_k}$.

A linear stability analysis of Equations (1)–(3) predicts the conditions for which a μ-pair of parametric fields starts to oscillate. By calculating the eigenvalues corresponding to Equations (1)–(3) linearized around the cw steady state solution, one obtains [49]

$$\lambda_\pm = -\gamma - g(\eta_{\mu00\mu} + \eta_{\bar\mu00\bar\mu}^*)|A_0|^2 \pm \sqrt{g^2|\eta_{\mu\bar\mu00}|^2|A_0|^4 - \left[\Delta_0 - D_2\mu^2 - ig(\eta_{\mu00\mu} - \eta_{\bar\mu00\bar\mu}^*)|A_0|^2\right]^2}, \quad (5)$$

where $D_2 \simeq -2\pi^2 c^3\beta''/L^2 n_0^3 = -(c/2n_0)D_1^2\beta''$ accounts for the group velocity dispersion at ω_0, with $\beta'' = \left.\frac{d^2 k}{d\omega^2}\right|_{\omega_0}$, and $n_0 = n(\omega_0)$ the refractive index at ω_0. Side modes start to oscillate, i.e., the zero solution for the parametric fields becomes unstable when the real part of an eigenvalue goes from negative to positive values. The coupling constants which appear in Equation (5) are: $\eta_{\mu\bar\mu00}$, which is the parametric gain related to cascaded SHG and OPO, whereby two photons at frequency ω_0 annihilate and two parametric photons at ω_μ and $\omega_{\bar\mu}$ are created, mediated by a SH photon; and $\eta_{\mu00\mu}$ ($\eta_{\bar\mu00\bar\mu}$), which is related to the sum frequency process between a parametric photon at ω_μ ($\omega_{\bar\mu}$) and the pump. The latter process is the most relevant nonlinear loss at the threshold (second term of r.h.s of Equation (5)), and provides a nonlinear phase shift (last term in the square brackets of r.h.s of Equation (5)). The lowest threshold occurs for a pair of parametric fields which starts to grow close to the minima of the sum frequency generation (SFG) efficiency.

A general expression for the dynamic equations for any number of interacting fields can be derived heuristically [49], yielding for each field A_μ, nearly resonant with the μ-th cavity mode,

$$\dot{A}_\mu = -(\gamma_\mu + i\Delta_\mu)A_\mu - g\sum_{\substack{\rho,\sigma \\ \nu = \rho+\sigma-\mu}} \eta_{\mu\nu\rho\sigma} A_\nu^* A_\rho A_\sigma + F_{\text{in}}, \quad (6)$$

where the summation over the indices ρ and σ goes over all the cavity resonant modes. The complex coupling constants are given by Equation (4), while the constraint over ν accounts for energy conservation. The coupled mode Equation (6) is formally analogous to the modal expansion for Kerr combs [60,61] and describes the whole comb dynamics. It is worth noting that the information provided by the linear stability analysis only holds for the very beginning of comb formation. Very quickly, a large number of cavity modes under the gain curve grow from noise. At the same time, they interact with each other through multiple nonlinear processes. These processes are not considered in the linear stability analysis, which intrinsically considers only three interacting modes. The long-term spectral configuration is thus the result of a complex interaction between many modes, over thousands of cavity round trips [52].

3. Time-Domain Model for Quadratic Combs

An alternative description of quadratic comb dynamics can be given in terms of time evolution of the slowly varying intracavity field envelopes. Let us define the envelopes $A(z,\tau)$ for the fundamental and $B(z,\tau)$ for the second harmonic electric fields in a resonator. Field dynamics can be described by an infinite dimensional map (Ikeda map) for the field amplitudes [50,51], which describes the evolution of cavity fields over the mth round trip, along with the boundary condition for the fields at the end of each round trip. The propagation equations for the fields $A_m(z,\tau)$ and $B_m(z,\tau)$ read as

$$\frac{\partial A_m}{\partial z} = \left[-\frac{\alpha_{c1}}{2} - i\frac{k_1''}{2}\frac{\partial^2}{\partial \tau^2}\right] A_m + i\kappa B_m A_m^* e^{-i\Delta kz}, \tag{7}$$

$$\frac{\partial B_m}{\partial z} = \left[-\frac{\alpha_{c2}}{2} - \Delta k'\frac{\partial}{\partial \tau} - i\frac{k_2''}{2}\frac{\partial^2}{\partial \tau^2}\right] B_m + i\kappa A_m^2 e^{i\Delta kz}, \tag{8}$$

where $z \in [0, L]$ is the position along the cavity round trip path; $\alpha_{c1,2}$ are propagation losses (hereafter, subscripts 1 and 2 denote fields at ω_0 and $2\omega_0$, respectively); $k_{1,2}'' = d^2k/d\omega^2|_{\omega_0, 2\omega_0}$ are the group velocity dispersion coefficients; $\Delta k' = dk/d\omega|_{2\omega_0} - dk/d\omega|_{\omega_0}$ is the corresponding group-velocity mismatch or temporal walk-off. The "fast-time" variable τ describes the temporal profiles of the fields in a reference frame moving with the group velocity of light at ω_0.

For the case of intracavity SHG, the fields at the beginning of the $(m+1)$th round trip are related to the fields at the end of the previous mth round trip according to the following cavity boundary conditions,

$$A_{m+1}(0,\tau) = \sqrt{1-\theta_1}\, A_m(L,\tau)\, e^{-i\delta_1} + \sqrt{\theta_1}\, A_{\text{in}} \tag{9}$$

$$B_{m+1}(0,\tau) = \sqrt{1-\theta_2}\, B_m(L,\tau)\, e^{-i\delta_2}, \tag{10}$$

where $\theta_{1,2}$ are power transmission coefficients at the coupling mirror, $\delta_1 \simeq (\omega_0 - \omega_{c1})t_R$ and $\delta_2 \simeq (2\omega_0 - \omega_{c2})t_R$ are the round trip phase detunings for the fields at ω_0 and $2\omega_0$, respectively, with ω_{c1} and ω_{c2} the frequencies of the respective nearest cavity resonance, and A_{in} is the external, constant driving field amplitude. It is worth noting that the Ikeda map of Equations (7)–(10) can describe different nonlinear systems (SHG or OPO, either singly or doubly resonant), by suitably choosing the boundary conditions. For a singly resonant cavity SHG, $\theta_2 = 1$, and the SH field resets at the beginning of each round trip, i.e., $B_{m+1}(0,\tau) = 0$.

For a relatively high-finesse resonator, the fundamental field evolves slowly during each round trip, and the infinite dimensional map may be averaged over one round trip length L. This averaging procedure yields a single mean field equation for the fundamental field amplitude [50],

$$t_R \frac{\partial A(t,\tau)}{\partial t} = \left[-\alpha_1 - i\delta_1 - iL\frac{k_1''}{2}\frac{\partial^2}{\partial \tau^2}\right] A - \rho A^* \left[A^2(t,\tau) \otimes I(\tau)\right] + \sqrt{\theta_1}\, A_{\text{in}}, \tag{11}$$

where t is a "slow time" variable, linked to the round trip index as $A(t = mt_R, \tau) = A_m(z = 0, \tau)$ [62–65], $\alpha_1 = (\alpha_{c1}L + \theta_1)/2$, $\rho = (\kappa L)^2$, $\otimes$ denotes convolution and the nonlinear response function $I(\tau) = \mathscr{F}^{-1}[\hat{I}(\Omega)]$, with $\hat{I}(\Omega) = [(1 - e^{-ix} - ix)/x^2]$, $x(\Omega) = \left[\Delta k + i\hat{k}(\Omega)\right] L$, and $\hat{k}(\Omega) = -\alpha_{c,2}/2 + i\left[\Delta k'\Omega + (k_2''/2)\Omega^2\right]$. Here, we define the direct and inverse Fourier transform operator as $\mathscr{F}[\cdot] = \int_{-\infty}^{\infty} \cdot e^{i\Omega\tau}\, d\tau$ and $\mathscr{F}^{-1}[\cdot] = (2\pi)^{-1} \int_{-\infty}^{\infty} \cdot e^{-i\Omega\tau}\, d\Omega$, respectively.

Similarly to the coupled mode equations in frequency domain, also the mean field Equation (11) exhibits an effective cubic nonlinearity, with a noninstantaneous response analogous to the delayed Raman response of cubic nonlinear media and other generalized nonlinear Schrödinger models.

Linear stability analysis of the cw solution of Equation (11) leads to the following expression for the eigenvalues [50],

$$\lambda_\pm = -\left(\alpha_1 + \rho P_0[\hat{I}(\Omega) + \hat{I}^*(-\Omega)]\right) \pm \sqrt{|\hat{I}(0)|^2 \rho^2 P_0^2 - \left(\delta_1 - \frac{k_1'' L}{2}\Omega^2 - i\rho P_0[\hat{I}(\Omega) - \hat{I}^*(-\Omega)]\right)^2}, \quad (12)$$

which, baring the notation, is substantially equivalent to Equation (5). Figure 4a shows the MI gain, $\Re[\lambda_+]$ profile as a function of the walk-off parameter $\Delta k'$. Clearly, there is no MI for zero walk-off, and MI appears for sufficiently large values of walk-off, revealing the fundamental role of group-velocity mismatch for the formation of quadratic optical frequency combs and related dissipative temporal patterns.

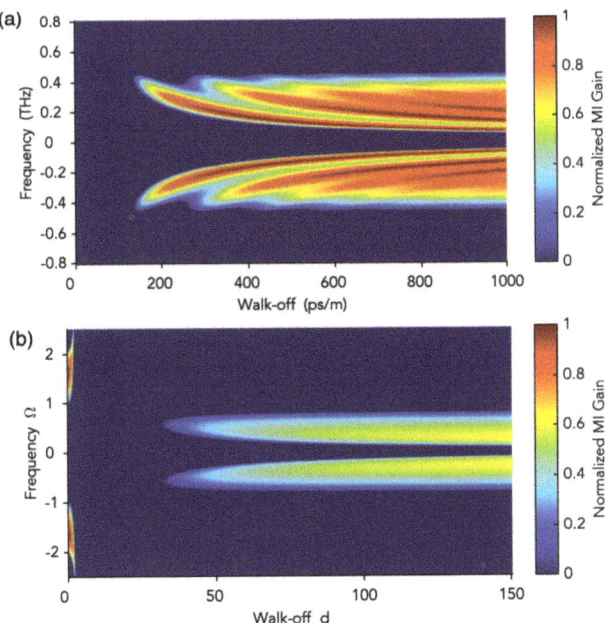

Figure 4. Modulation instability gain profiles as a function of temporal walk-off. (**a**) Singly resonant cavity SHG. (**b**) Doubly resonant cavity SHG (parameters are normalized according to Ref. [51]). Adapted with permission from [50,51]. Copyrighted by the American Physical Society.

Hansson et al. [52] demonstrated that the general system of coupled mode Equation (6) can be derived from the map of Equations (7)–(10). However, frequency domain coupled mode equations are not exactly equivalent to the time domain mean field Equation (11): the two approaches differ in the way the dispersion is averaged, although they provide almost equal results for the system of Ref. [48].

Theoretical models, in addition to providing useful insight into the physics of quadratic combs, can be a practical tool for simulating the comb dynamics, giving access to information not always available from the experiment. Both the frequency and time domain formalisms here described lend themselves to the numerical simulation of comb dynamics. Coupled mode Equation (6) is in general more time consuming than time domain approaches, unless it can be cast in a way where fast Fourier transform (FFT) algorithms can effectively reduce the computation time [66]. Numerical integration of the Ikeda map or the derived mean-field equation usually relies on split-step Fourier methods [67,68].

According to this method, propagation along each integration step is carried out in two steps. In a first step, the nonlinear and driving terms are propagated by means of a 4th-order Runge–Kutta method. The dispersive and absorption terms are propagated in a second step, where their propagation operator is evaluated in the Fourier domain, using an FFT algorithm. The simulation initiates by assuming a constant amplitude, input driving field A_{in} that describes the resonant pump laser. More importantly, in the first step a numerical white-noise background of one photon per mode must be added in order to seed the nonlinear processes which lead to the comb. Whereas the numerical integration of Ikeda map requires a spatial step size smaller than the cavity round trip length, the mean-field equation can be numerically integrated with temporal step sizes of the order of the round trip time, for the benefit of the computation time.

Figure 5 shows two spectra, (a) and (b), and the respective temporal patterns, (c) and (d), obtained by numerically integrating Equation (11). The simulations have been performed using the parameters from Ref. [48], in the case of quasi-phase matched SHG, for a constant input power of 2 and 7 W, respectively, and a small positive detuning. The simulated spectra are in good agreement with the experimental spectra shown in Figure 2b,c. For the moment, we cannot determine the temporal profile corresponding to a comb spectra. Hence, numerical simulations provide insights on the temporal feature of comb dynamics. We notice that the temporal pattern (c) associated to spectrum (a) has a stable periodic structure (also called Turing or roll pattern), which entails a strong phase coupling between the spectral modes, i.e., a mode-locked regime. Instead, the spectrum of Figure 5b, with secondary combs around the primary sidebands, corresponds to an irregular temporal pattern with no evidence of intermodal phase coupling. Moreover, it does not appear to reach a stationary regime. In both cases, the emission is not purely pulsed, as typically occurs for combs generated in femtosecond, mode-locked lasers, but the temporal patterns coexist with a flat background. The coexistence of a temporal pattern with a flat background is frequent for Kerr combs [69], as well as for combs generated in quantum cascade lasers [70,71]. In fact, in femtosecond laser combs the emission of short pulses is due to a particular phase relation between laser mode—i.e., all the modes have equal phases. However, in a wider sense, mode-locking only requires that a stable phase relation holds between all the mode fields. Finally, numerical simulations also reveal a slow drift of the temporal patterns (both at the fundamental and the SH fields) in the reference frame moving with the group velocity of the FF.

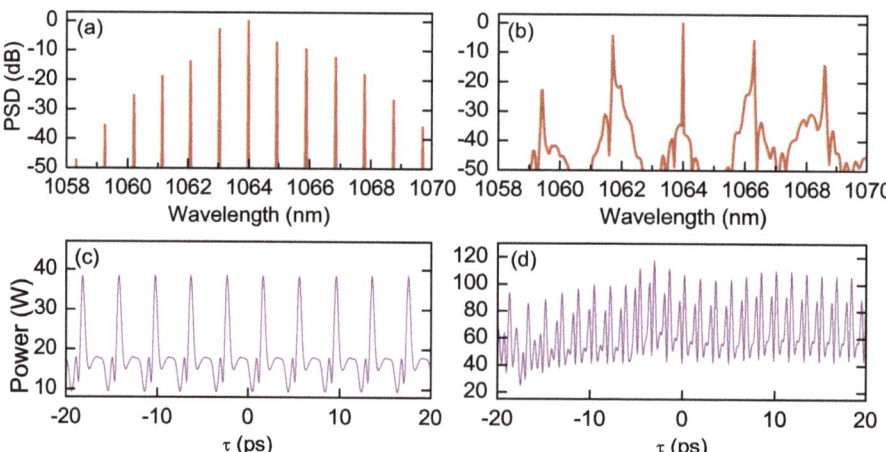

Figure 5. Numerical simulation of Equation (11), using the parameters of the system in Ref. [48]. (**a**) Input power 2 W, $\delta_1 = 0.001$. (**b**) Input power 7 W, $\delta_1 = 0.01$. (**c**,**d**) Details of the respective temporal patterns.

When $\theta_2 < 1$, the infinite dimensional map of Equations (7)–(10) describes the case of a doubly resonant optical cavity, where also second harmonic fields may resonate. Leo et al. theoretically analyzed this system [51] and derived a couple of two mean-field equations, which accurately model comb dynamics. These equations read, assuming phase-matched SHG,

$$t_R \frac{\partial A}{\partial t} = \left[-\alpha_1 - i\delta_1 - i\frac{k_1'' L}{2}\frac{\partial^2}{\partial \tau^2}\right] A + i\kappa L B A^* + \sqrt{\theta_1} A_{in}, \quad (13)$$

$$t_R \frac{\partial B}{\partial t} = \left[-\alpha_2 - i\delta_2 - \Delta k' L \frac{\partial}{\partial \tau} - i\frac{k_2'' L}{2}\frac{\partial^2}{\partial \tau^2}\right] B + i\kappa L A^2, \quad (14)$$

where α_2 is the cavity loss of the SH field.

Under realistic conditions, the two mean-field Equations (13) and (14) can be reduced to a single mean-field equation, analogously to Equation (11) for singly resonant cavity SHG. One obtains

$$t_R \frac{\partial A}{\partial t} = \left[-\alpha_1 - i\delta_1 - i\frac{k_1'' L}{2}\frac{\partial^2}{\partial \tau^2}\right] A - \rho A^* \left[A^2 \otimes J\right] + \sqrt{\theta_1} A_{in}, \quad (15)$$

where the Fourier transform of the kernel function J is

$$\hat{J}(\Omega) = \frac{1}{\alpha_2 + i\delta_2 - i\Delta k' L \Omega - i\frac{k_2'' L}{2}\Omega^2}. \quad (16)$$

A linear stability analysis of the cw solution (for both the Ikeda map and the mean-field approximations) reveals the significant role of temporal walk-off in enabling comb formation. However, in this case, MI gain may also occur for zero or relatively small values of the walk-off (Figure 4b).

4. Combs in Optical Parametric Oscillators

Degenerate optical parametric oscillation is the inverse process of cavity SHG, when the pump field A_{in} at the FF ω_0 is replaced by a pump field B_{in} at the SH frequency $2\omega_0$. Its dynamics can be described by an infinite dimensional map as well, where, in addition to Equations (7) and (8), the following boundary conditions hold for the fields at the beginning of each round trip,

$$A_{m+1}(0,\tau) = \sqrt{1-\theta_1}\, A_m(L,\tau)\, e^{-i\delta_1} \quad (17)$$

$$B_{m+1}(0,\tau) = B_{in}. \quad (18)$$

Here, we consider an OPO cavity where only the parametric field resonates. It is straightforward to extend the analysis to the case when the harmonic pump field also resonates.

Following the approach of Ref. [50], the infinite dimensional map can be combined into a single mean-field equation for the parametric field A, which reads, assuming $\Delta k = 0$ [54],

$$t_R \frac{\partial A(t,\tau)}{\partial t} = \left[-\alpha_1 - i\delta_1 - i\frac{L k_1''}{2}\frac{\partial^2}{\partial \tau^2}\right] A(t,\tau) - \mu^2 A^*(t,\tau)\left[A^2(t,\tau) \otimes I(\tau)\right] + i\mu B_{in} A^*(t,\tau), \quad (19)$$

where all the physical parameters and the kernel function I are the same as in Equation (11). We note that Equation (19) is similar to the corresponding mean-field equation for comb dynamics in cavity SHG, except for the parametric driving force (last term on the r.h.s.). Equation (19) has a trivial zero solution, $A_0 = 0$, and a nontrivial time independent solution, $A_0 = |A_0|e^{i\phi}$. From a linear stability analysis of the constant solution, we derived the following expression for the eigenvalues [54],

$$\lambda_\pm = -\left[\alpha_1 + \mu^2 |A_0|^2 \mathcal{I}_+(\Omega)\right] \pm \sqrt{(\alpha_1^2 + \delta_1^2) - [\delta_1 - D_2\Omega^2 - i\mu^2|A_0|^2 \mathcal{I}_-(\Omega)]^2}, \quad (20)$$

where $|A_0|^2 = [-\alpha_1 \pm \sqrt{\mu^2 B_{in}^2 - \delta_1^2}]/\mu^2 \hat{I}(0)$ is the squared modulus of the nontrivial solution and $\mathcal{I}_\pm(\Omega) = \hat{I}(\Omega) \pm \hat{I}^*(-\Omega)$. Similarly, for the zero solution the eigenvalues of the linearized system are

$$\lambda_\pm = -\alpha_1 \pm \sqrt{\mu^2 B_{in}^2 - (\delta_1 - D_2\Omega^2)^2}. \tag{21}$$

Both solutions exhibit MI gain for $\mathrm{Re}[\lambda_+] > 0$, which is shown in Figure 6a,b as a function of the cavity detuning. From Equation (20) it clearly appears that MI gain for the nontrivial solution depends both on walk-off $\Delta k'$, through $\mathcal{I}_\pm(\Omega)$, and GVD. As for singly resonant cavity SHG, MI only manifests itself for relatively high walk-off values, while it is absent for zero walk-off, as shown in Figure 6c. The instability of the zero solution, which is not expected in the usual dispersionless analysis of the OPO, does not depend on the walk-off, but it is rather induced by GVD. Actually, GVD is responsible for the unequal spacing between cavity resonances, so that they are asymmetrically displaced with respect to the degeneracy frequency ω_0, when the latter is perfectly resonant. Thus, GVD effectively favors parametric oscillations close to the degeneracy frequency. For normal dispersion, a positive detuning between the degeneracy frequency and the nearest cavity resonance can make symmetric an initially asymmetric pair of distant resonances, which now can more favourably oscillate than the degeneracy frequency ω_0. The larger the detuning, the more distant the symmetric resonances are. For small negative detunings no resonance pair can be symmetrically displaced around ω_0, and MI gain is maximum at the degeneracy frequency, decreasing as a function of the detuning amplitude. The same occurs in the case of anomalous dispersion, provided that the detuning sign is reversed.

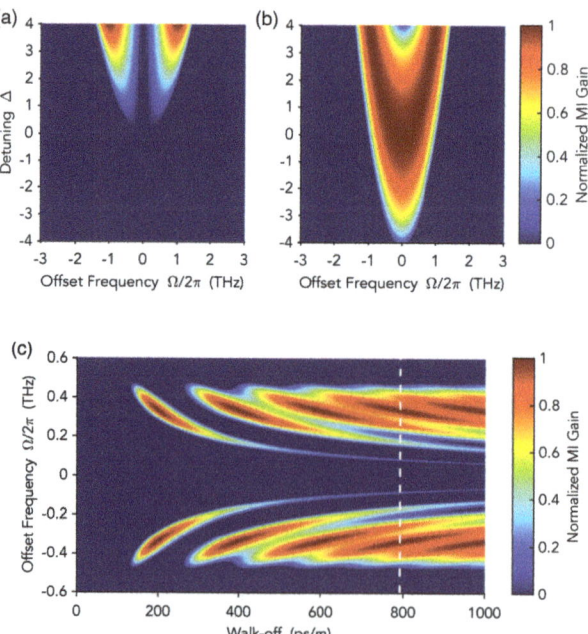

Figure 6. Optical frequency combs (OFC) in a degenerate OPO. (**a**,**b**) show the MI gain as a function of the normalized cavity detuning $\Delta = \delta_1/\alpha_1$, for the constant solution and the zero solution, respectively. (**c**) MI gain profiles as a function of the temporal walk-off. Adapted with permission from [54]. Copyrighted by the American Physical Society.

Frequency comb generation in an OPO has been demonstrated by using a nearly degenerate OPO pumped by a frequency doubled cw Nd:YAG laser (Figure 7). The OPO was based on a 15-mm-long periodically-poled 5%-MgO-doped lithium niobate crystal, with a grating period of $\Lambda = 6.92$ µm, enclosed in a bow-tie cavity resonating for the parametric wavelengths around 1064 nm, similar to that used for cavity SHG. The nonlinear crystal was located between two high-reflectivity spherical mirrors (with radius of curvature = 100 mm), while a flat high-reflectivity mirror was mounted on a piezoelectric actuator for cavity length control. A fourth, partially reflective flat mirror (R = 98%) allowed us to couple out the generated parametric radiation. The SH beam entered the OPO cavity from a first spherical mirror, passed through the nonlinear crystal, and left the cavity at the second spherical mirror. The FSR of the cavity was 505 MHz. We observed combs for pump powers higher than 85 mW (about three times the OPO threshold of 30 mW) and studied the effect of small cavity detunings on the comb spectra. Figure 8a–c show the experimental comb spectra recorded for $\Delta = -0.30, 0.00, 0.30$, respectively, with 300 mW of pump power. We found a good agreement with the corresponding spectra, shown in Figure 8d–f, calculated by numerically integrating the mean-field Equation (19). Experimental spectra for negative and zero detunings are very similar, displaying 1 FSR line spacing, whereas for the positive detuning the experimental spectrum consists of two pairs of widely spaced symmetric lines.

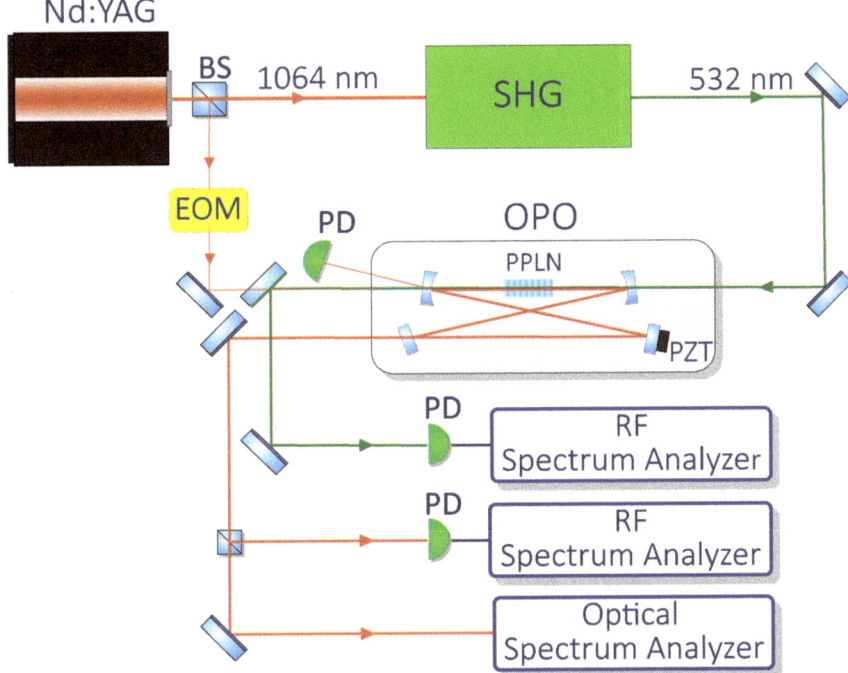

Figure 7. OFC in a degenerate OPO. Scheme of the experimental setup: beam splitter (BS), electro-optic phase modulator (EOM), periodically poled lithium niobate crystal (PPLN), piezoelectric actuator (PZT), photodiode (PD). Adapted with permission from [54]. Copyrighted by the American Physical Society.

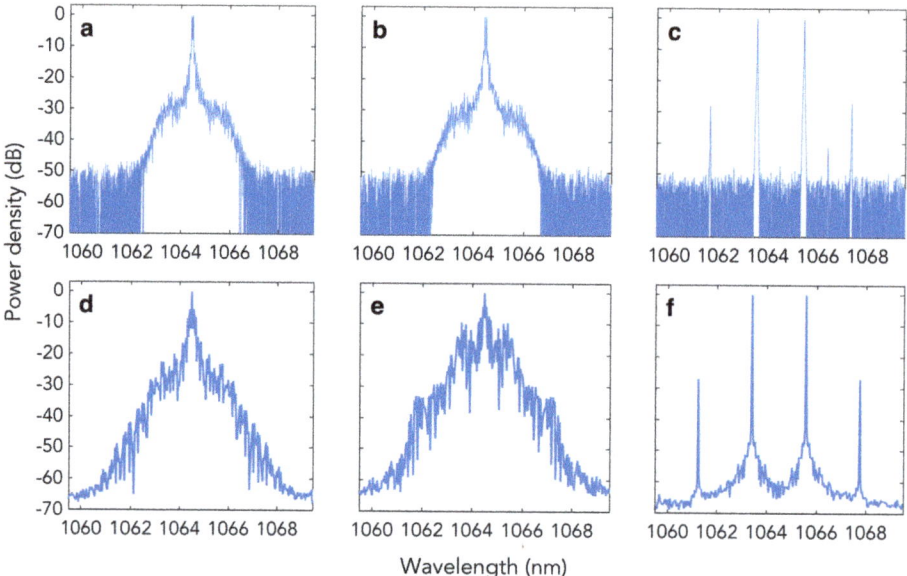

Figure 8. (**a**–**c**) Experimental OPO optical spectra for detunings $\Delta = -0.30, 0.00, 0.30$, respectively. (**d**–**f**) Corresponding numerically calculated spectra. From [54]. Copyrighted by the American Physical Society.

5. Single Envelope Equation

Models based on the two field envelopes, i.e., Equations (7)–(10) and their approximations hold as long as there is a single dominant nonlinear process and the combs are confined around two carrier frequencies. When the combs start to overlap, or multiple nonlinear processes play a prominent role, frequency comb generation may be studied by means of a more general model, based on a single-envelope equation combined with the boundary conditions that relate the fields between successive round trips and the input pump field [55],

$$\mathcal{F}[A^{m+1}(\tau, 0)] = \sqrt{\hat{\theta}(\Omega)} \mathcal{F}[A_{\text{in}}] + \sqrt{1 - \hat{\theta}(\Omega)} \, e^{i\phi_0} \mathcal{F}[A^m(\tau, L)] \tag{22}$$

$$\left[\partial_z - D\left(i\frac{\partial}{\partial \tau}\right) + \frac{\alpha_d}{2}\right] A^m(\tau, z) = i\rho_0 \left(1 + i\tau_{\text{sh}}\frac{\partial}{\partial \tau}\right) p_{\text{NL}}(\tau, z, A^m). \tag{23}$$

The boundary condition, Equation (22), is written in the Fourier domain, in order to account for the frequency dependence of the transmission coefficient θ at the input port of the resonator. It determines the intra-cavity field $A_{m+1}(\tau, z = 0)$ at the beginning of $(m+1)$th round trip in terms of the field at the end of the previous round trip $A_m(\tau, z = L)$ and the pump field A_{in}. Equation (23) is written in a reference frame moving at the group velocity at ω_0: p_{NL} is the broadband envelope of the nonlinear polarization $P_{\text{NL}} = P_{\text{NL}}^{(2)} + P_{\text{NL}}^{(3)} + \ldots = \epsilon(\chi^{(2)}E^2 + \chi^{(3)}E^3 + \ldots)$; $\rho_0 = \omega_0/2n_0 c \epsilon_0$; $\tau_{\text{sh}} = 1/\omega_0$ is the shock coefficient that describes the frequency dependence of the nonlinearity, and α_d is the distributed linear loss coefficient. Dispersion to all orders is included by the operator D,

$$D\left(i\frac{\partial}{\partial \tau}\right) = \sum_{l \geq 2} i\frac{\beta_l}{l!} \left(i\frac{\partial}{\partial \tau}\right)^l, \tag{24}$$

where $\beta_l = (d^l \beta / d\omega^l)_{\omega=\omega_0}$ are expansion coefficients of the propagation constant $\beta(\omega)$.

Figure 9 shows a spectrum obtained from the numerical simulation of Equations (22) and (23), when SHG and nondegenerate optical parametric oscillation are simultaneously quasi-phase matched in a radially poled, lithium niobate microresonator, pumped at 1850 nm (162 THz). In this case, the quasi-phase matching period for SHG (25.56 µm) simultaneously quasi-phase matches a nondegenerate OPO with idler (signal) at 56 THz (106 THz). The broadband power spectral density shows a generation of a multi-comb array, extending from the mid-infrared into the ultraviolet with a spacing of a single FSR (around 92 GHz). In addition to combs at the FF, SH, and third-harmonic (TH), two additional combs are generated around signal and idler frequencies. Moreover, several secondary combs appear between the FF and the SH and between the SH and the TH, respectively. These combs are generated by sum-frequency generation and difference frequency generation processes. For instance, the comb SC1 centered at 218 THz results from SFG between the idler and the FF, while SC3 (around 380 THz) results from SFG between the idler and the SH. On the other hand, DFG between the SH (TH) and the idler leads to a secondary comb SC2 (SC4) centered at 268 THz (430 THz).

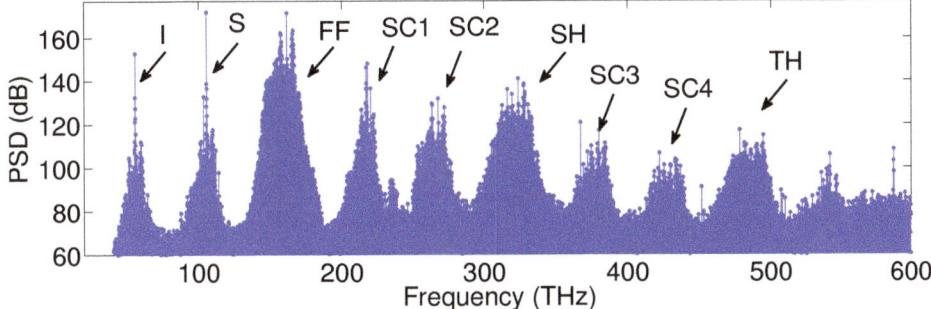

Figure 9. Numerical simulation of the single-envelope map when SHG and OPO processes are simultaneously phase matched in a lithium niobate microresonator pumped by 100 mW of cw power at 1850 nm. Reprinted with permission from [55] © The Optical Society.

6. Perspectives

Because of the intrinsically higher strength of the quadratic nonlinearity with respect to the third-order one, quadratic comb generation can be less demanding in terms of power density and cavity quality factor. Although quadratic combs have been generated in bulk cavities with moderate pump powers, their performance could increase if implemented in miniaturized devices, thus further extending and stimulating new applications [72,73]. Scaling the resonator to micrometric size may be beneficial for quadratic combs, allowing for a dramatic reduction of threshold power and a flexible management of the dispersion through a geometric design, allowing for a broader comb emission. As a matter of fact, direct generation of quadratic frequency combs has been very recently observed in chip-scale lithium niobate devices, such as periodically poled linear waveguide resonators [74,75], or exploiting naturally phase-matched SHG in whispering-gallery-mode resonators [76,77].

Several materials with second-order nonlinearity are suitable to be shaped into low-loss small-footprint resonators. Most of them have been used to generate Kerr combs [78–82], and, in some cases, secondary quadratic effects have been reported [78,79] or explicitly considered [83]. In contrast to Kerr combs, quadratic combs usually require more stringent conditions on phase matching and group velocity mismatch between different spectral components. Natural [84,85], cyclic [86], and quasi-[87,88] phase matching have been used in crystalline whispering-gallery-mode resonators. More recently, significant progress has been made in the fabrication of integrated, high-Q, lithium niobate microresonators for $\chi^{(2)}$ processes [89–92]. III-V materials provide an interesting photonic platform for second-order nonlinear optics, and different techniques have been devised to achieve phase matching [93], in particular for resonant structures [94–96]. It is worth noting that the well developed silicon platform can also be exploited for second-order nonlinear interaction. In fact, Timurdogan et al. demonstrated that a large "dressed" $\chi^{(2)}$ nonlinearity can be induced by breaking the crystalline center-symmetry of silicon when a direct-current field is applied across p-i-n junctions in ridge waveguides [97], enabling the implementation of quasi-phase matching schemes.

Unlike optical frequency combs in mode-locked lasers, parametrically generated combs do not usually correspond to a stable pulsed emission in the time domain. Different temporal regimes are possible, from chaotic to perfectly coherent states. The formation of temporal cavity solitons in a cw-pumped nonlinear resonator has attracted a particular interest in connection with parametrically generated combs [98]. Combs associated to a cavity soliton are broadband and highly coherent, which makes them ideal for low noise and metrological applications. In fact, cavity solitons are robust states which circulate indefinitely in a cavity, thanks to the double compensation between nonlinearity and chromatic dispersion and between cavity losses and cw driving. Recent theoretical works aim at identifying the dynamical regimes that exhibit soliton states or localized solutions in cavity SHG systems [99–102] or OPOs [103,104].

Finally, optical frequency combs are attracting a growing interest as sources of complex quantum states of light for high-dimensional quantum computation [26,105,106]. Second-order nonlinear optical systems are efficiently used for generation of quantum states of light: the classical correlations that establish in three-wave-mixing processes hold at the quantum level as well, leading, for instance, to generation of squeezed light or bipartite entanglement in an OPO. Tripartite, or quadripartite multicolor entanglement has been predicted in second-order nonlinear devices [107], in particular when multiple cascaded second-order nonlinear interactions occur, in traveling-wave or intracavity processes [108–110]. Interestingly, a recent study based on the three-wave model of Equations (1)–(3) predicts five-partite entanglement between one-octave-distant modes [111]. This result suggests that quadratic combs could exhibit multipartite entanglement between frequency modes, which are essential for scalable measurement-based quantum computing [112]. To fully explore these features, a general and complete analysis of the quantum dynamics of quadratic combs is needed [113].

Author Contributions: Designed and performed the experiments, I.R., S.M., M.P., P.M., and M.D.R.; theoretical modeling I.R, F.L., T.H., M.E., S.W., and M.D.R.; writing–original draft preparation, I.R. and M.D.R.; funding acquisition, M.E., P.D.N., S.W., and M.D.R.; all authors analyzed the data, discussed the results, read and edited the manuscript. All authors have read and agreed to the published version of the manuscript.

Funding: This research was funded by Ministero dell'Istruzione, dell'Università e della Ricerca (MIUR), PRIN 2015KEZNYM (NEMO); Ministero degli Affari Esteri e della Cooperazione Internazionale, project NOICE Joint Laboratory; European Union's Horizon 2020 research and innovation programme (Qombs Project, FET Flagship on Quantum Technologies grant no. 820419); the Rutherford Discovery Fellowships of the Royal Society of New Zealand and the Marsden Fund of the Royal Society of New Zealand. The work of S. W. is supported by the Ministry of Education and Science of the Russian Federation (Minobrnauka) (14.Y26.31.0017). T.H. acknowledges funding from the Swedish Research Council (Grant No. 2017-05309).

Conflicts of Interest: The authors declare no conflict of interest. The funders had no role in the design of the study; in the collection, analyses, or interpretation of data; in the writing of the manuscript, or in the decision to publish the results.

Abbreviations

The following abbreviations are used in this manuscript:

cw	Continuous wave
DFG	Difference frequency generation
FF	Fundamental frequency
FFT	Fast Fourier transform
FSR	Free spectral range
FWM	Four-wave mixing
GVD	Group velocity dispersion
MI	Modulation instability

OFC   Optical frequency comb
OPO   Optical parametric oscillator
SH   Second harmonic
SHG   Second harmonic generation
TH   Third harmonic

References

1. Jones, D.J.; Diddams, S.A.; Ranka, J.K.; Stentz, A.J.; Windeler, R.S.; Hall, J.L.; Cundiff, S.T. Carrier-envelope phase control of femtosecond mode-locked lasers and direct optical frequency synthesis. *Science* **2000**, *288*, 635–639. [CrossRef] [PubMed]
2. Holzwarth, R.; Udem, T.; Hänsch, T.W.; Knight, J.C.; Wadsworth, W.J.; Russell, P.S.J. Optical frequency synthesizer for precision spectroscopy. *Phys. Rev. Lett.* **2000**, *85*, 2264–2267. [CrossRef] [PubMed]
3. Hall, J.L. Nobel Lecture: Defining and measuring optical frequencies. *Rev. Mod. Phys.* **2006**, *78*, 1279–1295. [CrossRef]
4. Hansch, T.W. Nobel Lecture: Passion for precision. *Rev. Mod. Phys.* **2006**, *78*, 1297–1309. [CrossRef]
5. Newbury, N.R. Searching for applications with a fine-tooth comb. *Nat. Photonics* **2011**, *5*, 186–188. [CrossRef]
6. Predehl, K.; Grosche, G.; Raupach, S.M.F.; Droste, S.; Terra, O.; Alnis, J.; Legero, T.; Hansch, T.W.; Udem, T.; Holzwarth, R.; et al. A 920-Kilometer Optical Fiber Link for Frequency Metrology at the 19th Decimal Place. *Science* **2012**, *336*, 441–444. [CrossRef]
7. Clivati, C.; Cappellini, G.; Livi, L.F.; Poggiali, F.; de Cumis, M.S.; Mancini, M.; Pagano, G.; Frittelli, M.; Mura, A.; Costanzo, G.A.; et al. Measuring absolute frequencies beyond the GPS limit via long-haul optical frequency dissemination. *Opt. Express* **2016**, *24*, 11865–11875. [CrossRef]
8. Insero, G.; Borri, S.; Calonico, D.; Pastor, P.C.; Clivati, C.; D'Ambrosio, D.; De Natale, P.; Inguscio, M.; Levi, F.; Santambrogio, G. Measuring molecular frequencies in the 1–10 μm range at 11-digits accuracy. *Sci. Rep.* **2017**, *7*, 12780. [CrossRef]
9. Di Sarno, V.; Aiello, R.; De Rosa, M.; Ricciardi, I.; Mosca, S.; Notariale, G.; De Natale, P.; Santamaria, L.; Maddaloni, P. Lamb-dip spectroscopy of buffer-gas-cooled molecules. *Optica* **2019**, *6*, 436–441. [CrossRef]
10. Steinmetz, T.; Wilken, T.; Araujo-Hauck, C.; Holzwarth, R.; Haensch, T.W.; Pasquini, L.; Manescau, A.; D'Odorico, S.; Murphy, M.T.; Kentischer, T.; et al. Laser frequency combs for astronomical observations. *Science* **2008**, *321*, 1335–1337. [CrossRef]
11. McCracken, R.A.; Charsley, J.M.; Reid, D.T. A decade of astrocombs: recent advances in frequency combs for astronomy. *Opt. Express* **2017**, *25*, 15058–15078. [CrossRef]
12. Obrzud, E.; Rainer, M.; Harutyunyan, A.; Anderson, M.H.; Liu, J.; Geiselmann, M.; Chazelas, B.; Kundermann, S.; Lecomte, S.; Cecconi, M.; et al. A microphotonic astrocomb. *Nat. Photonics* **2018**, *13*, 31–35. [CrossRef]
13. Adler, F.; Thorpe, M.J.; Cossel, K.C.; Ye, J. Cavity-Enhanced Direct Frequency Comb Spectroscopy: Technology and Applications. *Annu. Rev. Anal. Chem.* **2010**, *3*, 175–205. [CrossRef] [PubMed]
14. Keilmann, F.; Gohle, C.; Holzwarth, R. Time-domain mid-infrared frequency-comb spectrometer. *Opt. Lett.* **2004**, *29*, 1542–1544. [CrossRef] [PubMed]
15. Picqué, N.; Hansch, T.W. Frequency comb spectroscopy. *Nat. Photonics* **2019**, *13*, 146–157. [CrossRef]
16. Schliesser, A.; Picqué, N.; Hänsch, T.W. Mid-infrared frequency combs. *Nat. Photonics* **2012**, *6*, 440–449. [CrossRef]
17. Rieker, G.B.; Giorgetta, F.R.; Swann, W.C.; Kofler, J.; Zolot, A.M.; Sinclair, L.C.; Baumann, E.; Cromer, C.; Petron, G.; Sweeney, C.; et al. Frequency-comb-based remote sensing of greenhouse gases over kilometer air paths. *Optica* **2014**, *1*, 290–298. [CrossRef]
18. Yu, M.; Okawachi, Y.; Griffith, A.G.; Picqué, N.; Lipson, M.; Gaeta, A.L. Silicon-chip-based mid-infrared dual-comb spectroscopy. *Nat. Commun.* **2018**, *9*, 1869. [CrossRef]
19. Pfeifle, J.; Brasch, V.; Lauermann, M.; Yu, Y.; Wegner, D.; Herr, T.; Hartinger, K.; Schindler, P.; Li, J.; Hillerkuss, D.; et al. Coherent terabit communications with microresonator Kerr frequency combs. *Nat. Photonics* **2014**, *8*, 375–380. [CrossRef]

20. Kemal, J.N.; Pfeifle, J.; Marin-Palomo, P.; Pascual, M.D.G.; Wolf, S.; Smyth, F.; Freude, W.; Koos, C. Multi-wavelength coherent transmission using an optical frequency comb as a local oscillator. *Opt. Express* **2016**, *24*, 25432–25445. [CrossRef]
21. Marin-Palomo, P.; Kemal, J.N.; Karpov, M.; Kordts, A.; Pfeifle, J.; Pfeiffer, M.H.P.; Trocha, P.; Wolf, S.; Brasch, V.; Anderson, M.H.; et al. Microresonator-based solitons for massively parallel coherent optical communications. *Nature* **2017**, *546*, 274–279. [CrossRef] [PubMed]
22. Roslund, J.; de Araújo, R.M.; Jiang, S.; Fabre, C.; Treps, N. Wavelength-multiplexed quantum networks with ultrafast frequency combs. *Nat. Photonics* **2014**, *8*, 109–112. [CrossRef]
23. Dutt, A.; Luke, K.; Manipatruni, S.; Gaeta, A.L.; Nussenzveig, P.; Lipson, M. On-Chip Optical Squeezing. *Phys. Rev. Appl.* **2015**, *3*, 044005. [CrossRef]
24. Reimer, C.; Kues, M.; Roztocki, P.; Wetzel, B.; Grazioso, F.; Little, B.E.; Chu, S.T.; Johnston, T.; Bromberg, Y.; Caspani, L.; et al. Generation of multiphoton entangled quantum states by means of integrated frequency combs. *Science* **2016**, *351*, 1176–1180. [CrossRef] [PubMed]
25. Imany, P.; Jaramillo-Villegas, J.A.; Odele, O.D.; Kyunghun, H.A.N.; Leaird, D.E.; Lukens, J.M.; Lougovski, P.; Minghao, Q.I.; Weiner, A.M. 50-GHz-spaced comb of high-dimensional frequency-bin entangled photons from an on-chip silicon nitride microresonator. *Opt. Express* **2018**, *26*, 1825–1840. [CrossRef] [PubMed]
26. Kues, M.; Reimer, C.; Lukens, J.M.; Munro, W.J.; Weiner, A.M.; Moss, D.J.; Morandotti, R. Quantum optical microcombs. *Nat. Photonics* **2019**, *13*, 170–179. [CrossRef]
27. Del'Haye, P.; Schliesser, A.; Arcizet, O.; Wilken, T.; Holzwarth, R.; Kippenberg, T.J. Optical frequency comb generation from a monolithic microresonator. *Nature* **2007**, *450*, 1214–1217. [CrossRef]
28. Kippenberg, T.J.; Holzwarth, R.; Diddams, S.A. Microresonator-based optical frequency combs. *Science* **2011**, *332*, 555–559. [CrossRef]
29. Pasquazi, A.; Peccianti, M.; Razzari, L.; Moss, D.J.; Coen, S.; Erkintalo, M.; Chembo, Y.K.; Hansson, T.; Wabnitz, S.; Del'Haye, P.; et al. Micro-combs: A novel generation of optical sources. *Phys. Rep.* **2018**, *729*, 1–81. [CrossRef]
30. Gaeta, A.L.; Lipson, M.; Kippenberg, T.J. Photonic-chip-based frequency combs. *Nat. Photonics* **2019**, *13*, 158–169. [CrossRef]
31. Maddaloni, P.; Malara, P.; Gagliardi, G.; De Natale, P. Mid-infrared fibre-based optical comb. *New J. Phys.* **2006**, *8*, 262. [CrossRef]
32. Sun, J.H.; Gale, B.J.S.; Reid, D.T. Composite frequency comb spanning 0.4–2.4 μm from a phase-controlled femtosecond Ti:sapphire laser and synchronously pumped optical parametric oscillator. *Opt. Lett.* **2007**, *32*, 1414–1416. [CrossRef] [PubMed]
33. Wong, S.T.; Plettner, T.; Vodopyanov, K.L.; Urbanek, K.; Digonnet, M.; Byer, R.L. Self-phase-locked degenerate femtosecond optical parametric oscillator. *Opt. Lett.* **2008**, *33*, 1896–1898. [CrossRef] [PubMed]
34. Gambetta, A.; Ramponi, R.; Marangoni, M. Mid-infrared optical combs from a compact amplified Er-doped fiber oscillator. *Opt. Lett.* **2008**, *33*, 2671–2673. [CrossRef]
35. Adler, F.; Cossel, K.C.; Thorpe, M.J.; Hartl, I.; Fermann, M.E.; Ye, J. Phase-stabilized, 15 W frequency comb at 2.8–4.8 μm. *Opt. Lett.* **2009**, *34*, 1330–1332. [CrossRef]
36. Galli, I.; Cappelli, F.; Cancio, P.; Giusfredi, G.; Mazzotti, D.; Bartalini, S.; De Natale, P. High-coherence mid-infrared frequency comb. *Opt. Express* **2013**, *21*, 28877–28885. [CrossRef]
37. Diddams, S.A.; Ma, L.S.S.; Ye, J.; Hall, J.L. Broadband optical frequency comb generation with a phase-modulated parametric oscillator. *Opt. Lett.* **1999**, *24*, 1747–1749. [CrossRef]
38. Kourogi, M.; Nakagawa, K.; Ohtsu, M. Wide-span optical frequency comb generator for accurate optical frequency difference measurement. *IEEE J. Quantum Electron.* **1993**, *29*, 2693–2701. [CrossRef]
39. Ulvila, V.; Phillips, C.R.; Halonen, L.L.; Vainio, M. Frequency comb generation by a continuous-wave-pumped optical parametric oscillator based on cascading quadratic nonlinearities. *Opt. Lett.* **2013**, *38*, 4281–4284. [CrossRef]
40. Ulvila, V.; Phillips, C.R.; Halonen, L.L.; Vainio, M. High-power mid-infrared frequency comb from a continuous-wave-pumped bulk optical parametric oscillator. *Opt. Express* **2014**, *22*, 10535–10543. [CrossRef]
41. Ostrovskii, L.A. Self-action of Light in Crystals. *JETP Lett.* **1967**, *5*, 272–275.
42. Desalvo, R.; Hagan, D.J.; Sheik-Bahae, M.; Stegeman, G.; Van Stryland, E.W.; Vanherzeele, H. Self-focusing and self-defocusing by cascaded second-order effects in KTP. *Opt. Lett.* **1992**, *17*, 28–30. [CrossRef] [PubMed]

43. Stegeman, G.I. $\chi^{(2)}$ cascading: Nonlinear phase shifts. *Quantum Semiclass. Opt.* **1999**, *9*, 139–153. [CrossRef]
44. Schiller, S.; Byer, R.L. Quadruply resonant optical parametric oscillation in a monolithic total-internal-reflection resonator. *J. Opt. Soc. Am. B* **1993**, *10*, 1696–1707. [CrossRef]
45. Schiller, S.; Breitenbach, G.; Paschotta, R.R.; Mlynek, J. Subharmonic-pumped continuous-wave parametric oscillator. *Appl. Phys. Lett.* **1996**, *68*, 3374–3376. [CrossRef]
46. Schneider, K.; Schiller, S. Multiple conversion and optical limiting in a subharmonic-pumped parametric oscillator. *Opt. Lett.* **1997**, *22*, 363–365. [CrossRef] [PubMed]
47. White, A.G.; Lam, P.K.; Taubman, M.S.; Marte, M.A.M.; Schiller, S.; McClelland, D.E.; Bachor, H.A. Classical and quantum signatures of competing $\chi^{(2)}$ nonlinearities. *Phys. Rev. A* **1997**, *55*, 4511–4515. [CrossRef]
48. Ricciardi, I.; Mosca, S.; Parisi, M.; Maddaloni, P.; Santamaria, L.; De Natale, P.; De Rosa, M. Frequency comb generation in quadratic nonlinear media. *Phys. Rev. A* **2015**, *91*, 063839. [CrossRef]
49. Mosca, S.; Ricciardi, I.; Parisi, M.; Maddaloni, P.; Santamaria, L.; De Natale, P.; De Rosa, M. Direct generation of optical frequency combs in $\chi^{(2)}$ nonlinear cavities. *Nanophotonics* **2016**, *5*, 316–331. [CrossRef]
50. Leo, F.; Hansson, T.; Ricciardi, I.; De Rosa, M.; Coen, S.; Wabnitz, S.; Erkintalo, M. Walk-off-induced modulation instability, temporal pattern formation, and frequency comb generation in cavity-enhanced second-harmonic generation. *Phys. Rev. Lett.* **2016**, *116*, 033901. [CrossRef]
51. Leo, F.; Hansson, T.; Ricciardi, I.; De Rosa, M.; Coen, S.; Wabnitz, S.; Erkintalo, M. Frequency-comb formation in doubly resonant second-harmonic generation. *Phys. Rev. A* **2016**, *93*, 043831. [CrossRef]
52. Hansson, T.; Leo, F.; Erkintalo, M.; Coen, S.; Ricciardi, I.; De Rosa, M.; Wabnitz, S. Singly resonant second-harmonic-generation frequency combs. *Phys. Rev. A* **2017**, *95*, 013805. [CrossRef]
53. Zakharov, V.E.; Ostrovsky, L.A. Modulation instability: The beginning. *Phys. D Nonlinear Phenom.* **2009**, *238*, 540–548. [CrossRef]
54. Mosca, S.; Parisi, M.; Ricciardi, I.; Leo, F.; Hansson, T.; Erkintalo, M.; Maddaloni, P.; De Natale, P.; Wabnitz, S.; De Rosa, M. Modulation Instability Induced Frequency Comb Generation in a Continuously Pumped Optical Parametric Oscillator. *Phys. Rev. Lett.* **2018**, *121*, 093903. [CrossRef]
55. Hansson, T.; Leo, F.; Erkintalo, M.; Anthony, J.; Coen, S.; Ricciardi, I.; De Rosa, M.; Wabnitz, S. Single envelope equation modeling of multi-octave comb arrays in microresonators with quadratic and cubic nonlinearities. *J. Opt. Soc. Am. B* **2016**, *33*, 1207–1215. [CrossRef]
56. Drever, R.W.P.; Hall, J.L.; Kowalski, F.V.; Hough, J.; Ford, G.M.; Munley, A.J.; Ward, H. Laser phase and frequency stabilization using an optical resonator. *Appl. Phys. B* **1983**, *31*, 97–105. [CrossRef]
57. Ricciardi, I.; De Rosa, M.; Rocco, A.; Ferraro, P.; De Natale, P. Cavity-enhanced generation of 6 W cw second-harmonic power at 532 nm in periodically-poled MgO:LiTaO$_3$. *Opt. Express* **2010**, *18*, 10985–10994. [CrossRef]
58. De Rosa, M.; Conti, L.; Cerdonio, M.; Pinard, M.; Marin, F. Experimental Measurement of the Dynamic Photothermal Effect in Fabry-Perot Cavities for Gravitational Wave Detectors. *Phys. Rev. Lett.* **2002**, *89*, 237402. [CrossRef]
59. Carmon, T.; Yang, L.; Vahala, K.J. Dynamical thermal behavior and thermal self-stability of microcavities. *Opt. Express* **2004**, *12*, 4742–4750. [CrossRef]
60. Chembo, Y.K.; Yu, N. Modal expansion approach to optical-frequency-comb generation with monolithic whispering-gallery-mode resonators. *Phys. Rev. A* **2010**, *82*, 033801. [CrossRef]
61. Chembo, Y.K.; Strekalov, D.V.; Yu, N. Spectrum and dynamics of optical frequency combs generated with monolithic whispering gallery mode resonators. *Phys. Rev. Lett.* **2010**, *104*, 103902. [CrossRef] [PubMed]
62. Haelterman, M.; Trillo, S.; Wabnitz, S. Dissipative modulation instability in a nonlinear dispersive ring cavity. *Opt. Commun.* **1992**, *91*, 401–407. [CrossRef]
63. Leo, F.; Coen, S.; Kockaert, P.; Gorza, S.P.; Emplit, P.; Haelterman, M. Temporal cavity solitons in one-dimensional Kerr media as bits in an all-optical buffer. *Nat. Photonics* **2010**, *4*, 471–476. [CrossRef]
64. Coen, S.; Randle, H.G.; Sylvestre, T.; Erkintalo, M. Modeling of octave-spanning Kerr frequency combs using a generalized mean-field Lugiato–Lefever model. *Opt. Lett.* **2013**, *38*, 37–39. [CrossRef]
65. Coen, S.; Erkintalo, M. Universal scaling laws of Kerr frequency combs. *Opt. Lett.* **2013**, *38*, 1790–1792. [CrossRef]
66. Hansson, T.; Modotto, D.; Wabnitz, S. On the numerical simulation of Kerr frequency combs using coupled mode equations. *Opt. Commun.* **2014**, *312*, 134–136. [CrossRef]

67. Agrawal, G.P. *Nonlinear Fiber Optics*, 3rd ed.; Academic Press: San Diego, CA, USA, 2001.
68. Weideman, J.; Herbst, B. Split-step methods for the solution of the nonlinear Schrödinger equation. *SIAM* **1986**, *23*, 485–507. [CrossRef]
69. Godey, C.; Balakireva, I.V.; Coillet, A.; Chembo, Y.K. Stability analysis of the spatiotemporal Lugiato-Lefever model for Kerr optical frequency combs in the anomalous and normal dispersion regimes. *Phys. Rev. A* **2014**, *89*, 063814. [CrossRef]
70. Khurgin, J.B.; Dikmelik, Y.; Hugi, A.; Faist, J. Coherent frequency combs produced by self frequency modulation in quantum cascade lasers. *Appl. Phys. Lett.* **2014**, *104*, 081118. [CrossRef]
71. Cappelli, F.; Consolino, L.; Campo, G.; Galli, I.; Mazzotti, D.; Campa, A.; de Cumis, M.S.; Pastor, P.C.; Eramo, R.; Rösch, M.; et al. Retrieval of phase relation and emission profile of quantum cascade laser frequency combs. *Nat. Photonics* **2019**, *13*, 562–568. [CrossRef]
72. Breunig, I. Three-wave mixing in whispering gallery resonators. *Laser Photonics Rev.* **2016**, *10*, 569–587. [CrossRef]
73. Strekalov, D.V.; Marquardt, C.; Matsko, A.B.; Schwefel, H.G.L.; Leuchs, G. Nonlinear and quantum optics with whispering gallery resonators. *J. Opt.* **2016**, *18*, 123002. [CrossRef]
74. Ikuta, R.; Asano, M.; Tani, R.; Yamamoto, T.; Imoto, N. Frequency comb generation in a quadratic nonlinear waveguide resonator. *Opt. Express* **2018**, *26*, 15551–15558. [CrossRef] [PubMed]
75. Stefszky, M.; Ulvila, V.; Abdallah, Z.; Silberhorn, C.; Vainio, M. Towards optical-frequency-comb generation in continuous-wave-pumped titanium-indiffused lithium-niobate waveguide resonators. *Phys. Rev. A* **2018**, *98*, 053850. [CrossRef]
76. Hendry, I.; Trainor, L.S.; Xu, Y.; Coen, S.; Murdoch, S.G.; Schwefel, H.G.L.; Erkintalo, M. Experimental observation of internally-pumped parametric oscillation and quadratic comb generation in a $\chi^{(2)}$ whispering-gallery-mode microresonator. *arXiv* **2019**, arXiv:1912.02804.
77. Szabados, J.; Puzyrev, D.N.; Minet, Y.; Reis, L.; Buse, K.; Villois, A.; Skryabin, D.V.; Breunig, I. Frequency comb generation via cascaded second-order nonlinearities in microresonators. *arXiv* **2019**, arXiv:1912.00945.
78. Levy, J.S.; Foster, M.A.; Gaeta, A.L.; Lipson, M. Harmonic generation in silicon nitride ring resonators. *Opt. Express* **2011**, *19*, 11415–11421. [CrossRef]
79. Jung, H.; Stoll, R.; Guo, X.; Fischer, D.; Tang, H.X. Green, red, and IR frequency comb line generation from single IR pump in AlN microring resonator. *Optica* **2014**, *1*, 396–399. [CrossRef]
80. Pu, M.; Ottaviano, L.; Semenova, E.; Yvind, K. Efficient frequency comb generation in AlGaAs-on-insulator. *Optica* **2016**, *3*, 823–826. [CrossRef]
81. Wang, C.; Zhang, M.; Yu, M.; Zhu, R.; Hu, H.; Lončar, M. Monolithic lithium niobate photonic circuits for Kerr frequency comb generation and modulation. *Nat. Commun.* **2019**, *10*, 978. [CrossRef]
82. Zhang, M.; Buscaino, B.; Wang, C.; Shams-Ansari, A.; Reimer, C.; Zhu, R.; Kahn, J.M.; Lončar, M. Broadband electro-optic frequency comb generation in a lithium niobate microring resonator. *Nature* **2019**, *568*, 373–377. [CrossRef]
83. Xue, X.; Leo, F.; Xuan, Y.; Jaramillo-Villegas, J.A.; Wang, P.H.; Leaird, D.E.; Erkintalo, M.; Qi, M.; Weiner, A.M. Second-harmonic-assisted four-wave mixing in chip-based microresonator frequency comb generation. *Light Sci. Appl.* **2016**, *6*, e16253. [CrossRef]
84. Fürst, J.U.; Strekalov, D.V.; Elser, D.; Aiello, A.; Andersen, U.L.; Marquardt, C.; Leuchs, G. Low-Threshold Optical Parametric Oscillations in a Whispering Gallery Mode Resonator. *Phys. Rev. Lett.* **2010**, *105*, 263904. [CrossRef]
85. Fürst, J.U.; Strekalov, D.V.; Elser, D.; Lassen, M.; Andersen, U.L.; Marquardt, C.; Leuchs, G. Naturally Phase-Matched Second-Harmonic Generation in a Whispering-Gallery-Mode Resonator. *Phys. Rev. Lett.* **2010**, *104*, 153901. [CrossRef]
86. Lin, G.; Fürst, J.U.; Strekalov, D.V.; Yu, N. Wide-range cyclic phase matching and second harmonic generation in whispering gallery resonators. *Appl. Phys. Lett.* **2013**, *103*, 181107. [CrossRef]
87. Meisenheimer, S.K.; Fürst, J.U.; Werner, C.; Beckmann, T.; Buse, K.; Breunig, I. Broadband infrared spectroscopy using optical parametric oscillation in a radially-poled whispering gallery resonator. *Opt. Express* **2015**, *23*, 24042–24047. [CrossRef]

88. Mohageg, M.; Strekalov, D.; Savchenkov, A.; Matsko, A.; Ilchenko, V.; Maleki, L. Calligraphic poling of Lithium Niobate. *Opt. Express* **2005**, *13*, 3408–3419. [CrossRef]
89. Guarino, A.; Poberaj, G.; Rezzonico, D.; Degl'Innocenti, R.; Günter, P. Electro-optically tunable microring resonators in lithium niobate. *Nat. Photonics* **2007**, *1*, 407–410. [CrossRef]
90. Wang, C.; Burek, M.J.; Lin, Z.; Atikian, H.A.; Venkataraman, V.; Huang, I.C.; Stark, P.; Lončar, M. Integrated high quality factor lithium niobate microdisk resonators. *Opt. Express* **2014**, *22*, 30924–30933. [CrossRef]
91. Liang, H.; Lin, Q.; Luo, R.; Jiang, W.C.; Zhang, X.C.; Sun, X. Nonlinear optical oscillation dynamics in high-Q lithium niobate microresonators. *Opt. Express* **2017**, *25*, 13504–13516.
92. Wu, R.; Zhang, J.; Yao, N.; Fang, W.; Qiao, L.; Chai, Z.; Lin, J.; Cheng, Y. Lithium niobate micro-disk resonators of quality factors above 10^7. *Opt. Lett.* **2018**, *43*, 4116–4119. [CrossRef]
93. Helmy, A.S.; Abolghasem, P.; Stewart Aitchison, J.; Bijlani, B.J.; Han, J.; Holmes, B.M.; Hutchings, D.C.; Younis, U.; Wagner, S.J. Recent advances in phase matching of second-order nonlinearities in monolithic semiconductor waveguides. *Laser Photonics Rev.* **2010**, *5*, 272–286. [CrossRef]
94. Kuo, P.S.; Bravo-Abad, J.; Solomon, G.S. Second-harmonic generation using $\bar{4}$-quasi-phasematching in a GaAs whispering-gallery-mode microcavity. *Nat. Commun.* **2014**, *5*, 3109. [CrossRef]
95. Mariani, S.; Andronico, A.; Lemaître, A.; Favero, I.; Ducci, S.; Leo, G. Second-harmonic generation in AlGaAs microdisks in the telecom range. *Opt. Lett.* **2014**, *39*, 3062–3064. [CrossRef]
96. Parisi, M.; Morais, N.; Ricciardi, I.; Mosca, S.; Hansson, T.; Wabnitz, S.; Leo, G.; De Rosa, M. AlGaAs waveguide microresonators for efficient generation of quadratic frequency combs. *J. Opt. Soc. Am. B* **2017**, *34*, 1842–1847. [CrossRef]
97. Timurdogan, E.; Poulton, C.V.; Byrd, M.J.; Watts, M.R. Electric field-induced second-order nonlinear optical effects in silicon waveguides. *Nat. Photonics* **2017**, *11*, 200–206. [CrossRef]
98. Herr, T.; Brasch, V.; Jost, J.D.; Wang, C.Y.; Kondratiev, N.M.; Gorodetsky, M.L.; Kippenberg, T.J. Temporal solitons in optical microresonators. *Nat. Photonics* **2014**, *8*, 145–152. [CrossRef]
99. Hansson, T.; Parra-Rivas, P.; Bernard, M.; Leo, F.; Gelens, L.; Wabnitz, S. Quadratic soliton combs in doubly resonant second-harmonic generation. *Opt. Lett.* **2018**, *43*, 6033–6036. [CrossRef]
100. Villois, A.; Skryabin, D.V. Soliton and quasi-soliton frequency combs due to second harmonic generation in microresonators. *Opt. Express* **2019**, *27*, 7098–7107. [CrossRef]
101. Erkintalo, M.; Li, Z.; Parra-Rivas, P.; Leo, F. Dynamics of Kerr-like Optical Frequency Combs Generated via Phase-mismatched Second-harmonic Generation. In Proceedings of the 2019 Conference on Lasers and Electro-Optics Europe and European Quantum Electronics Conference, Munich, Germany, 23–27 June 2019.
102. Lobanov, V.E.; Kondratiev, N.M.; Shitikov, A.E.; Bilenko, I.A. Two-color flat-top solitonic pulses in $\chi^{(2)}$ optical microresonators via second-harmonic generation. *Phys. Rev. A* **2020**, *101*, 013831. [CrossRef]
103. Villois, A.; Kondratiev, N.; Breunig, I.; Puzyrev, D.N.; Skryabin, D.V. Frequency combs in a microring optical parametric oscillator. *Opt. Lett.* **2019**, *44*, 4443–4446. [CrossRef]
104. Parra-Rivas, P.; Gelens, L.; Leo, F. Localized structures in dispersive and doubly resonant optical parametric oscillators. *Phys. Rev. E* **2019**, *100*, 032219. [CrossRef]
105. Menicucci, N.C.; Flammia, S.T.; Pfister, O. One-Way Quantum Computing in the Optical Frequency Comb. *Phys. Rev. Lett.* **2008**, *101*, 130501. [CrossRef]
106. Pfister, O. Continuous-variable quantum computing in the quantum optical frequency comb. *J. Phys. B At. Mol. Opt. Phys.* **2020**, *53*, 012001. [CrossRef]
107. Villar, A.S.; Martinelli, M.; Fabre, C.; Nussenzveig, P. Direct production of tripartite pump-signal-idler entanglement in the above-threshold optical parametric oscillator. *Phys. Rev. Lett.* **2006**, *97*, 140504. [CrossRef]
108. Pfister, O.; Feng, S.; Jennings, G.; Pooser, R.C.; Xie, D. Multipartite continuous-variable entanglement from concurrent nonlinearities. *Phys. Rev. A* **2004**, *70*, 020302. [CrossRef]
109. Guo, J.; Zou, H.; Zhai, Z.; Zhang, J.; Gao, J. Generation of continuous-variable tripartite entanglement using cascaded nonlinearities. *Phys. Rev. A* **2005**, *71*, 034305. [CrossRef]
110. Pennarun, C.; Bradley, A.S.; Olsen, M.K. Tripartite entanglement and threshold properties of coupled intracavity down-conversion and sum-frequency generation. *Phys. Rev. A* **2007**, *76*, 063812. [CrossRef]

111. He, G.; Sun, Y.; Hu, L.; Zhang, R.; Chen, X.; Wang, J. Five-partite entanglement generation between two optical frequency combs in a quasi-periodic $\chi^{(2)}$ nonlinear optical crystal. *Sci. Rep.* **2017**, *7*, 9054. [CrossRef] [PubMed]
112. Raussendorf, R.; Browne, D.E.; Briegel, H.J. Measurement-based quantum computation on cluster states. *Phys. Rev. A* **2003**, *68*, 022312. [CrossRef]
113. Chembo, Y.K. Quantum dynamics of Kerr optical frequency combs below and above threshold: Spontaneous four-wave mixing, entanglement, and squeezed states of light. *Phys. Rev. A* **2016**, *93*, 033820. [CrossRef]

© 2020 by the authors. Licensee MDPI, Basel, Switzerland. This article is an open access article distributed under the terms and conditions of the Creative Commons Attribution (CC BY) license (http://creativecommons.org/licenses/by/4.0/).

Review

Nonlinear Optics in Microspherical Resonators

Gabriele Frigenti [1,2,3,†], **Daniele Farnesi** [2,†], **Gualtiero Nunzi Conti** [1,2] **and Silvia Soria** [2,*]

1 Centro Fermi—Museo Storico della Fisica e Centro Studi e Ricerche "Enrico Fermi", Compendio del Viminale, Piazza del Viminale 1, 00184 Roma, Italy; g.frigenti@ifac.cnr.it (G.F.); g.nunziconti@ifac.cnr.it (G.N.C.)
2 CNR-IFAC, Istituto di Fisica Applicata "Nello Carrara", Consiglio Nazionale delle Ricerche, via Madonna del Piano 10, I50019 Sesto Fiorentino (FI), Italy; d.farnesi@ifac.cnr.it
3 Laboratorio Europeo di Spettroscopia Nonlineare (LENS) - Università degli Studi di Firenze, via Nello Carrara 1, I50019 Sesto Fiorentino (FI), Italy
* Correspondence: s.soria@ifac.cnr.it
† These authors contributed equally to this work.

Received: 18 February 2020; Accepted: 10 March 2020; Published: 13 March 2020

Abstract: Nonlinear frequency generation requires high intensity density which is usually achieved with pulsed laser sources, anomalous dispersion, high nonlinear coefficients or long interaction lengths. Whispering gallery mode microresonators (WGMRs) are photonic devices that enhance nonlinear interactions and can be exploited for continuous wave (CW) nonlinear frequency conversion, due to their capability of confine light for long time periods in a very small volume, even though in the normal dispersion regime. All signals must be resonant with the cavity. Here, we present a review of nonlinear optical processes in glass microspherical cavities, hollow and solid.

Keywords: kerr nonlinearity; whispering gallery mode; optical resonators; stimulated brillouin scattering; optomechanical oscillations

1. Introduction

Optical resonators have been gaining a lot of interest in recent decades in all branches of modern optics, both linear and nonlinear optics [1]. Among these resonators whispering gallery modes resonators (WGMR) have shown high mode stability and high quality factors, up to 10^{11}. Their fabrication is rather simple and inexpensive. WGMR are total internal reflection resonators and they were first introduced by Lord Rayleigh for sound waves propagating close to the dome wall in St. Paul's cathedral, London [2]. The same phenomenology can be applied to the optical domain, and it was analyzed in depth by Mie and Debye. The geometry of the resonator determines the volume and field distribution of the modes. This kind of monolithic resonators are excellent platforms for fundamental and applied studies of nonlinear interactions between light and matter mainly due to their long photon lifetimes (long temporal confinement) and their small mode volumes (spatial confinement). Temporal and spatial confinement have made possible optical frequency conversion with low-power continuous wave (CW) lasers with powers ranging from micro-watts to milliwatts. However, the high circulating intensities inside a WGMR are not a sufficient condition for efficient harmonic generation, parametric and hyper-parametric oscillations: these phenomena require fulfilling phase and mode matching and energy conservation conditions [3,4].

This review will describe the applications in nonlinear optics of silica microspherical WGMR, both solid and hollow microcavities. Figure 1 shows an illustration of the three types of WGMR reviewed in this paper. These 3D WGMR show very high quality factor Q, their fabrication is easy and regarding specifically nonlinear frequency generation, their very dense mode spectra eases the phase-matching processes required for parametric and hyper-parametric interactions. Microspheres are the first monolithic WGMR that were investigated theoretically and can be obtained by melting amorphous materials such as silica [5]. Hollow microspherical WGMR or the so-called microbubbles (MBR) [6] are also an important family of WGMR which can be tuned beyond a free spectral range (FSR) by changing their radius, either via piezo-electric [7] or gas pressure stress [8,9]. Tuning the FSR and the generated combs can be very important for practical applications such as gas sensing or molecular spectroscopy. However, the size of a solid silica microsphere or microbottles is very difficult to change once it is fabricated. A successful way to tune solid WGMR is by coating them with functional materials, such as nanoparticles [10,11]. Hybrid microspheres were also used for Kerr switching [12,13].

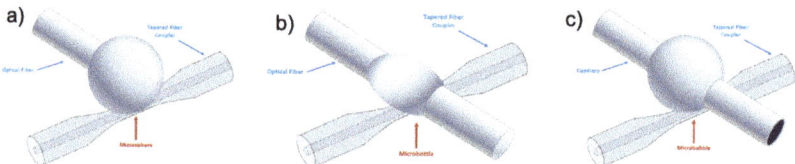

Figure 1. Schematic drawing of microspherical WGMR: (**a**) Microsphere, (**b**) Microbottle and (**c**) Microbubble. The coupling tapered fiber is also sketched in this picture.

Silica glass is a centrosymmetric material; therefore, second order nonlinear interactions are forbidden. Here, the elemental nonlinear interaction is due to the third-order susceptibility χ^3 effects, in which four photons are coupled. Previous work in this area, however, has been focused on toroidal WGMRs, where most of the excitable modes are constrained to be the equatorial ones [14], or on highly nonlinear materials [3,15]. Efficient generation of visible light via third-order sum-frequency generation (TSFG) or four-wave mixing (FWM), and third-harmonic generation (THG) in silica microspherical WGMR have been explored. MBR were also studied in the THz domain [16] or for generating two-photon fluorescence (TPF) of the filling liquid [17]. Detailed description of their properties can be found in several books and reviews [1,18,19].

In this review, several nonlinear effects in silica microspheres will be illustrated (see Figure 2). Generally, nonlinear processes are classified into parametric (hyper-parametric) and non-parametric processes. Parametrical processes are predominant for non or near resonant interactions where the initial and final quantum states are the same, which means that there is no real absorption of photons by the material. Since these processes involve only virtual energy levels, their lifetimes are extremely short (less than a femtosecond). On the other hand, non-parametric processes involve real energy levels with different initial and final quantum states. In this case, there is energy transfer from the photons to the host medium with a relative longer lifetime and it is predominant for resonant interactions. Harmonic generation, TSFG, FWM and coherent anti-Stokes Raman spectroscopy (CARS) are parametric interactions whereas stimulated Raman scattering (SRS) and stimulated Brilluoin scattering (SBS) are non-parametric processes (see Figure 2b).

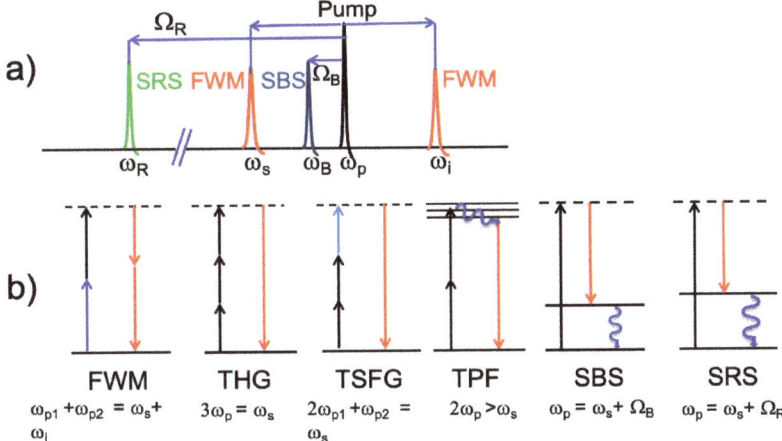

Figure 2. (a) Illustration of the infrared spectra observed when nonlinear phenomena are generated. (b) Schematic representation of the energy levels of different nonlinear processes (from left to right): FWM, THG, TSFG, TPF, SBS and SRS. Real energy levels are denoted by solid lines (non-parametric processes) while virtual states are indicated by dashed lines.

We will start from tunable optical harmonic generation with extremely narrow linewidth. In the most general case of TSFG, three different waves interact with a nonlinear medium to generate a fourth wave of different frequency ($\omega_{TSFG} = \omega_1 + \omega_2 + \omega_3$). If the three input frequencies are degenerate, the result will be the THG; in this case the energy conservation requires $\omega_{THG} = 3\omega_1$. The additional phase-matching condition requires $n(\omega_{THG}) = n(\omega_p)$, which in general can be fulfilled, because linear and nonlinear dispersion can be compensated for by the dense distribution of degenerate whispering gallery modes (WGMs) with different polar number and decreasing effective index $n_{eff} = m/kR$ [1]. In a WGMR, an additional boundary condition leads to a strict value for the resonant frequency, which may be in conflict with the strict energy conservation, meaning that ω_{TSFG} or ω_{THG} is out of resonance.

Stimulated Raman scattering (SRS) is a pure gain process and, therefore, the phase-matching condition is established automatically. Stimulated anti-Stokes Raman scattering (SARS) has been observed in single-component microdroplets with strong Raman gain [20] and from minority species in multi-component microdroplets with low Raman gain [21]; in the latter case, external seeding at the Stokes frequency was used to enhance the SARS signal. However, SARS does require phase-matching, to be efficiently generated and it is thresholdless. WGMRs provide ultra-low thresholds [22] for SRS generation which is a huge advantage. In consequence, one can find extensive literature about SRS generation in WGMRs, but not SARS. SARS and CARS are four-wave mixing (FWM) processes, where vibrational transition frequencies match the beating frequencies between the involved waves, usually, the pump and the anti-Stokes (or Stokes) waves. Cavity resonant enhanced SARS generation, multi-order SARS, and third-order nonlinear processes in silica WGMRs at any dispersion value [23,24] were presented by Farnesi et al. [25]. The linewidth of the cavity-enhanced SARS emission is similar to that of SRS and the pump, and it is a high-quality mode [26]. SBS is an inelastic scattering process like SRS but it is a coherent interaction of light photons and acoustic phonons instead of light photons and light phonons (SRS). The photon-phonon interaction is enhanced due to the overlap of both waves inside the WGMR, which acts as a dual photonic-phononic cavity, named also as phoXonic cavity [27,28]. Another important difference between SBS and SRS are the gain bandwidth and gain coefficient. SBS has one of the largest gain coefficients but a narrower gain bandwidth compared to SRS that constrains WGMR geometries, since the Brillouin frequency shift should match the free spectral range (FSR) of the resonator. It has been demonstrated that this constrain can be bypassed by using higher-order modes [18,29]. All these nonlinear phenomena can occur simultaneously despite their different lifetimes. It has been observed in conjunction THG,SRS and TSFG; or SBS and FWM.

2. Kerr Effects

Silica glass is a centrosymmetric material with an odd-order nonlinear polarization which is directly proportional to the nonlinear third-order susceptibility $\chi_{(3)}$ of the hosting material. The $\chi_{(3)}$ is a fourth order tensor with 81 elements. The third-order nonlinear polarization is responsible for several nonlinear phenomena, such as THG, FWM, TSFG, the optical Kerr effect and CARS.

2.1. Third-Order Sum-Frequency Generation

The nonlinear polarization for third-order sum-frequency generation (TSFG) [30] can be written as:

$$\mathbf{P}^{NLS}(\mathbf{r},\omega) = \chi^{(3)} \mathbf{E}_a(\mathbf{r},\omega_a) \mathbf{E}_b(\mathbf{r},\omega_b) \mathbf{E}_c(\mathbf{r},\omega_c) \tag{1}$$

where $E(\omega_i)$ is the electric field amplitude at frequency ω_i. This equation shows that in the third-order approximation, the radiation at the new frequency $\omega = \omega_a + \omega_b + \omega_c$ (energy conservation) can be generated by an intense field containing ω_a, ω_b, and ω_c. This general case, or third-order sum-frequency generation (TSFG) has been studied in WGM structures, in particular in liquid droplets, starting from 1989 [31–33]. TSFG is a weak process; even when conditions are optimized, the emission is only 10^{-4} times the typical intensity of SRS. A model for TSFG in spherical dielectric microresonators can be based on the work of Chew et al. [34] for emission from a polarization source within a sphere. A particular case of TSFG is the third-harmonic generation (THG) in which the three input frequencies are degenerate and so $\omega_{TGH} = 3\omega_1$.

The polarization in Equation (1) can be written as

$$\mathbf{P}^{NLS}(\mathbf{r},\omega) = D \sum_{jkl} \chi^{(3)}_{ijkl} E_a(\mathbf{r},\omega_a) E_b(\mathbf{r},\omega_b) E_c(\mathbf{r},\omega_c) \tag{2}$$

where j,k,l are the three orthogonal coordinate directions and D is the number of distinct permutation of ω_a, ω_b and ω_c. Being silica an isotropic material, only three independent elements can be considered:

$$\chi_{ijkl} = \chi_{1122} \delta_{ij} \delta_{kj} + \chi_{1212} \delta_{ik} \delta_{jl} + \chi_{1212} \delta_{il} \delta_{jk} \tag{3}$$

In spherical coordinates (r, θ, ψ), for the transverse electric field (TE) in a sphere, the radial component is zero and $\chi^{(3)}_{1111} = \chi^{(3)}_{1122} + \chi^{(3)}_{1212} + \chi^{(3)}_{1221}$.
In the simplest case of THG, $\chi^{(3)}_{1122} = \chi^{(3)}_{1212} = \chi^{(3)}_{1221}$, and so:

$$\chi^{(3)}_{ijkl} = \chi^{(3)}_{1122}(\delta_{ij}\delta_{kj} + \delta_{ik}\delta_{jl} + \delta_{il}\delta_{jk}) \tag{4}$$

and the θ component of the polarization is:

$$P^{NLS}_\theta = 3\chi^3_{1122}(E^3_{a\theta} + E_{a\theta} E^2_{a\phi}) \tag{5}$$

The radiations generated by the polarization $\mathbf{P}^{NLS}(\mathbf{r}',\omega)$, where $\mathbf{r}'$ is the source position, induce additional fields which have to satisfy the boundary conditions at the surface of the sphere. The solution is a combination of spherical Bessel and Hankel functions, j_n and h^1_n, respectively.

If the three waves generating the TSFG are standing waves, then the output is a standing wave too. The fields (TE) can be written as:

$$\mathbf{E_s}(\mathbf{r},\omega_s) = A g_n j_n(k_t r) \left[\mathbf{Y}_{nnm}(\theta,\phi) + \mathbf{Y}^*_{nnm}(\theta,\phi) \right] / 2 \tag{6}$$

where A is the amplitude factor proportional to $\sqrt{I_s}$. The field components are labeled n_s and m_s, where s is a, b or c.

Obtaining the total power at ω requires the integration over all ω:

$$P^T_{n_3 m_3} = \int_0^\infty P_{n_3 m_3}(\omega) d\omega \qquad (7)$$

The TSFG power is proportional to the spatial overlap integrals as well as to a frequency overlap integral; the former ones are calculated by integrating the product of the fields of the TSFG mode and of the three generating modes over the sphere volume. In addition to the energy and momentum conservation, in WGMR the pump and the generated frequency must be resonant. When these three conditions are fulfilled, the high quality factor enhances the interaction. The three conditions are quite difficult to be fulfilled simultaneously. The intermodal dispersion of the different spatial WGMs can be used, however, to obtain highly efficient frequency conversion. The inset of Figure 3 shows a picture of the microsphere with the TH signal with the characteristic upper and lower green lobes along the polar direction. As expected, TH signal is codirectional with the pump. Figure 3 shows the emission spectrum of THG at 519.6 nm when pumping with 1556.9 nm.

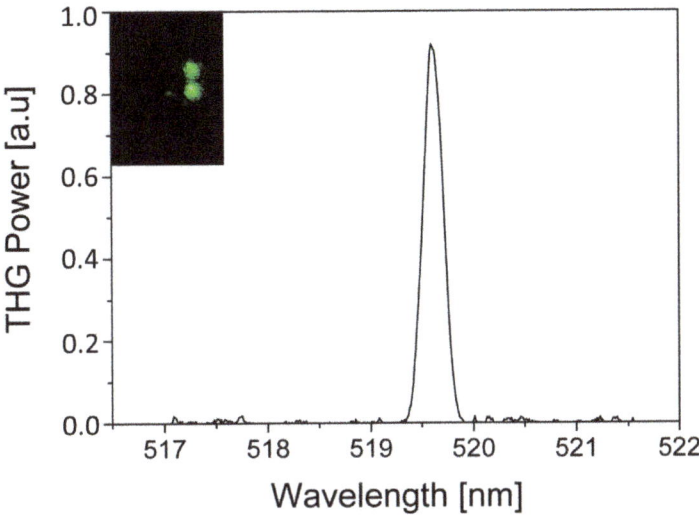

Figure 3. Emission spectrum indicating third-harmonic generation at 519.6 nm when pumping at 1556.9 nm, whereas the inset picture was taken during the spectral measurements. Reproduced with modifications from Ref. [35].

Asano et al. [36] observed THG in silica microbottle resonators. This particular resonator showed interface and surface effects that allowed the simultaneous generation of THG and second harmonic generation (SHG). In this work, the pump powers are over 200 mW, quite high compared to previous works in other types of WGMR. [25]. A way to lower the launched pump power into the microresonator is coating its surface. Dominguez et al. [37] used a similar strategy for second harmonic generation, coating the silica microspheres with a crystal violet monolayer. Chen et al. [38] have published very recently impressive results in terms of efficiency. The authors have coated silica microspheres with a thin layer of 4-[4-diethylamino(styril)]pyridium (DSP) molecules. DSP has a high third-order nonlinear coefficient and the efficiency of THG is 4 order of magnitude higher than the reported in bare silica microspheres. The authors also observed multiemissions due to TSFG.

2.2. Four-Wave Mixing

Third-order four-wave mixing (FWM) is a hyper-parametric oscillation where two pump photons ω_{pump} generate a signal ω_S and an idler photon ω_I. It requires two conditions to be satisfied: the momentum conservation, which is intrinsically satisfied in WGM resonators [39], and the energy conservation, which is not a priori satisfied since the separation between adjacent modes $\nu_{FSR} = |\nu_m - \nu_{m+1}|$ can vary due to the material and cavity dispersion. Indeed, only in recent works this process has been observed by coupling a CW laser into microcavities exploiting the Kerr nonlinearity to enable cascaded four-wave mixing. The resonances of the WGMR will also impose that the new generated frequencies will be discrete, creating a frequency comb.

The comb generation can occur in two different ways: as a Type I (or natively mode spaced comb), with sidebands separated by one free spectral range (FSR), and a Type II (or multimode spaced comb), with sidebands separated from the pump by several FSRs. Cascaded FWM preserves the initial spacing to higher-order emerging sidebands thank to the conservation of the energy in the parametric processes [40].

The Kerr comb formation starts by generation of the first symmetrical lines generated in a degenerate FWM process when the parametric gain overcomes the loss of the cavity. The separation of the new lines from the pump depends on the dispersion and the pump power. The threshold of the parametric frequency conversion [24], at which the gain of the excited sidebands is equal to the cavity decay rate, is:

$$P_{th} = \frac{\kappa^2 n_0^2 V_{eff}}{8\eta \omega_0 c n_2} \tag{8}$$

and the gain of the sidebands, for $P_{pump} > P_{th}$, can be written as: [41]

$$G = \sqrt{\kappa^2 (\frac{P_{abs}}{P_{th}})^2 - 4\left(\omega_0 - \omega_p + \mu^2 D_2 - \kappa \frac{P_{abs}}{P_{th}}\right)^2} \tag{9}$$

where P_{abs} is the power absorbed by the cavity. At the threshold ($G = \kappa$), from Equation (9) we obtain:

$$\sqrt{\frac{f^2}{|a_0|^2 - 1}} - d_2 \mu_{th}^2 + |a_0|^2 - \sqrt{|a_0|^4 - 1} = 0 \tag{10}$$

The theory of generation of frequency combs in silica microspheres has been described in a 2010 article by Chembo et al. [42].

The first experimental demonstration of FWM in silica microspheres was done by Kippenberg et al. [43]. The authors showed the generation of a pair of signal-idler photons created by two pump photons separated by one FSR with an emission ratio close to unity. Frequency combs in microspheres were first demonstrated by Agha et al. [44] where the first theoretical model was established. The same group published a broader comb in microspheres [45], but their theoretical model only predicted FWM and modulation instabilities in resonators with anomalous dispersion. However, nonlinear hyper-parametrical oscillations have been also achieved in the regime of normal dispersion [23,25,39,46]. This occurrence is due to the cavity boundary conditions that introduce an additional degree of freedom: the frequency detuning of the pump from the eigenmode of the nonlinear resonator [23,47]. Zhang et al. [10] proposed hybrid silica microspheres for generation of tunable Kerr and Raman-Kerr combs. The authors have coated the polar cap of the microsphere with iron oxide nanoparticles. Since the WGM are excited at the equator of the microspheres, far away from the iron oxide nanoparticles, the high quality factor Q was not spoiled. The authors fed the control light into the microsphere through the fiber stem (see Figure 4). This control light was absorbed by the iron oxide nanoparticles and due to a strong photothermal effect, the comb was tuned. The achieved tuning of the Kerr comb was about 0.8 nm whereas the Raman-Kerr comb was tuned about 2.67 nm. The proposed photothermal tuning is an all-optical method and has less disadvantages than the mechanical methods [48] which present mechanical interferences and need cryogenic temperatures. However, the tuning range achieved by mechanical methods was about 450 GHz at 10 K for a microsphere of 40 µm diameter.

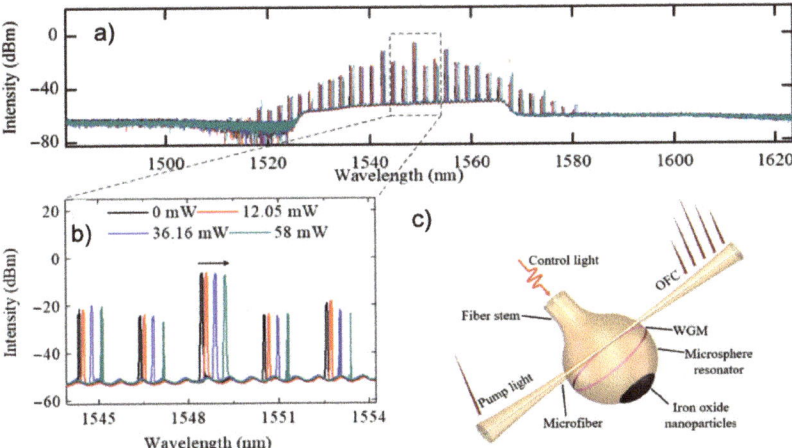

Figure 4. (a) Generated Kerr combs in a hybrid microsphere for different control light powers, (b) zoom-in of the spectra of panel (a,c) Sketch of experimental set-up showing an iron oxide nanoparticle coated silica microsphere of 248 µm diameter for photothermal tuning of generated optical frequency combs (OFC). The pink ring is the excited whispering gallery mode and the black polar cap represents the coated area with iron oxide nanoparticles. Reproduced with modifications from Ref. [10].

For microbubbles, the first demonstration of cascaded FWM was done by Li et al. [49]. Broader combs in microbubbles were demonstrated by Farnesi et al. [50]. The authors here realized a "Type I" or natively mode spaced comb with sidebands separated by one FSR (see Figure 5a) and a "Type II" or multimode spaced comb with sidebands separated by several FSR (see Figure 5b). Figure 5c shows the FWM pairs in the vicinity of the pump at 1.55264 nm in the backward direction. In this case, the FWM pairs are separated by one FSR. At 14 THz from the pump, centered at 1508 nm, an anti-Stokes comb was observed. In this case, the intensity of the anti-stokes component is high enough to generate its own parametrical oscillation, with a separation smaller than the FSR of the cavity. MBRs are spheroidal WGMR with quite dense spectral characteristics with two nearly equidistant mode families characterized by the same azimuthal but different vertical quantum number. The presence of these two mode families gives the different frequency spacing and the asymmetric spectrum (see Figure 5d [50]).

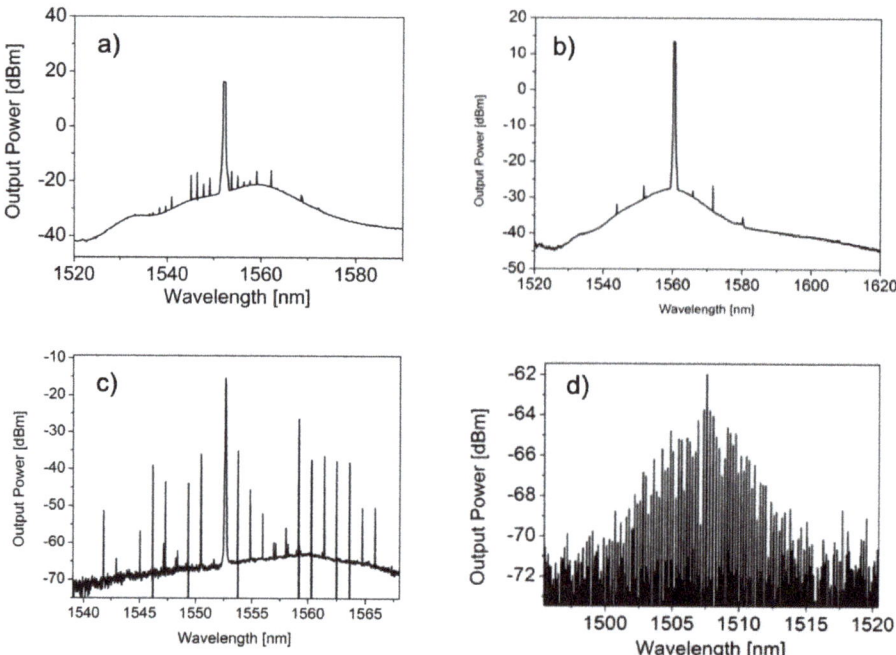

Figure 5. Experimental spectra of: (**a**) a Type I comb, with a frequency offset of 1 FSR, (**b**) a Type II comb, with a frequency offset of 5 FSR, (**c**) FWM in the vicinity of the pump spaced by azimuthal FSR and (**d**) Modulation intensity around the anti-stokes component at 1508 nm, with a frequency offset of 2 vertical FSR (2X0, 12 nm) measured for a microbubble of 475 μm diameter. Reproduced with modifications from Adapted with permission from Ref. [50] © The Optical Society.

Yang et al. [51,52] have experimentally measured frequency combs in the visible (pump wavelength centered at 765 nm) by engineering the dispersion through wall thickness of the microbubbles and degenerate FWM in hollow microbottles.

2.3. Stimulated Raman Scattering

The inelastic scattering of a photon with an optical phonon, which originates from a finite response time of the third-order nonlinear polarization of the material, is called the Raman scattering effect. Spontaneous Raman scattering occurs when a monochromatic light beam propagates in a material like silica. Some of the photons are transferred to new frequencies. The scattered photons may lose (Stokes shift) or gain (anti-Stokes shift) energy.

The left diagram in Figure 2b represents the absorption of a pump photon with energy with the consequent excitation of a molecule from the ground state (G) into a higher virtual energy state (V). The energy difference between the ground and the excited level is equal to pump photon's energy. In a second step, the molecule falls to an intermediate level (I), which is generated by its own periodical oscillations or rotations. This decay is accompanied by a Stokes photon emission. The destruction of the pump and the generation of the Stokes photon happen simultaneously because V is a virtual state. The energy difference between the pump and the Stokes photon is equal to the difference between the energy levels I and G; the remaining energy is the vibrational energy delivered to the molecule.

In contrast to other kinds of nonlinear phenomena where the molecule returns to its ground level after the interaction, an energy transfer between a photon and a molecule takes place here. Raman scattering is a pure gain process and depends on particular material resonances. In crystalline media, these resonances show a very narrow bandwidth. On the other hand, in amorphous silica the molecular vibration modes are overlapped with each other and create a continuum [53]. Contrary to spontaneous emission, SRS can transform a large part of power into a new frequency-shifted wave with the intensity growing exponentially with the propagation distance in the nonlinear (NL) material. Highly efficient scattering can occur as a result of the stimulated version under excitation by an intense laser beam; 10% or more of the energy of the incident laser beam can be converted.

Generally, in WGM resonators the lasing threshold occurs when the cavity round-trip gain equals round-trip loss. The intra-cavity gain coefficient is related to the bulk Raman gain coefficient g_b (in silica the maximum is 6.5×10^{14} m/W at 1550 nm) through the equation:

$$g_R = \frac{c^2}{C(\Gamma) 2 n^2 V_{eff}} g_b \tag{11}$$

where V_{eff} is the effective modal volume and $C(\Gamma)$ is the modal coupling. The threshold pump power can be derived by the gain coefficient, taking into account the power build-up factor in the resonator:

$$P_{th} = \frac{\pi^2 n^2 V_{eff}}{\lambda_p \lambda_R C(\Gamma) g_R Q^2} \tag{12}$$

Thus, the threshold follows an inverse dependence on the squared quality factor Q of the cavity. This explains how an increase in Q will cause a two-fold benefit in terms of reducing cavity round-trip losses as well as of increasing the Raman gain, due to the Raman gain dependence on the pump intensity.

Contrary to the parametric effects, Raman scattering is intrinsically phase-matched over the energy levels of the molecule. In other words, it is a pure gain process.

The SRS can be summarized in two parts: (1) the molecular vibrations modulate the refractive index of the medium at the resonant frequency ω_v and the frequency sidebands are induced in the laser field. (2) the Stokes field at frequency $\omega_S = \omega_L - \omega_v$ beats with the laser field to produce the modulation of the total intensity which excites the molecular oscillation at frequency $\omega_L - \omega_S = \omega_v$. In this way the two processes reinforce one another. The Raman emission can be thought as a down-conversion of a pump photon and phonon associated with the vibrational mode of the molecule. The anti-Stokes wave is generated together with the Stokes one, through FWM in which two pump photons annihilate themselves to produce Stokes and anti-Stokes photon can occur if $2\omega_L = \omega_a + \omega_s$, providing the total momentum conservation. This leads to the phase-matching condition $\Delta k = 2k(\omega_L) - k(\omega_a) - k(\omega_s) = 0$ where $k(\omega)$ is the propagation constant. When the phase mismatch Δk is large, Stokes emission experiences gain whereas the anti-Stokes experiences loss. For a perfect phase-matching, the anti-Stokes wave is strongly coupled to the Stokes one, preventing the growth of the latter.

In 2003 a microcavity-based cascaded Raman laser was demonstrated in WGM silica microspheres with sub-milliwatt pump power [54]. In cascaded Raman oscillation, the Raman signals serve to secondary pump field and generate higher-order Raman waves. As the pump power is increased, the first Stokes line extracts power from the pump until it becomes strong enough to seed the generation of a next Stokes line. This process can continue to generate more Raman peaks. The cascade process can be modeled as coupled harmonic oscillators with the pump and the Raman fields by including higher-order coupling terms(see for instance, [55]) The SRS and cascaded SRS in the infrared region occurs as standing waves because the Raman gain that amplifies the waves is the same for waves traveling in either the forward or the backward direction. In the presence of these phenomena, we have also observed TSFG in the visible, obtaining multicolor emission (red, orange, yellow, and green) by tuning the pump wavelength. Figure 6 shows the spectra measured for each different color and the corresponding microscope picture of the microsphere.

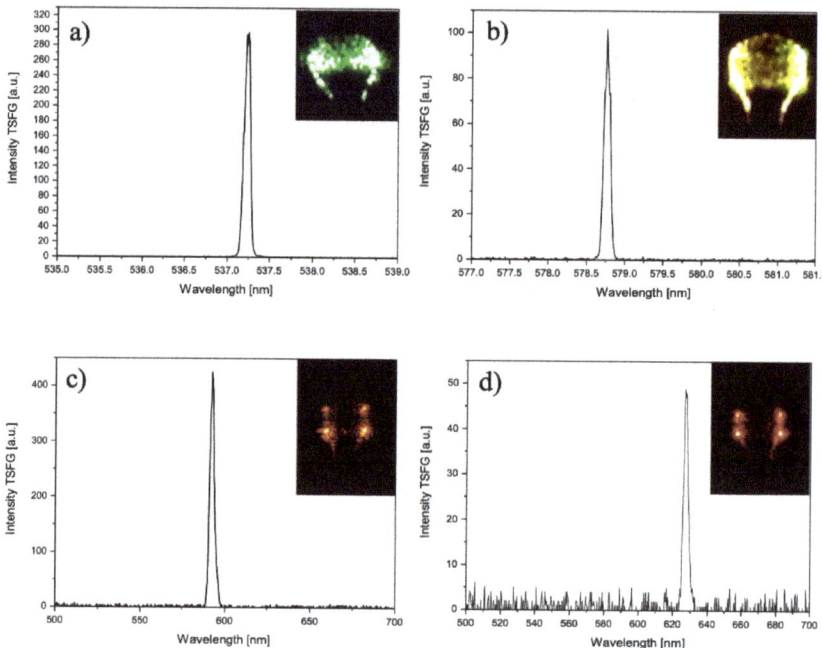

Figure 6. Various spectra obtained with different pump wavelengths showing TSFG standing waves generated among cascaded Raman lines with emission at: (**a**) 537.2 nm; (**b**) 578.8 nm, (**c**) 592 nm, and (**d**) 625 nm. Reproduced with modifications from Ref. [35].

In these cases, the pump was high enough to generate several orders of Stokes Raman lines and Raman combs. TSFG and THG can also occur simultaneously. Figure 7 shows the picture of a multicolor emission, one at 519.2 nm and one at 625 nm, corresponding to the THG signal of the pump laser and the TSFG, respectively.

As mentioned before, the red standing wave is the result of TSFG whereas the green traveling wave is TH signal. To fulfill the phase-matching condition and dispersion compensation, we must excite higher-order polar modes ($l- | m | > 1$). High polar order modes can be excited by placing the coupling taper far from the equatorial plane where the intensity peaks of these modes are located [1]. These modes also improve mode matching which is another requirement for efficient nonlinear frequency generation. The overlap of the WGM eigenfunctions corresponds to the power of TSFG and THG. In the latter case, the total power is proportional to the overlap of the TH field with the cubic power of the pump field.

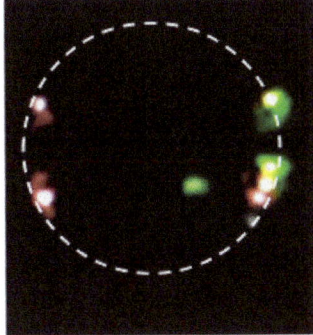

Figure 7. Optical picture of the microsphere showing the TSFG standing wave in the red and the traveling wave in the green of the THG.

WGMR Raman lasers can be used in sensing applications. WGMR-based lasers have a very narrow linewidth, they are dopant-free and they can attain high detection resolution down to single nanoparticles [56]. However, biochemical sensors need to work at telecom wavelengths where water shows a very high absorption [57,58]. A way to overcome such limitation is to use laser pumps in the visible or near infrared (NIR) [59,60]. Figure 8 shows the SRS line at 807 nm when a microsphere of about 50 µm diameter is pumped at 778 nm.

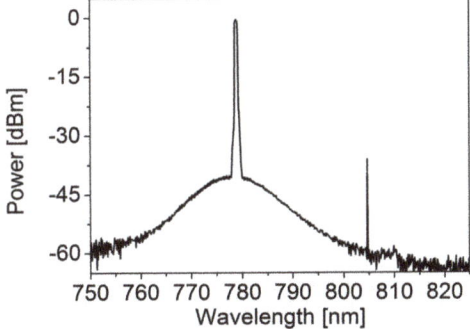

Figure 8. Experimental spectra of cascaded Raman lasing in a microsphere of 50 µm diameter. The laser pump is centered at about 778 nm, the Raman line is centered at about 807 nm.

Stimulated anti-Stokes Raman scattering (SARS) requires phase-matching, to be efficiently generated, and it is thresholdless. The nonlinear polarization of the anti-Stokes wave, P_{as}^{NL}, is defined by the relation:

$$P_{as}^{NL} = \chi^{(3)} E_P^2 E_S^* e^{i(2k_P - k_S)z} \tag{13}$$

where E_P and E_S are the amplitude of the pump and Stokes waves, respectively; χ^3 is the third-order nonlinear susceptibility, and k_i are the propagation vectors for the pump and Stokes waves [20,21]. Farnesi et al. [25] measured SARS in solid microspherical WGMR in the normal dispersion regime [23,39,46], contrary to well-known theoretical models [44,45], which predicted modulation instabilities and FWM in WGMR only with anomalous dispersion. The cavity boundary conditions introduce an extra degree of freedom, namely the frequency detuning between the pump and the eigenmode of the nonlinear WGMR [23,47]. As stated in [25], there was a negative shift of about 30 MHz of the resonant frequency due to the Kerr effect plus a larger thermo-optic frequency shift. However, the FSR of the resonator showed just a slight change [23]. The linear dispersion also changes slightly since it is calculated as the variation of the FSR [40]. Similar results have been achieved by Soltani et al. [61]. The group have been able to enhance by a factor of 4 the results of Farnesi et al. [25] using hybrid microspheres. In this particular case, silica microspheres have been functionalized with a layer of gold nanorods in polyethilenglycol (PEG). Farnesi et al. [25] have measured some unusual features, namely strong anti-Stokes components and extraordinarily symmetric spectra. Usually in SRS, the Stokes waves are exponentially enhanced, whereas the anti-Stokes waves are exponentially absorbed [62]. Anti-Stokes Raman components are coupled to the Stokes Raman ones [62,63], independently of the magnitude of dispersion. As a result of the coupling and of an effectively phase-matched hyper-parametric process, the anti-Stokes wave grows along the microsphere directly proportioned to the Stokes wave. When there is perfect phase-matching condition, each eigensolution is an equal combination of Stokes and anti-Stokes components with a power ratio of one. When the phase mismatching conditions deviates from zero gradually, we are in an intermediate case where the anti-Stokes/Stokes power ratio is given by the following equation:

$$\frac{P_a}{P_S} = \left| \frac{\gamma q P}{2\gamma q P + \beta_2 \Omega^2} \right|^2 \tag{14}$$

where P is the cavity build-up pump power, $\Omega/2\pi$ is the frequency shift between the pump and the first Raman order, β_2 is the linear dispersion, γ is the nonlinearity coefficient and $q = (1-\alpha) + \alpha\chi^3 = 0.82 + i0.25$ is a complex number that depends on the Raman susceptibility of silica and on the fractional contribution of the electronic susceptibility to the total nonlinear index [26,64].

When there is no phase match, the Stokes and anti-Stokes components are effectively decoupled.

This expression is valid for both regimes, normal and anomalous, but the linear dispersion $\Delta k = \beta_2 \Omega^2$ must be large. The values obtained from Equation (14) are in close agreement with the experimental ones, given the uncertainties in β_2 and γ, as it can be seen in Figure 9.

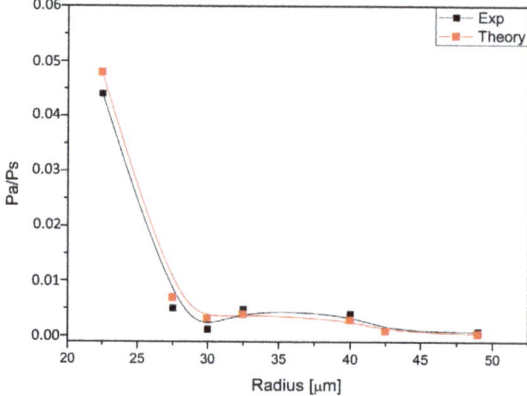

Figure 9. P_a/P_S ratio: experimental (solid black squares) and calculated values (solid red circles) The lines are a guide to the eye.

2.4. Stimulated Brillouin Scattering

SBS is also an inelastic scattering process but it results from the coherent interaction of light photons and acoustic phonons instead of optical phonons. The WGMR acts as a dual photonic-phononic cavity due to the overlap of both waves inside the resonator. SBS is automatically phase-matched because it is a pure gain process, like SRS. The SBS gain coefficient is one of the largest but with a small gain bandwidth [62]. The narrow bandwidth places very stringent conditions on the geometry of resonators since it would require a Brillouin frequency shift that equals the free spectral range (FSR) of the WGMR. [36,65,66]. We need to excite high order modes [29] with vertical FSR smaller than the fundamental FSR [50] in order to bypass such a stringent condition. SBS can further reduce its threshold power when it is resonantly enhanced, and it can be as low as some micro-watts [65]. The threshold power is directly proportional to the mode volume [67]

$$P_{th} = \frac{\pi^2 n^2 V_{eff}}{\lambda_p \lambda_s B g_b Q_p Q_s} \frac{1}{1 + \frac{Q_m \lambda_m}{2\pi r}} \tag{15}$$

where the subscripts p, s, m refer to pump, Stokes and mechanical modes, and B is the mode overlap. SBS has been demonstrated in silica [67,68] and tellurite [69] microspheres, microbottles [70] and microbubbles [29,71,72]. Hollow and solid spherical WGMR have been used to generate both backward and forward SBS. As with SRS, cascaded SBS can also be generated in WGMR. Even and odd orders are observed in both directions, but showing different lasing efficiency (even (odd) orders are more efficient in forward (backward) direction) [72]. Scattering can occur in forward direction with frequencies in the MHz–GHz range whereas backward are in the GHz range. The SBS frequency in silica glass is about 11 GHz and it scales with the optical one, with a bandwidth in the range of 20–60 MHz at telecom frequencies. In our experiments, the free spectral range (FSR) of our MBR is 141 GHz (diameter about 475 µm) and 105 GHz (diameter about 675 µm). Therefore, we can obtain SBS only by using high order modes, in that case the vertical FSR is much less than the FSR [50]. Frequency combs and SBS can coexist in both forward and backward directions, as Figure 10. SBS efficiency in microbubbles can be so high to allow degenerated FWM from the Brillouin laser line. Figure 10 shows in forward direction, cascaded FWM from the second order Brillouin laser line (1.54522 nm) for a pump wavelength centered at 1.54504 nm

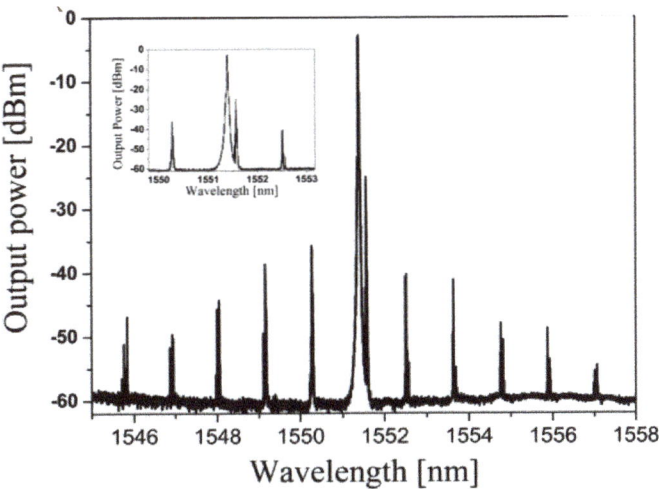

Figure 10. Native Kerr comb and second order SBS in forward direction: pump power 72 mW at 1551,344 nm. (Inset: zoom of the spectrum showing the 2nd order SBS laser line and the first pair of FWM lines) Reproduced with modifications from Ref. [73].

3. Kerr Switching in Hybrid Resonators

Inorganic materials still show weak nonlinearity, slow dynamics and the difficulty of discrimination between thermal and Kerr nonlinearity at room temperature limits their performance. Significant advantages can be obtained if organic-inorganic hybrid systems can be used [74,75]. π-conjugated polymers are extremely suitable nonlinear optical materials which show structural flexibility, relative ease of preparation, high χ^3 values and high photostability. Hybrid polyfluorene derivatives-silica WGMR have been demonstrated as very good candidates for all-optical switching [12,13] where two beams are present, namely a pump beam that switches a probe [76,77]. The electronic Kerr effect is almost instantaneous (picosecond timescales) and due to the enhancement of the WGMR combined with a strong third-order nonlinearity, the intensities used are well below the damage threshold of the conjugated polymer. The material refractive index n and the absorption coefficient α depend on the light intensity I in the material according to the equations $n = n_0 + n_2 I + n_4 I^2 + \ldots$ (with the nonlinear refractive index $n_2 \approx \Re(\chi^3)$, $n_4 \approx \Re(\chi^5)$, where χ^3 and χ^5 are the third- and fifth- order nonlinear optical susceptibilities) and $\alpha = \alpha_0 + \beta I$ (with the nonlinear absorption coefficient $\beta \approx \Im(\chi^3)$). All-optical switching for a probe signal I_{probe}, which is resonant with the microsphere, can be realized using a resonant pump beam I_{pump}, which affects the coated cavity resonance position by changing the refractive index of the coating in the corresponding wavelength range [78].

If the χ^3 and χ^5 are caused by fast electronic Kerr nonlinearity, then, as mentioned before, the nonlinear switching is on a picosecond time scale, which is the most desired situation for the optical switching. However, thermal nonlinearities can restrict the use of the hybrid devices because the spectral response is sensitive to the input power of the probe signal as well [76]. In that case, the light-induced changes in the refractive index can be described phenomenologically by $n_2 = (\delta n / \delta T) T_L$, where T_L is the laser-induced change of the temperature of the nonlinear medium, the corresponding switching time being about 10^{-3}–10^{-5} s. In other words, the thermal switching of a nonlinear medium for the case of a standard Ti-sapphire laser should be the same for the pulsed or CW mode operation regime. Moreover, an intrinsically weak but highly localized probe beam can also participate in this type of the light-induced WGM switching.

Murzina et al. [13] reported on the all-optical switching of WGM in silica microspheres with two types of coatings, an active one based on a Kerr polymeric material (polyfluorene derivative, PF(o)n) [12] and an inert polymer based on an anionic copolymer made of methacrylic acid and methyl methacrylate (Eudragit® L100) [79]. The authors modeled the overlap of the coupled optical field with the polymer layer and verified the role of the probe field experimentally for both polymer coatings.

Figure 11 shows an sketch of the set-up, an optical picture of a microspherical WGMR, with a diameter of about 250 µm and the taper; and the control test with a bare WGMR. A Tunics Plus was used as probe beam, which is a semiconductor external-cavity laser tunable in the spectral range of 1.55–1.6 µm and with 300 kHz linewidth. The pump probe was a Mira 900-f Ti-Sapphire (Coherent). The laser probe was coupled into the WGMR through a homemade tapered fiber whereas the laser pump was coupled into a multimode fiber that illuminated a hemisphere of the WGMR. To make a polymer coverage, the dip coating technique was used. The authors obtained layers of about 100 nm thickness for PF(o) coatings and of about 50 nm thickness for Eudragit coatings. The Q values were higher than 10^8 for bare microspheres and higher than 10^6 after polymer coating. To attain adequate solubility in common organic solvents and mesogenic behavior, the polyfluorene derivative, PF(o)n, was functionalized at the C9 position of the fluorine ring with two pendant octyl chains [80]. It shows a maximum absorption at λ_{abs} = 379 nm, and the measured n_6 is about 2×10^{-10} cm^2/W and β coefficient 7×10^{-7} cm/W [8,14]. As the first step, pump-induced effects on the barre WGMR in were studied. No shift was seen as the averaged pump power was increased up to 30 mW. Thus, we may assume that the Kerr nonlinearity and the thermal nonlinear effects of pure fused silica were negligible for these pump-and-probe levels.

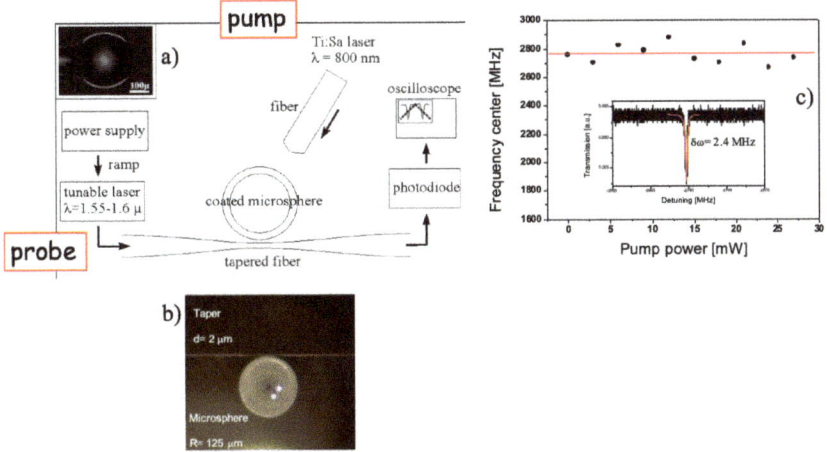

Figure 11. (a) Experimental pump-and-probe set-up. Left hand side inset: optical image of the WGMR. (b) Picture of the microsphere and the coupling taper; (c) Frequency center of the WGMR resonance versus pump power. The red line is a guide to the eye. Inset: Zoom of a typical resonance for a bare microsphere, the red line is a Lorentzian fit with a FWHM of about 2.4 MHz. Reproduced with modifications from [12].

Figure 12a shows the WGM spectrum measured for the probe wavelength of 1600 nm and for two different pump powers for the mode-locked regime of the Ti:Sapphire laser. The Ti:Sapphire was tuned at 775 nm to generate the two-photon absorption (TPA) in the PF(o)n coating of the hybrid WGMR. Thermal nonlinearities can be ruled out since no broadening of the resonance, hysteresis or asymmetries could be observed in the transmission spectra.

The results of the pump-and-probe experiments are shown in Figure 12b, where a frequency shift of 2 GHz is obtained in pulsed regime for an average pump power of 35 mW at 775 nm for a probe of 1600 nm. To discriminate the thermal shift from the Kerr shift, we have performed measurements in CW and pulsed regime for the same average pump powers, similarly to a previous work [12]. For the same wavelength and average pump power, the authors obtained a much lower spectral shift of 250 MHz in the CW regime (Figure 12b). In here, we have also tested the influence of the wavelength of the pump beam. Figure 12b also shows a frequency shift of 200 MHz obtained in the mode-locked regime for an average pump power up to 21 mW at 825 nm. The detuning is of the same order of magnitude as the CW regime. We have chosen 825 nm as a pump beam because two-photon absorption (TPA) is not feasible; and because it is also far from the second harmonic of the probe beam. In that case the pump beam acts as a spectrally broad thermal source only.

Figure 12c shows the frequency shift versus pump power for λ_{probe} = 1558 nm at two different laser regimes for the PF(o) coated WGMR. At λ_{pump} = 775 nm a clear quadratic dependency can be observed, whereas at λ_{pump} = 825 nm the dependency is linear and the detuning is of the same order of magnitude as the CW regime, indicating again that in absence of TPA, the pump acts as a thermal source. It can also be observed that in the case of λ_{probe} = 1558 nm the magnitude of the shift is greater than in the case of λ_{probe} = 1600 nm, for both regimes, pulsed and CW. Figure 12d shows an almost null red-shift up to 20 mW of pump power for λ_{probe} = 1558 nm for a WGMR coated with Eudragit® L100, an inert polymer.

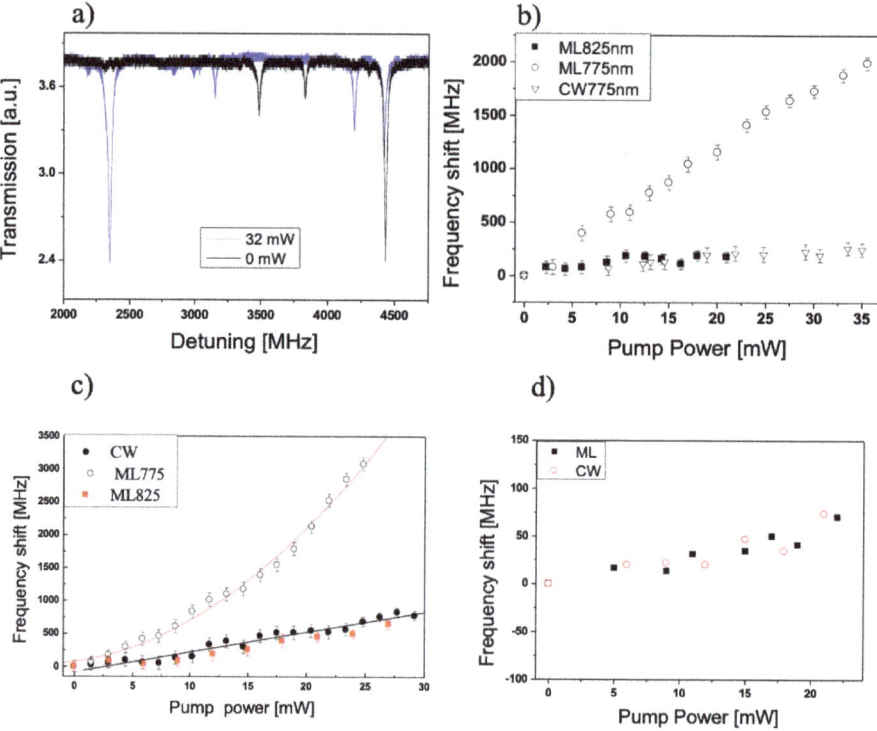

Figure 12. (**a**) Typical WGM spectra measured for a polymer-coated microsphere for two different pump powers: (black) pump laser off and (blue) 32 mW. λ_{pump} = 775 nm, λ_{probe} = 1600 nm. (**b**) Pump power dependence of the detuning of WGM in PF(o)n coated microspheres for mode-locked regime of the Ti-sapphire pump laser: 825 nm (filled squares) and 775 nm (empty circles); and CW at 775 nm (empty downside triangles). (**c**) The probe wavelength is λ_{probe} = 1558 nm, switching for CW (filled circles), mode-locked at 775 nm (empty circles) and mode-locked at 825 nm. (**d**) Eudragit coated microspheres for mode-locked regime of the Ti-sapphire pump laser for two different regimes: CW (empty circles) and pulsed (filled squares). The probe wavelength is λ_{probe} = 1558 nm. Reproduced with modifications from [13].

4. Two-Photon Fluorescence

Two-Photon Fluorescence (TPF) is a validated technique for imaging and detection of labeled biological material such as peptides [81] and steroids [82,83]. TPF has numerous advantages over conventional one photon fluorescence (OPF), being the most important one the large Stokes shift between emission and excitation. The large energy gap between the fields lowers the background noise. Other advantages are the reduced static photo-bleaching of dyes due to the absorption's quadratic dependence on the intensity; and the wide range of fluorescent dyes with high quantum yields and molar extinction in the visible. However, TPF requires high photon density flux, reached by tightly focusing the laser light. To avoid tight focusing and achieve the needed intensities, much research has resorted to enhancing photonic platforms such as resonators [84]. MBR are the most suitable WGMR for TPF, being hollow they can be filled with liquids. For the TPF demonstration Pastell et al. [17] filled MBR with a 10^{-3} and 10^{-4} solution of fluorescein and 10^{-6} solution of Rhodamine 6G.

The authors used a modified confocal microscope for coupling the light into the microbubble resonator [85], an inverted light microscope (Nikon Eclipse TE2000U) up-graded to a multi-modal imaging system even though it can be used either as a bright-field microscope or as a phase contrast microscope. The microscope was pumped by a Mira 900-f Ti: Sapphire (repetition rate 76 MHz and 150fs pulse duration, Coherent) with an average power of 1.2 W. It has an average power of 1.2 W and it can be tuned over a wavelength range of 690–950 nm within which falls the two-photon absorption spectra of many fluorophores [86]. The wavelength in the experiments was set to 800 nm. The raster scan of the beam was stopped and it was coupled into the MBR by focusing the laser beam tangential to the bubble wall with two different dry objectives, namely 4X and 10X and 0.5 NA. The excitation light was filtered by a dichroic mirror (FF720-SDi01, Semrock) and a BG39 Schott filter. We tested first the bubble with the fluorescein filling that was imaged with a 4X dry objective using a CCD camera to see the complete WGM at the equator. Figure 13 shows the TPF band around the equator and the TPF partially coupled back to the MBR wall. In this case, the MBR was filled with fluorescein and imaged with a 4X dry objective using a CCD camera to see the complete WGM at the equator.

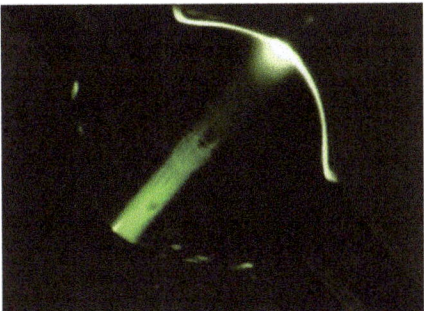

Figure 13. Fluorescence image of the MBR filled with 10^{-3} solution of fluorescein, showing the TPF band and lobes, and the fluorescence coupled back into the MBR wall . Reproduced with modifications from [17].

The two-photon nature of the emitted signal was validated by checking its dependence on the excitation laser power. Figure 14 shows an optical image of a MBR for an incident power of about 190 mW (the power at the focal plane is just the 30 of the incident power), its intensity plot (center); and a logarithmic representation of the TPEF signal from a Rhodamine 6G filled microbubble versus incident laser power at the focal plane. A linear fit to the data has slope close to 2, ensuring the quadratic dependence of the obtained signal. The two lobes that correspond to the WGM are clearly seen.

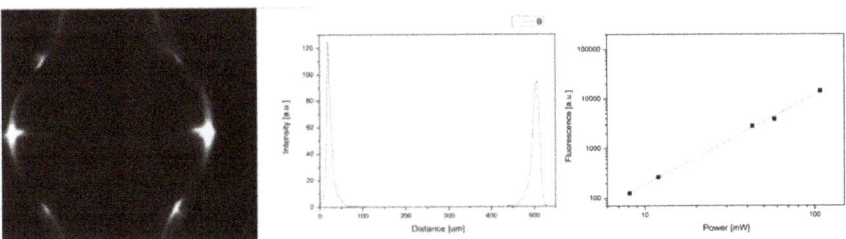

Figure 14. Fluorescence image of the MBR filled with 10^{-6} solution of Rhodamine 6G, showing the TPF lobes, the corresponding intensity plot (center) and the TPF signal versus the pump laser power in log-log scale. The red line is the linear fit with slope close to 2. The probe wavelength is λ_{pump} = 800 nm. Reproduced with modifications from [17].

5. Conclusions

In this paper, we present an overview of the latest advances in the area of nonlinear frequency generation in WGMR. The review is limited to microspherical solid and hollow WGMR and Kerr phenomena. The Kerr effects we reviewed are third-harmonic generation, third-order sum-frequency generation, frequency combs, Kerr switching and Two-Photon Fluorescence. Stimulated Raman Scattering and Stimulated Brillouin Scattering and combination of other nonlinear phenomena such as FWM are discussed in separated subsections. WGMR are excellent platforms to understand how light, sound and matter interact. The ultimate goal of the nonlinear research in photonic devices is few or single-photon interactions to pave the way for quantum compact photonic devices at room temperature.

Author Contributions: Conceptualization, S.S.; writing—original draft preparation, S.S.; writing—review and editing, G.F., D.F., G.N.C.; figure preparation, D.F. Figures 7 and 8 are original unpublished data from our labs, the picture and data were taken by D.F., the experiment was conceived by G.N.C. and the data analyzed by S.S. All authors have read and agreed to the published version of the manuscript.

Funding: This research was partially funded by Centro Fermi project MiFo.

Acknowledgments: Franco Cosi is gratefully acknowledged for his technical support.

Conflicts of Interest: The authors declare no conflict of interest. The funders had no role in the design of the study; in the collection, analyses, or interpretation of data; in the writing of the manuscript, or in the decision to publish the results.

References

1. Chiasera, A.; Dumeige, Y.; Feron, P.; Ferrari, M.; Jestin, Y.; Nunzi Conti, G.; Pelli, S.; Soria, S.; Righini, G. Spherical whispering-gallery-mode microresonators. *Laser Phot. Rev.* **2010**, *4*, 457–482. [CrossRef]
2. Rayleigh, L. The problem of the whispering gallery. *Philos. Mag.* **1910**, *20*, 1001. [CrossRef]
3. Qian, S.X.; Chang, R.K. Multi-order Stokes emission from micrometer-size microdroplets. *Phys. Rev. Lett.* **1986**, *56*, 926–929. [CrossRef] [PubMed]
4. Dumeige, Y.; Trebaol, S.; Ghisa, L.; Nguyen, T.K.N.; Tavernier, H.; Féron, P. Determination of coupling regime of high Q resonators and optical gain of highly selective amplifiers. *JOSA B* **2008**, *25*, 2073–2080. [CrossRef]
5. Ilchenko, V.; Gorodetsky, M.; Vyatchanin, S. Coupling and tunability of optical whispering gallery modes—A basis for coordinatemeter. *Opt. Commun.* **1994**, *107*, 41–48. [CrossRef]
6. Sumetsky, M.; Dulashko, Y.; Windele, R.S. Super free spectral range tunable optical microbubble resonator. *Opt. Lett.* **2010**, *35*, 1866–1868. [CrossRef] [PubMed]
7. Shu, F.; Zhang, P.; Qian, Y.; Wang, Z.; Wan, S.; Zou, C.L.; Guo, G.; Dong, C. A mechanically tuned comb in a dispersion-engineered silica microbubble resonator. *Sci. China-Phys. Mech. Astron.* **2020**, *63*, 254211. [CrossRef]
8. Henze, R.; Seifert, T.; Ward, J.; Benson, O. Tuning whispering gallery modes using internal aerostatic pressure. *Opt. Lett.* **2011**, *36*, 4536–4538. [CrossRef]
9. Madugani, R.; Yang, Y.; Le, V.H.; Ward, J.M.; Nic Chormaic, S. Linear Laser Tuning Using a Pressure-Sensitive Microbubble Resonator. *IEEE Photonics Technol. Lett.* **2016**, *28*, 1134–1137. [CrossRef]
10. Zhu, S.; Shi, L.; Ren, L.; Zhao, Y.; Jiang, B.; Xiao, B.; Zhang, X. Controllable Kerr and Raman-Kerr frequency combs in functionalized microsphere resonators. *Nanophotonics* **2019**, *8*, 2321–2329. [CrossRef]
11. Zhu, S.; Jiang, B.; Xiao, B.; Shi, L.; Zhang, X. Tunable Brillouin and Raman microlasers using hybrid microbottle resonators. *Nanophotonics* **2019**, *8*, 931–940. [CrossRef]
12. Razdolskiy, I.; Berneschi, S.; Nunzi Conti, G.; Pelli, S.; Murzina, T.; Righini, G.; Soria, S. Hybrid microspheres for nonlinear Kerr switching devices. *Opt. Express* **2011**, *10*, 9523–9528. [CrossRef] [PubMed]
13. Murzina, T.V.; Berneschi, S.; Barucci, A.; Cosi, F.; Nunzi Conti, G.; Razdolskiy, I.; Soria, S. Kerr versus thermal non-linear effects studied by hybrid whispering gallery mode resonators. *Opt. Mater. Express* **2012**, *2*, 1088–1094. [CrossRef]
14. Carmon, T.; Vahala, K. Visible continuous emission from a silica microphotonic device by third-harmonic generation. *Nat. Phys.* **2007**, *3*, 430. [CrossRef]
15. Lin, H.B.; Campillo, A. CW nonlinear optics in droplet microcavities displaying enhanced gain. *Phys. Rev. Lett.* **1994**, *73*, 2440–2443. [CrossRef]

16. Vogt, D.; Leonhardt, R. Terahertz whispering gallery mode bubble resonator. *Optica* **2017**, *4*, 809–812. [CrossRef]
17. Pastells, C.; Marco, P.; Merino, D.; Loza-Alvarez, P.; Quercioli, F.; Pasquardini, L.; Lunelli, L.; Pederzolli, C.; Daldoso, N.; Farnesi, D.; et al. Two photon versus one photon fluorescence excitation in whispering gallery mode microresonators. *J. Lumin.* **2016**, *170*, 860–865. [CrossRef]
18. Lin, G.; Coillet, A.; Chembo, Y.K. Nonliner phtonics with high Q whsipering gallery mode resonators. *Adv. Opt. Photonics* **2017**, *9*, 828–890. [CrossRef]
19. Righini, G.; Dumeige, Y.; Feron, P.; Ferrari, M.; Nunzi Conti, G.; Soria, S.; Ristiić, D. Whispering gallery mode microresonators: Fundamentals and applications. *Riv. Del Nuovo Cimento* **2011**, *34*, 435.
20. Leach, D.H.; Chang, R.K.; Acker, W.P. Stimulated anti-Stokes Raman scattering in microdroplets. *Opt. Lett* **1992**, *7*, 387–389. [CrossRef]
21. Roman, V.E.; Popp, J.; Fields, M.H.; Kiefer, W. Minority species detection in aerosols by stimulated anti-Stokes-Raman scattering and external seeding. *Appl. Opt.* **1999**, *38*, 1418–1422. [CrossRef] [PubMed]
22. Spillane, S.; Kippenberg, T.; Vahala, K. Ultralow-threshold Raman laser using a spherical dielectric microcavity. *Nature* **2002**, *415*, 621. [CrossRef] [PubMed]
23. Matsko, A.B.; Savchenkov, A.A.; Maleki, L. Normal group-velocity dispersion Kerr frequency comb. *Opt. Lett.* **2012**, *37*, 43–45. [CrossRef] [PubMed]
24. Matsko, A.; Strekalov, D.; Ilchenko, V.; Maleki, L. Optical hyper-parametric oscillations in a whispering gallery mode resonator: Threshold and phase diffusion. *Phys. Rev. A* **2005**, *71*, 033804. [CrossRef]
25. Farnesi, D.; Cosi, F.; Trono, C.; Righini, G.C.; G. Nunzi Conti.; Soria, S. Stimulated anti-Stokes Raman scattering resonantly enhanced in silica microspheres. *Opt. Lett.* **2014**, *39*, 5993. [CrossRef] [PubMed]
26. Roos, P.A.; Meng, L.S.; Murphy, S.K.; Carlsten, J.L. Approaching quantum-limited cw anti-Stokes conversion through cavity-enhanced Raman-resonant four-wave mixing. *J. Opt. Soc. Am. B* **2004**, *21*, 357–363. [CrossRef]
27. Maldovan, M.; Thomas, E. Simultaneous localization of photons and phonons in two-dimensional periodic structures. *Appl. Phys. Lett.* **2006**, *88*, 251907. [CrossRef]
28. Rolland, Q.; Oudich, M.; El-Jallal, S.; Dupont, S.; Pennec, Y.; Gazalet, J.; Kastelik, J.C.; Lévêque, G.; Djafari-Rouhani, B. Acousto-optic couplings in two-dimensional phoxonic crystal cavities. *Appl. Phys. Lett.* **2012**, *101*, 061109. [CrossRef]
29. Lu, Q.; Liu, S.; Wu, X.; Liu, L.; Xu, L. Stimulated Brillouin laser and frequency comb generation in high Q microbubble resonators. *Opt. Lett.* **2016**, *41*, 1736–1739. [CrossRef]
30. Hill, S.C.; Leach, D.H.; Chang, R.K. Third-order sum-frequency generation in droplets: Model with numerical results for third-harmonic generation. *J. Opt. Soc. Am. B* **1993**, *10*, 16–33. [CrossRef]
31. Acker, W.P.; Leach, D.H.; Chang, R.K. Third-order optical sum-frequency generation in micrometer-sized liquid droplets. *Opt. Lett.* **1989**, *14*, 402–404. [CrossRef] [PubMed]
32. Leach, D.H.; Acker, W.P.; Chang, R.K. The effect of the phase velocity and spatial overlap of spherical resonances on sum-frequency generation in droplets. *Opt. Lett.* **1990**, *15*, 894–896. [CrossRef] [PubMed]
33. Leach, D.H.; Chang, R.K.; Acker, W.P.; Hill, S.C. Third order sum-frequency generation in droplets: Experimental results. *J. Opt. Soc. Am. B* **1993**, *10*, 34–45. [CrossRef]
34. Chew, H.; Sculley, M.; Kerker, M.; McNulty, P.J.; Cooke, D.D. Raman and fluorescent scattering by molecules embedded in small particles: Results for coherent optical processes. *J. Opt. Soc. Am.* **1978**, *68*, 1686–1689. [CrossRef]
35. Farnesi, D.; Berneschi, S.; Barucci, A.; Righini, G.; Soria, S.; Nunzi Conti, G. Optical frequency conversion in silica-whispering-gallery-mode microspherical resonators. *Phys. Rev. Lett.* **2014**, *112*, 093901. [CrossRef]
36. M.Asano.; Komori, S.; Ikuta, R.; N. Imoto, S.O.; Yamamoto, T. Visible light emission from a silica microbottle resonotar by second and third harmonic generation. *Opt. Lett.* **2016**, *41*, 5793–5796. [CrossRef]
37. Kozyreff, G.; Dominguez-Juarez, J.L.; Martorell, J. Nonlinear optics in spheres: From second harmonic scattering to quasi-phase matched generation in whispering gallery modes. *Laser Photonics Rev.* **2011**, *5*, 737–749. [CrossRef]
38. Chen, J.H.; Shen, X.; Tang, S.J.; Cao, Q.T.; Gong, Q.; Xiao, Y.F. Microcavity Nonlinear Optics with an organically functionalized surface. *Phys. Rev. Lett.* **2019**, *123*, 173902. [CrossRef]
39. Savchenkov, A.A.; Matsko, A.B.; Ilchenko, V.S.; Solomatine, I.; Seidel, D.; Maleki, L. Tunable Optical Frequency Comb with a Crystalline Whispering Gallery Mode Resonator. *Phys. Rev. Lett.* **2008**, *101*, 093902. [CrossRef]
40. Del'Haye, P. Optical Frequency Comb Generation in Monolithic Resonators. Ph.D. Thesis, Ludwig-Maximilians-Universitaet, München, Germany, 2011.

41. Kippenberg, T.; Spillane, S.; Vahala, K. Kerr-nonlinearity optical parametric oscillation in an ultrahigh-Q toroid microcavity. *Phys. Rev. Lett.* **2004**, *93*, 083904. [CrossRef]
42. Chembo, Y.K.; Strekalov, D.V.; Yu, N. Spectrum and dynamics of optical frequency combs generated with monolithic whispering gallery mode resonators. *Phys. Rev. Lett.* **2010**, *104*, 103902. [CrossRef] [PubMed]
43. Kippenberg, T.J.; Spillane, S.M.; Vahala, K.J. Demonstration of ultra-high-Q small mode volume toroid microcavities on a chip. *Appl. Phys. Lett.* **2004**, *85*, 6113–6115. [CrossRef]
44. Agha, I.H.; Okawachi, Y.; Foster, M.A.; Sharping, J.E.; Gaeta, A.L. Four-wave-mixing parametric oscillations in dispersion-compensated high-Q silica microspheres. *Phys. Rev. A* **2007**, *76*, 043837. [CrossRef]
45. Agha, I.H.; Okawachi, Y.; Gaeta, A.L. Theoretical and experimental investigation of broadband cascaded four-wave mixing in high-Q microspheres. *Opt. Express* **2009**, *17*, 16209. [CrossRef] [PubMed]
46. Liang, W.; Savchenkov, A.A.; Ilchenko, V.S.; Eliyahu, D.; Seidel, D.; Matsko, A.B.; Maleki, L. Generation of a coherent near-infrared Kerr frequency comb in a monolithic microresonator with normal GVD. *Opt. Lett.* **2014**, *39*, 2920. [CrossRef] [PubMed]
47. Coen, S.; Haelterman, M. Modulational Instability Induced by Cavity Boundary Conditions in a Normally Dispersive Optical Fiber. *Phys. Rev. Lett.* **1997**, *79*, 4139. [CrossRef]
48. Dinyari, K.; Barbour, R.; Golter, D.A.; Wang, H. Mechanical tuning of whispering gallery modes over a 0.5Thz tuning range with MHz resolution in silica microsphere at cryogenic temperatures. *Opt. Express* **2011**, *19*, 17966–17972. [CrossRef]
49. Li, M.; X.Wu.; L.Liu.; L.Xu. Kerr Parametric oscillations and frequency comb generation from dispersion compensated silica micro-bubble resonators. *Opt. Express* **2013**, *21*, 16908–16913. [CrossRef]
50. Farnesi, D.; Barucci, A.; Righini, G.C.; Nunzi Conti, G.; Soria, S. Generation of broadband hyperparametric oscillations in silica microbubbles. *Opt. Lett.* **2015**, *40*, 4508–4511. [CrossRef]
51. Yang, Y.; Jiang, X.; Kasumie, S.; Zhao, G.; Xu, L.; Ward, J.; Yang, L.; Nic Chormaic, S. Four-wave mixing parametric oscillation and frequency comb generation at visible wavelengths in a silica microbubble resonator. *Opt. Lett.* **2016**, *41*, 5266–5269. [CrossRef]
52. Yang, Y.; Ooka, Y.; Thompson, R.M.; Ward, J.M.; Nic Chormaic, S. Degenerate four-wave-mixing in a silica, hollow bottle microresonator. *Opt. Lett.* **2016**, *41*, 575–579. [CrossRef] [PubMed]
53. Shuker, R.; Gammon, R.W. Raman-Scattering Selection-Rule Breaking and the Density of States in Amorphous Materials. *Phys. Rev. Lett.* **1970**, *25*, 222. [CrossRef]
54. Min, B.; Kippenberg, T.; Vahala, K. A compact fiber-compatible cascaded Raman laser. *Opt. Lett.* **2003**, *28*, 17. [CrossRef] [PubMed]
55. Kippenberg, T.J.; Spillane, S.M.; Min, B.K.; Vahala, K.J. Theoretical and experimental study of stimulated and cascaded Raman scattering in ultrahigh-Q optical microcavities. *IEEE J. Sel. Top. Quantum Electron.* **2004**, *10*, 1219–1228. [CrossRef]
56. Ozdemir, S.K.; Zhu, J.; Yang, X.; Peng, B.; Yilmaz, H.; He, L.; Monifi, F.; Huang, S.; Long, G.; Yang, L. Highly sensitive detection of nanoparticles with a self-referenced and self-heterodyned whispering-gallery Raman microlaser. *Proc. Natl. Acad. Sci. USA* **2014**, *111*, E3836. [CrossRef]
57. Soria, S.; Berneschi, S.; Brenci, M.; Cosi, F.; Nunzi Conti, G.; Pelli, S.; Righini, G.C. Optical Microspherical Resonators for Biomedical Sensing. *Sensors* **2011**, *11*, 785–805. [CrossRef]
58. Righini, G.C.; Soria, S. Biosensing by WGM micro spherical resonators. *Sensors* **2016**, *16*, 905. [CrossRef]
59. Chistiakova, M.V.; Armani, A.M. Cascaded Raman microlaser in air and in buffer. *Opt. Lett.* **2012**, *37*, 4068–4070. [CrossRef]
60. Ooka, Y.; Yang, Y.; Ward, J.M.; Nic Chormaic, S. Raman Lasing in a hollow bottle like microresonator. *Appl. Phys. Express* **2015**, *8*, 092001. [CrossRef]
61. Soltani, S.; Diep, V.; Zeto, R.; Armani, A. Stimulated anti-Stokes Raman emission generated by gold nanorod coated optical resonators. *ACS Photonics* **2018**, *5*, 3550–3556. [CrossRef]
62. Boyd, R. *Nonlinear Optics*; Academic Press: Cambridge, MA, USA, 2008.
63. Shen, Y.R. *The Principles of Nonlinear Optics*; John Wiley and Sons, Inc.: Hoboken, NJ, USA, 2003.
64. Stolen, R.H.; Gordon, J.P.; J.Tomlinson, W.; Haus, H.A. Raman response function of silica-core fibers. *J. Opt. Soc. Am. B* **1989**, *6*, 1159–1166. [CrossRef]
65. Grudinin, I.; Matsko, A.; Maleki, L. Brillouin lasing with a CaF2 whispering gallery mode resonator. *Phys. Rev. Lett.* **2009**, *102*, 043902. [CrossRef] [PubMed]

66. Eggleton, B.; Poulton, G.; Rant, P. inducing and harnessing stimulated Brillouin scattering in photonic integrated circuits. *Adv. Opt. Photonics* **2013**, *5*, 536–587. [CrossRef]
67. Tomes, M.; Carmon, T. Photonic micro-electromechanical systems vibrating at X-band (11GHz) rates. *Phys. Rev. Lett.* **2009**, *102*, 113601. [CrossRef] [PubMed]
68. Bahl, G.; Zehnpfennig, J.; Tomes, M.; Carmon, T. Stimulated optomechanical excitation of surface acoustic waves in a microdevice. *Nat. Commun.* **2011**, *2*, 403. [CrossRef] [PubMed]
69. Guo, C.; Che, K.; Zhang, P.; Wu, J.; Huang, Y.; Xu, H.; Cai, Z. Low-threshold stimulated Brilluoin scattering in high Q whispering gallery mode tellurite microspheres. *Opt. Express* **2015**, *23*, 32261–32266. [CrossRef]
70. Asano, M.; Takeuchi, Y.; Ozdemir, S.; Ikuta, R.; Yang, L.; Imoto, N.; Yamamoto, T. Stimulated Brillouin scattering and Brillouin coupled four wave mixing in a silica microbottle resonator. *Opt. Express* **2016**, *24*, 12082–12092. [CrossRef]
71. Bahl, G.; Kim, K.; Lee, W.; Liu, J.; Fan, X.; Carmon, T. Brillouin cavity optomechanicswith microfluid devices. *Nat. Commun.* **2013**, *4*, 1994. [CrossRef]
72. Farnesi, D.; Righini, G.; Nunzi Conti, G.; Soria, S. Efficient frequency generation in phoxonic cavities based on hollow whispering gallery mode resonators. *Sci. Rep.* **2017**, *7*, 44198. [CrossRef]
73. Rosello-Mecho, X.; Farnesi, D.; Frigenti, G.; Barucci, A.; RAtto, F.; Fernandez-Bienes, A.; Garcia-Fernandez, T.; Delgado-Pinar, M.; Andrés, M.; Nunzi Conti, G.; Soria, S. Parametrical Optomechanical Oscillations in PhoXonic Whispering Gallery Mode Resonators. *Sci. Rep.* **2019**, *9*, 7163. [CrossRef]
74. Yamaguchi, K.; Niimi, T.; Haraguchi, M.; Okamoto, T.; Fukui, M. Self-Modulation Scattering intensity from a silica microsphere coated with a sol-gel film doped with J-aggregates. *Jpn. J. Appl. Phys.* **2006**, *45*, 6750. [CrossRef]
75. Yamaguchi, K.; Fujii, M.; Haraguchi, M.; Okamoto, T.; Fukui, M. Nonlinear trimer resonators for compact ultrafast switching,. *Opt. Express* **2009**, *17*, 23204–23212. [CrossRef] [PubMed]
76. Tapalian, H.; Laine, J.P.; Lane, P. Thermo-optical switches using coated microsphere resonators. *IEEE Photonics Technol. Lett.* **2002**, *14*, 1118–1120. [CrossRef]
77. Roy, S.; Prasad, M.; Topolancik, J.; Vollmer, F. All-optical switch with bacteriorhodopsin protein coated Microcavities and its application to low power computing circuits. *J. Appl. Phys.* **2010**, *107*, 053115. [CrossRef]
78. Poellinger, M.; Rauschenbeutel, A. All-optical signal processing at ultra los powers in bottle microresonators using the Kerr effect. *Opt. Express* **2010**, *18*, 17764–17775. [CrossRef]
79. Soria, S.; F.Baldini.; Berneschi, S.; Cosi, F.; Giannetti, A.; Nunzi Conti, G.; Pelli, S.; Righini, G. High-Q polymer-coated microspheres for immunosensing applications. *Opt. Express* **2009**, *17*, 14694. [CrossRef]
80. Chinelatto, L.; Barrio, J.D.; Pinol, M.; Oriol, L.; Matranga, M.; Santo, M.D.; Barberi, R. Oligofluorenes blue emitters for cholesteric liquid crystals. *J. Photochem. Photobiol. A* **2010**, *210*, 130–139. [CrossRef]
81. Selle, A.; Kappel, C.; Bader, M.A.; Marowsky, G.; Winkler, K.; Alexiev, U. Picosecond pulse induced two photon fluorescence enhancement in biological material by application of grating waveguide structures. *Opt. Lett.* **2005**, *30*, 1683–1685. [CrossRef]
82. Thayil, A.; Muriano, A.; Salvador, P.; Galve, R.; Marco, M.; Zalvidea, D.; Loza-Alvarez, P.; Katchalski, T.; Grinvald, E.; Friesem, A.; Soria, S. Nonlinear immunofluorescent assay for androgenic hormones based on grating waveguide structures. *Opt. Express* **2008**, *16*, 13315–13322. [CrossRef]
83. Muriano, A.; Thayil, A.; Salvador, P.; Galve, R.; Loza-Alvarez, P.; Soria, S.; Marco, M. Two-photon fluorescent immunosensor for androgenic hormones using resonant grating waveguide structures. *Sens. Actuators B* **2012**, *174*, 394–401. [CrossRef]
84. Cohoon, G.A.; Khieu, K.; Norwood, R. Observation of two photon fluorescence of Rhodamine 6G in microbubble resonators. *Opt. Lett.* **2014**, *39*, 3098–3101. [CrossRef] [PubMed]
85. Martin, L.; Haro-Gonzalez, P.; Martin, I.R.; Navarro-Urrios, D.; Alonso, D.; Perez-Rodriguez, C.; Jaque, D.; Capujr, N.E. Whispering gallery modes in glass microspheres: Optimizing of pumping in a modified confocal microscope. *Opt. Lett.* **2011**, *36*, 615–617. [CrossRef] [PubMed]
86. Bestvater, F.; Spiess, E.; Stobrawa, G.; Hacker, M.; Feurer, T.; Porwol, T.; Berchner-Pfannschmidt, U.; Wotzlaw, C.; Acker, H. Two-photon fluorescence absorption and emission spectra of dyes relevant for cell imaging. *J. Microsc.* **2002**, *208*, 108–115. [CrossRef] [PubMed]

© 2020 by the authors. Licensee MDPI, Basel, Switzerland. This article is an open access article distributed under the terms and conditions of the Creative Commons Attribution (CC BY) license (http://creativecommons.org/licenses/by/4.0/).

Review

Nonlinear Optics in Dielectric Guided-Mode Resonant Structures and Resonant Metasurfaces

Varun Raghunathan *, Jayanta Deka, Sruti Menon, Rabindra Biswas and Lal Krishna A.S

ECE Department, Indian Institute of Science, Bangalore 560012, India; deka@iisc.ac.in (J.D.); sruti@iisc.ac.in (S.M.); rabindrab@iisc.ac.in (R.B.); lalkrishna@iisc.ac.in (L.K.A.S.)
* Correspondence: varunr@iisc.ac.in

Received: 21 March 2020; Accepted: 9 April 2020; Published: 24 April 2020

Abstract: Nonlinear optics is an important area of photonics research for realizing active optical functionalities such as light emission, frequency conversion, and ultrafast optical switching for applications in optical communication, material processing, precision measurements, spectroscopic sensing and label-free biological imaging. An emerging topic in nonlinear optics research is to realize high efficiency optical functionalities in ultra-small, sub-wavelength length scale structures by leveraging interesting optical resonances in surface relief metasurfaces. Such artificial surfaces can be engineered to support high quality factor resonances for enhanced nonlinear optical interaction by leveraging interesting physical mechanisms. The aim of this review article is to give an overview of the emerging field of nonlinear optics in dielectric based sub-wavelength periodic structures to realize efficient harmonic generators, wavelength mixers, optical switches etc. Dielectric metasurfaces support the realization of high quality-factor resonances with electric field concentrated either inside or in the vicinity of the dielectric media, while at the same time operate at high optical intensities without damage. The periodic dielectric structures considered here are broadly classified into guided-mode resonant structures and resonant metasurfaces. The basic physical mechanisms behind guided-mode resonances, electromagnetically-induced transparency like resonances and bound-states in continuum resonances in periodic photonic structures are discussed. Various nonlinear optical processes studied in such structures with example implementations are also reviewed. Finally, some future directions of interest in terms of realizing large-area metasurfaces, techniques for enhancing the efficiency of the nonlinear processes, heterogenous integration, and extension to non-conventional wavelength ranges in the ultra-violet and infrared region are discussed.

Keywords: optical resonances; nonlinear optics; harmonic generation; four-wave mixing; optical switching; sub-wavelength gratings; Mie scattering; Fano resonances; guided-mode resonance

1. Introduction

The field of nonlinear optics encompasses the study of the nonlinear interaction of incident light with material at sufficiently high optical intensity levels resulting in the generation of harmonics, conversion of frequencies from one band to another, ultra-fast switching etc. [1–3]. The materials of interest for realizing nonlinear optical devices are typically in bulk form with interaction lengths in the millimeter to even kilometer range to allow efficient build-up of the nonlinear signal of interest. With the emergence of accurate nanofabrication techniques, there is interest in exploring nonlinear optical effects within the purview of nano-photonics with structural dimensions comparable to or much less than the incident light wavelength [4]. At such length scales, interesting regimes for studying nonlinear optics emerge, in which the resonant optical interaction due to frequency selective light scattering or frequency selective light coupling into and out of the structures becomes significant [5]. The resonant effects lead to a build-up of electric field inside or in the vicinity of the structure, resulting in enhancement of the nonlinear optical effects being studied. The structures of interest for such studies can be broadly divided into metallic structures which support plasmonic-type resonances, [6] and dielectric structures which support Mie scattering-type resonances [7]. Plasmonic sub-wavelength structures support field localization at its periphery due to localized surface-plasmon resonances with multipolar electric-type characteristics. This can be contrasted with dielectric structures which support both electric- and magnetic-type resonances with field concentrated either inside or outside the structure. The plasmonic resonances generally result in higher field enhancement when compared to the dielectric resonances, however at the price of higher absorption losses due to free-carrier absorption and reduced optical damage thresholds. Plasmonic structures have been extensively explored for nonlinear optical studies over the past decades, with many good review articles written in this topic [8–10]. In recent times, there has been a resurgence of interest in studying dielectric periodic structures, with the focus on realizing low-loss, sub-wavelength artificially engineered surfaces, also popularly called as metasurfaces for shaping the amplitude, phase and polarization properties of light to realize ultra-thin lenses, polarizers, holograms etc. [11–13]. This has also triggered interest in artificially engineered dielectric surfaces consisting of one or two- dimensional grating structures for resonant nonlinear optical studies [14]. In this context, high refractive index materials are particularly explored due to the enhanced field concentration which can be achieved when compared to medium- or low-index materials. Materials such as Silicon, Germanium, and Gallium Arsenide which possess high refractive indices are also amenable to large area nanofabrication due to their existing use in complementary metal-oxide semiconductor (CMOS) compatible microelectronics and integrated optoelectronic applications [15,16]. There is also interest in exploring interesting physical mechanisms to create the optical resonances in such structures, for example using guided-mode resonances [17], electromagnetically-induced transparency (EIT) like resonances [18] and bound-states in continuum resonances [19] to enhance local electric fields and consequently amplify the nonlinear optical effects.

This review article is aimed at discussing some of the recent research efforts pursued in the study of resonant nonlinear optical effects in high-index contrast dielectric-based arrayed structures. First, a comparison of light scattering from isolated dielectric particles with that of resonant transmission or reflection spectra obtained with arrays of such dielectric elements in the form of one-dimensional and two-dimensional grating-like structures is presented. The benefits of using high-index dielectric particles when compared to low- or medium-index particles in terms of the robustness of the scattering features are discussed. Various physical mechanisms responsible for the resonant features observed in periodically arranged sub-wavelength dielectric structures are also discussed. In this context, guided-mode resonance phenomenon due to frequency selective in- and out-coupling of the incident light into transverse propagating waveguide modes of the high index structure are discussed. Guided-mode resonance structures in the form of fully-etched high contrast gratings, and partially-etched zero contrast gratings are studied. EIT-like resonance phenomenon due to the coupling of bright excitable modes with dark unexcitable modes is discussed. The recent research into bound-state in continuum which leads to discrete resonant states in a continuum of states to realize high quality factor resonances using asymmetric structures or off-axis excitation is also briefly discussed. Subsequently, various nonlinear optical phenomena and structures studied in dielectric metasurface platform are discussed. This includes second- and third- harmonic generation processes, wave-mixing processes in the form of sum-frequency and four-wave mixing, higher harmonic generation processes, ultra-fast optical switching utilizing electronic nonlinearities, photon acceleration processes in harmonic generation, and hybrid metasurfaces for nonlinear optical applications. Finally, the main conclusions of this review article with some future directions of interest in terms of large-area metasurfaces, techniques for enhancing the efficiency of the nonlinear processes, heterogenous integration, and extension to non-conventional wavelength ranges in the ultra-violet and infrared region are presented. There are few other detailed review articles published previously in the area of all-dielectric metasurfaces [12,13], resonant grating structures [17], and their nonlinear optical application [4,10,14,20].

2. Design Considerations for Resonant Dielectric Grating Structures

In this section, the resonant enhancement of the incident field in the dielectric sub-wavelength structures, which can be designed with an understanding of how the electromagnetic wave interacts with isolated entities and arrays of repetitive units is discussed. Furthermore, the role of refractive index of the dielectric structure in determining the extent of field enhancement within the structure is discussed. High index materials such as silicon and germanium are found to particularly result in well defined, high quality factor resonant spectral features.

2.1. Isolated Particle Versus Array

The interaction of electromagnetic fields with isolated sub-wavelength particles results in interesting linear light scattering which can be decomposed into basic electric and magnetic scattering modes. Plasmonic structures which rely on isolated, oligomeric and periodic array arrangement of metal sub-wavelength sized particles have been studied extensively over the last five decades [21]. Such structures rely on localized surface-plasmon resonances for field enhancement and find applications in fluorescence [22,23], Raman scattering [24–26], and nonlinear optical studies [8–10]. Plasmonic resonances result in multi-fold field enhancement with only electric multipolar characteristics and field localization outside the structure in order to satisfy the required boundary conditions. In similar lines to light scattering from metallic nanostructure, size-dependent resonant light scattering from dielectric nanostructures has also been studied [27]. The field enhancement in dielectric sub-wavelength structures is not as large as in plasmonic structures, however the field concentration inside the dielectric, the high damage thresholds and the robust scattering spectra has triggered interest in studying light scattering from dielectric particles in isolated and array form [28]. Figure 1a,b shows the simulated scattering spectrum from an isolated silicon cylindrical nanowire with varying diameter. Such isolated wires can be grown using chemical synthesis techniques and suspended in solution form [29] or can also be defined using nano-lithography techniques [30]. Experimentally reported study of scattering spectra from silicon nanowires of varying widths is also shown in Figure 1c (from [31]). In this study, the size-dependent visible resonant scattering from the nanowires results in vivid colors obtained using dark-field imaging. The experimental and simulated scattering spectra in Figure 1 are found to be in good agreement. Figure 2a shows the simulated scattering cross-section from an isolated silicon nanosphere of 150 nm diameter, which is further decomposed into characteristic electric and magnetic resonances. The scattering spectra for a sub-wavelength spherical particle, expressed as scattering efficiency can be expanded as a superposition of characteristic electric and magnetic resonant modes as follows: [7]

$$Q_{scatter} = \frac{2}{x^2} \sum_{i=1}^{\infty} (2i+1)\left(|a_i|^2 + |b_i|^2\right) \qquad (1)$$

$$\text{where,} \quad a_i = \frac{m^2 j_i(mx)[x j_i(x)]' - \mu_1 j_i(x)[mx j_i(mx)]'}{m^2 j_i(mx)[x h_i(x)]' - \mu_1 h_i(x)[mx j_i(mx)]'} \qquad (2)$$

$$b_i = \frac{\mu_1 j_i(mx)[x j_i(x)]' - j_i(x)[mx j_i(mx)]'}{\mu_1 j_i(mx)[x h_i(x)]' - h_i(x)[mx j_i(mx)]'} \qquad (3)$$

where a_i, b_i are the electric and magnetic mode coefficients respectively, which are expanded in terms of Bessel, Hankel, Ricatti-Bessel and Ricatti-Hankel functions, $x = ka$ refers to the modified dimension parameter, and $m = \frac{\sqrt{\varepsilon_1 \mu_1}}{\sqrt{\varepsilon_{host} \mu_{host}}}$ refers to the contrast parameter [7]. Simplified forms of the scattering expansion for specific structures can be found in ref. [32]. Typical field profile obtained close to the electric/magnetic dipolar and quadrapolar resonances for isolated spherical particle are also shown in Figure 2a.

The experimental demonstration of tunability of the scattering spectrum based on dielectric particle size is shown in Figure 2b (from [33]). For certain particle diameter, strong influence from magnetic dipole mode is observed (denoted by md in Figure 2b). The study of magnetic resonances in dielectric structures, in particular magnetic dipole modes has been of particular interest for the resonant enhancement of nonlinear optical effects [34] and can potentially be used to enhance light-matter interaction in materials with allowed magnetic transitions [35]. Figure 3 shows the scattering spectra for a silicon isolated sub-wavelength disk. The scattering spectra from sub-wavelength dielectric disks resemble that of sub-wavelength spheres with analytical models available for decomposition into magnetic and electric modes. Sub-wavelength disks are structures which can be fabricated using standard electron-beam lithography and etching processes, and are best suited for large areas, reproducible scaling for practical photonic device applications [36]. These are often studied in isolated, closely spaced arrays, and in collective oligomeric forms [37].

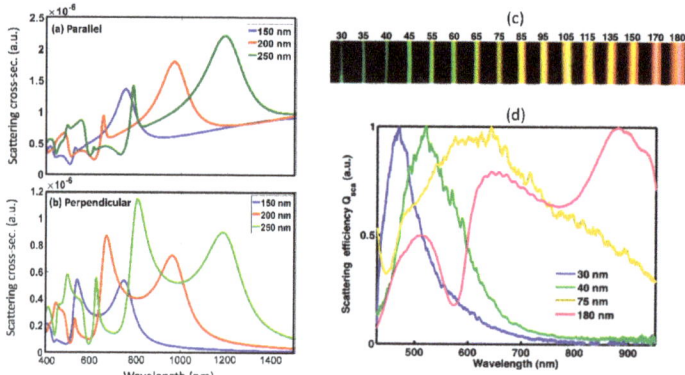

Figure 1. The simulated scattering cross section (in arbitrary units–a.u.) for silicon nanowires of varying diameter with incident light polarization oriented: (**a**) parallel and (**b**) perpendicular to the nanowire. (**c**) Experimentally obtained dark field images of nanowires showing light scattering for various width. (**d**) Experimentally obtained scattering spectra of nanowires of varying width. (Figures c and d are reproduced with permission from ref. [31]).

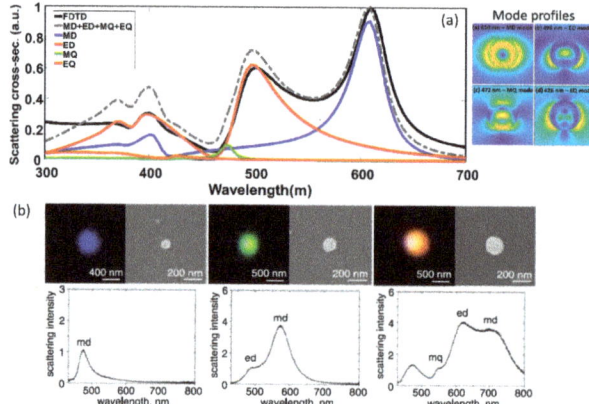

Figure 2. (**a**) Simulated scattering cross section (in arbitrary units–a.u.) for silicon nanospheres of fixed diameter of 150 nm (shown in black) and the result of decomposing the scattering spectra into magnetic dipole—MD (blue), electric dipole—ED (red), magnetic quadrupole—MQ (green) and electric quadrupole—EQ (brown). The sum of the MD, ED, MQ and EQ spectra is also shown (grey dashed). The field profiles for the MD, ED, MQ and EQ modes are also shown. (**b**) Experimentally obtained dark field scattering images and spectra for varying diameters of silicon sub-wavelength nanoparticles. (Figure b reproduced with permission from ref. [33]).

The arrangement of individual scatterers into a periodic array of dielectric nanowire one-dimensional (1D) grating structures [38] or spherical [39], cylindrical [34,40] two-dimensional (2D) grating structures has been of interest to tailor the overall transmission or reflection spectra at the resonance wavelengths. Even though such transmit- or reflect-arrays are well known in the microwave frequency range [41], at optical frequencies such structures have been realized only recently with advancement in precision nano-fabrication techniques, such as electron-beam, optical interference, stepper-based lithography, nanoimprint and self-assembly techniques [42]. Furthermore, the recent research interest in the area of surface-relief sub-wavelength features to realize metasurfaces has also led to a resurgence of interest in guided-mode resonance structures and resonant metasurfaces for sensing and nonlinear optical applications. The consequence of scaling from an isolated sub-wavelength cylinder to hexagonal array of 2D cylinders is illustrated in the transmission contour map of Figure 4. For these simulations, rigorous coupled-wave analysis (RCWA) method was used to simulate the two-dimensional array using S4, electromagnetic solver package [43]. The pitch of the hexagonal array is increased from 0.8 to 2.4 micrometer, keeping the dimensions of the individual cylinders same as in Figure 3. For large separation between the cylinders, the individual disks do not interact with each other and the transmission spectra shows poor contrast and remains unchanged. The hexagonal array also acts as a higher order diffraction grating with significant energy being directed to non-zero diffraction orders. The transmission spectra used to plot the contour in Figure 4 do not separate the different diffraction orders and hence does not show this effect. With a reduction in separation distance, there is a more prominent interaction between the cylinders, which shows up as characteristic high contrast resonant features in the spectra. These high contrast resonant features are found to be in close vicinity to the peaks in the scattering spectra observed for the isolated cylinder case, as shown in Figure 3. This clearly shows that, though scattering from individual dielectric objects is the underlying reason for the frequency selective interaction with the electromagnetic field, the collective effect due to the array is important in determining the overall resonant spectral features and the associated field enhancement. Isolated sub-wavelength metallic and dielectric particles have found their own niche applications in in-vivo cell imaging [44], super-resolution microscopy [45] etc. For practical applications in photonic devices, and in particular in nonlinear optics, arrayed sub-wavelength objects and their characteristic spectral resonances are found to be more relevant when compared to isolated ones. However, it needs to be mentioned that the ability to achieve high-quality factor resonances in isolated structures utilizing bound-states in continuum mechanism, also called as super-cavity resonance is another promising direction of recent research in recent times [46].

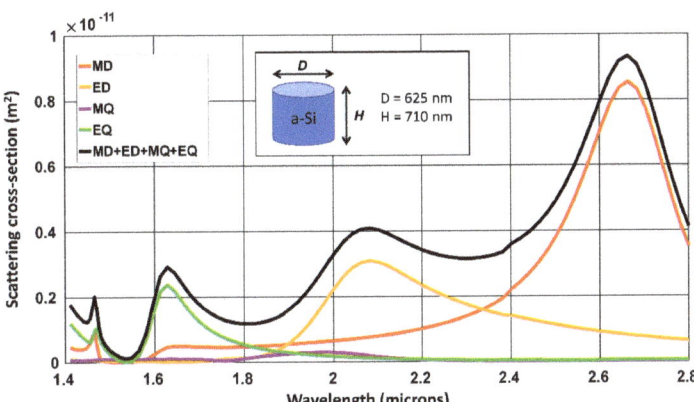

Figure 3. Scattering spectra from an isolated sub-wavelength cylinder separated into: magnetic dipole —MD (red), electric dipole—ED (orange), magnetic quadrapole—MQ (purple) and electric quadrapole—EQ (green). The sum of the MD, ED, MQ and EQ scattering spectra is shown in black.

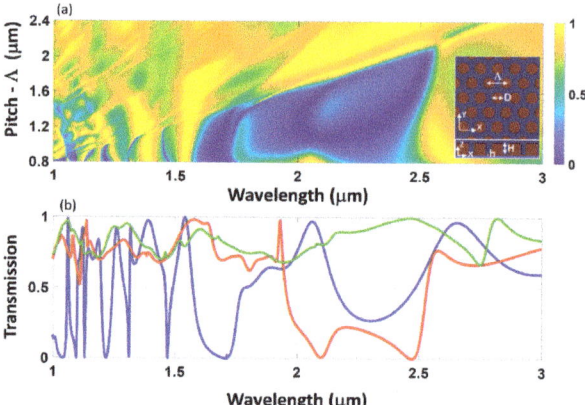

Figure 4. (**a**) Transmission spectra contour map for varying array pitch (Λ). The top view and side view profile of the hexagonal arrangement of sub-wavelength cylinders shown in the inset. The height and diameter of the structure are: H = 710 nm, h = 0 nm, and D = 625 nm. (**b**) Selected transmission spectra for pitch, Λ = 0.8 µm (blue curve), 1.2 µm (red curve) and 1.6 µm (green curve) are shown.

2.2. Effect of Refractive Index Contrast on Resonant Interaction

The refractive index of the sub-wavelength dielectric particle plays a critical role in determining the light scattering strength and hence its resonance characteristics. Figure 5 shows the scattering efficiency spectra for varying refractive indices for isolated dielectric sub-wavelength disk. A constant refractive index across the spectral range of interest and lossless dielectric medium are assumed in these simulations. The dimensions of the disk are same as in Figure 3. It is found that the contrast or sharpness of the resonances increases with increasing refractive index. This is associated with the higher quality factor of the resonance and stronger field concentration in the dielectric structure with higher refractive index. Furthermore, the scattering spectra is found to shift to longer wavelengths with increasing refractive index. This can be associated with the increased optical path length with increasing refractive index of the dielectric structure and its comparison relative to the wavelength range over which scattering is observed.

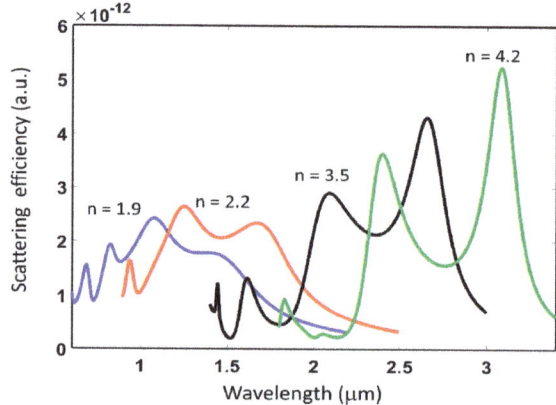

Figure 5. Scattering efficiency spectrum from isolated sub-wavelength dielectric sub-wavelength disk with dimensions same as in Figure 3 as a function of varying refractive index. Refractive index of 1.9 (blue curve), 2.2 (red curve), 3.5 (black curve) and n = 4.2 (green curve) are shown. The refractive index is assumed to be constant across the spectral range shown for each curve.

A detailed comparison of the scattering spectra of plasmonic and dielectric sub-wavelength structures as a function of increasing dielectric constant is reported in ref. [47].

In the case of 1D sub-wavelength periodic grating structures, the role of the refractive index of the grating material on the resonant spectra are shown in Figure 6 [48,49]. A schematic view of the 1D grating structures simulated is shown in Figure 6a. The transmission spectra contour maps of fully-etched, high-index contrast grating structure of silicon and medium-index contrast grating structure of silicon nitride are compared for varying grating heights. The spectra are shown for incident transverse electric (TE) polarization, oriented parallel to the gratings. It is found that, for silicon high contrast grating structures, the transmission spectra show prominent resonance features [50], as evident from the checkerboard type patterns in Figure 6b. In contrast to this, the silicon nitride medium contrast gratings show poor contrast resonance features, as shown in Figure 6c with a reduced wavelength range over which the resonance features are observed. This is a direct consequence of the reduced spectral window between the zeroth order diffraction and first order diffraction into the glass substrate for the case of silicon nitride when compared to silicon. The wavelength range across which zeroth order diffraction occurs, also called as the dual-mode resonance is denoted by the black arrows in each of the contour maps is found to be reduced for the silicon nitride structures. Thus, high index contrast periodic structures are generally better suited for engineering optical resonances for nonlinear optical applications. However, they invariably end up being highly lossy in the visible and near infrared wavelength region due to enhanced absorption above their energy bandgap. The medium index contrast materials such as silicon nitride, titanium oxide etc. do offer certain benefits in terms of extended low-loss transmission window when compared to high-index contrast structures. For example, in the case of silicon nitride, the low-loss optical window extends from close to 300 nm to 5 µm. This is particularly beneficial to realize high quality resonances both at the fundamental excitation and nonlinear signal wavelengths. Thus, there is still interest in research into alternate structures based on guided-mode resonances, partially etched zero contrast grating structures etc. to obtain prominent resonances using medium index contrast material systems [48].

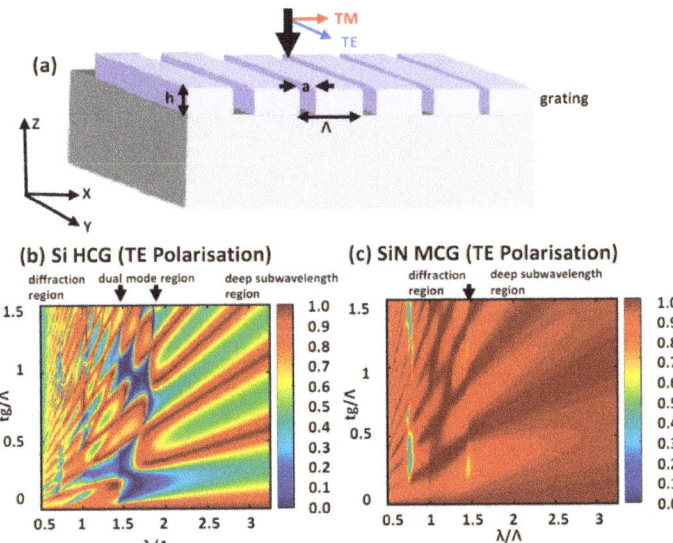

Figure 6. (**a**) Schematic of fully-etched dielectric one-dimensional grating structure. The simulated transmission spectra for (**b**) silicon high contrast grating and (**c**) silicon nitride medium contrast grating as a function of varying height for fixed incident TE polarization. The dimensions and wavelength are normalized by the pitch of the grating structure. (Figure is reproduced with permission from ref. [49]).

3. Physical Mechanisms behind Resonances in Arrayed Structures

In this section, we discuss the underlying physical mechanisms that lead to resonances in the arrayed structures. Though the resonances in isolated and arrayed structures can be accurately modelled and designed based on scattering expansion or coupled-wave analysis methods as discussed in Section 2, these techniques do not give much physical insight into the working of the arrayed resonant structures and how incident light field interacts with the structures. Thus, it is instructive to describe the underlying working mechanism of the arrayed structures used for resonant nonlinear optical studies. Here, the resonance phenomenon studied are broadly classified as guided-mode resonances, EIT- like resonances and bound-states-in-continuum resonances. Though this list is not exhaustive [17,51], a majority of the resonant structures studied fall into these categories.

3.1. Guided-Mode Resonances

Guided-mode resonances arise in dielectric grating-waveguide coupled structures due to the evanescent diffraction orders from the grating coupling the incident electromagnetic wave into guided modes of the waveguiding layer. In other words, the optical resonances can also be described by the wavelength selective in- and out-coupling of electromagnetic wave into the waveguide through interaction with the grating structure. The waveguide can be either a separate high index layer located close to the grating structures or can be the grating structure itself which is fully or partially etched. Few examples of such structures which support guided-mode resonances in various waveguide-grating arrangement are shown in Figure 7. Here, we briefly outline some of the properties of periodic grating structures and demonstrated applications. Recent review articles provide a comprehensive overview of the recent advances in resonant waveguide grating structures [17,52].

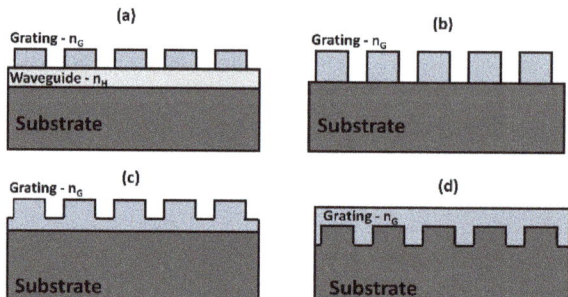

Figure 7. Cross-section view of various guided-mode resonance based grating structures with waveguide and grating layer refractive indices n_H and n_G respectively. (**a**) The grating and waveguide are made of different materials with the waveguide of higher index below the fully-etched grating structure. (**b**) Fully etched grating structures which can act as the effective waveguide. (**c**) Partially etched grating structures with the waveguide layer made of same material as the etched gratings. (**d**) Substrate grating structures is coated with a high-index waveguiding layer on top.

Wood anomaly in diffraction gratings has been studied in the past to explain the occurrence of sharp spectral orders in the diffracted light [53]. In particular, resonance type anomalies are explained based on leaky guided-modes supported in the waveguide grating structures. A basic waveguide-grating model used to describe the diffraction effect from the periodically index modulated structure leading to coupling of waveguide modes is shown in Figure 8a [38,54]. This model is adapted from the seminal paper by R. Magnusson et.al. [38], which explained the working mechanism and applications of the guided-mode resonant structures. Under weak perturbation, the supported guided mode resonances can be understood based on the frequency-selective excitation of the modes of an effective waveguide formed by the periodically index modulated structure with the grating providing the required phase matching to couple the free-space incident light to the waveguide mode. Typical resonant filter characteristics obtained at normal incidence is shown in shown in Figure 8b [38]. Under weak index modulation, the filter reflection spectra are found to be strongly angle sensitive, as shown in Figure 8b. This offers a mechanism to angle-tune the filter characteristics. Such resonant response from waveguide-grating structures have been utilized as band-pass and band-stop filters working across different electromagnetic spectral regions. They have also been proposed to be used as intra-cavity narrow-band high-reflection mirrors, photorefractive tunable filters, and as electro-optic switches [38]. Guided-mode resonances have also found applications as bio-sensors in immune-assays in which local reaction between the functionalized antibody and the target antigen leads to a shift in the guided-mode resonance position [48,50,52,55]. Such assays have been proposed to be used in bright-field (colorimetric), fluorescence and also in wide-field imaging modes. In the context of nonlinear optics, the guided-mode resonance grating structure have been utilized as the nonlinear media in which the nonlinear optical interaction is enhanced due to the local electric-field built-up in the structure close to the resonance wavelengths. Some of these examples are discussed in Section 4 below.

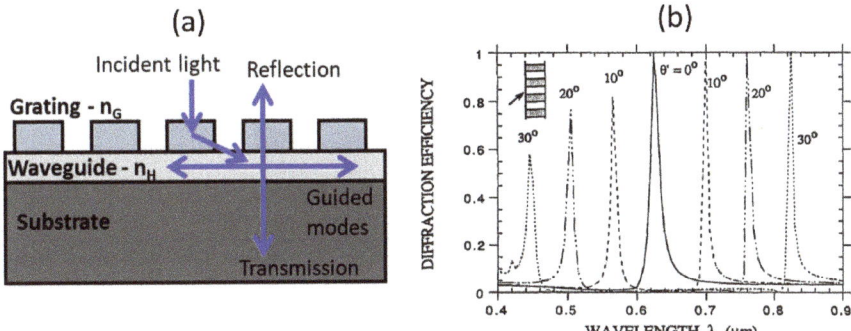

Figure 8. (a) A cross-section schematic of the guided-mode resonance structure showing the resonant coupling of incident light into the waveguide region through interaction with the grating structure. (b) Simulated filter response for the guided-mode resonance structure for parameters: pitch = 330 nm, height = 330 nm, dielectric constant difference (normalized), $\Delta \varepsilon / \varepsilon_{avg} = 0.05$ and center wavelength of 547 nm and dependence of the filter response on the angle of incidence. (Figure b is reproduced with permission from ref. [38]).

Fully etched high index contrast gratings, also termed as high-contrast gratings [50] are also used to create high quality factor resonances. A cross-section schematic view of such a structure is shown in Figure 7b above. The resonances in these structures can be understood either based on guided mode resonances with the grating coupling light into strongly-modulated effective waveguide created by the air/silicon periodic structure or based on optical modes supported in the periodic air-high index structure along the longitudinal propagation direction (i.e., along the optical axis) with energy exchanged with the interface due to reflection [56]. The resonant characteristics achieved in the high contrast grating structures have been classified as crossing and anti-crossing resonances based on the phase difference between the interacting longitudinal modes being odd and even multiple of π respectively [50,56]. The classification of the resonance spectra as crossing and anti-crossing features is shown in Figure 9. The anti-crossing resonances exhibit highly asymmetric Fano-like line shape with strong-coupling between the interacting modes. This results in strong field enhancement within the structure as shown in Figure 9c with high quality factor. In contrast, the crossing resonances exhibit more symmetric spectral shape, however with lower quality factor and reduced field enhancement, as shown in Figure 9d. The high contrast gratings have been used to build both narrow-band and broad-band reflectors with reflectivity greater than 95% for use as end-facet mirrors in vertical cavity surface-emitting lasers (VCSELs) [50]. They have also been used as bio-sensors based on refractive index changes altering the resonance characteristics [55]. The strong field-concentration has also been utilized for Raman scattering enhancement with suitable metal nanostructures patterned in regions of strong dielectric field concentration [57]. In the context of nonlinear optics, the strong nonlinear optical properties of the high index materials, such as Aluminum Gallium Arsenide and Silicon are leveraged in enhancing second-harmonic generation and four-wave mixing due to strong field enhancement near anti-crossing resonances [58,59]. This is further discussed in Section 4 below.

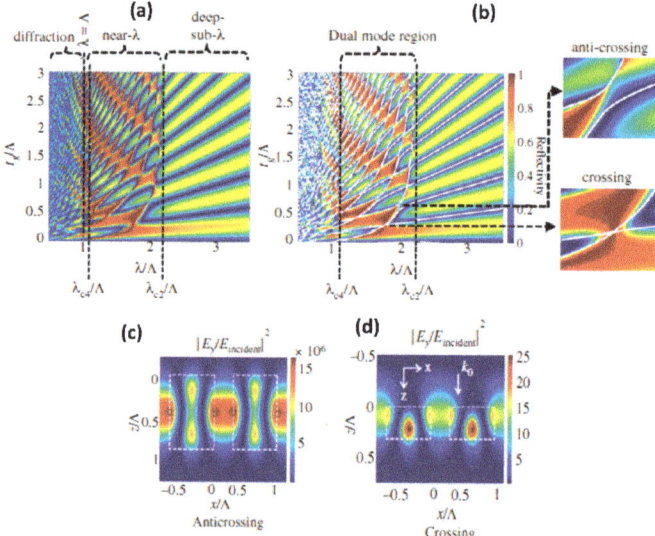

Figure 9. (**a**) Reflection spectra contour for free-standing silicon high contrast gratings with dimensions: $n_{grating}$ = 3.48, duty cycle = 70% and thickness varied. The various regions of operation of the grating are also shown. (**b**) The reflection spectra overlapped with the solutions to the eigen-mode equations of the resonant modes (white curves). The overlap regions of the white curves result in anti-crossing and crossing type resonance. (**c**) Field intensity profile at anti-crossing resonance. (**d**) Field intensity profile at crossing resonance. (Figures are reproduced with permission from ref. [50]).

In comparison to fully-etched, high contrast grating structures, partially etched structures, also termed as zero-contrast gratings [60,61], offer an additional degree of freedom for designing the grating characteristics based on the chosen etch depth. Such structures can achieve robust spectral characteristics comparable to, if not better than that of high contrast grating structures. As an example, the realization of broadband reflective filter using zero-contrast gratings is shown in Figure 10 [60]. The structures shown here have been optimized using an inverse design approach in which particle swarm search algorithm is utilized to quickly search a wide design space with the performance optimized based on achieving a desired value of figure-of-merit. In this example, the figure-of-merit is chosen as follows [60]:

$$Figure\ of\ merit = \left\{ \frac{1}{M} \sum_{i=1}^{M} \left[R_{desired}(\lambda_i) - R_{design}(\lambda_i) \right] \right\}^{1/2} \qquad (4)$$

The figure of merit compares the root-mean square error between the desired and designed reflection spectra with the goal of minimizing the difference between the two through optimized design. With multiple iterations this process helps optimize the device design to achieve the designed spectral response as close to the desired one as possible. A comparison of broadband reflector performance of zero-contrast gratings with high contrast gratings can also be found in ref. [60]. Such inverse design approaches are most promising to search a wide design space and at the same time achieve close to the ideal response for the resonant structure. Another emerging direction is the use of neural networks for optical metasurface design optimization [62,63] to achieve optimal specifications with often non-intuitive, but effective meta-atom shapes which can be fabricated with present-day advanced nanolithography tools.

The angle sensitivity of the guided-mode resonance filter is another useful characteristic for filter response tuning or for wide field of view application. This is also found to be a strong function of the effective index of the waveguide-grating structure, with higher index resulting in stronger field confinement in the unit-cells and hence less angular sensitivity [50]. In this context, high contrast gratings with highly confined field profiles within the etched structures are found to be more angle insensitive than the partially etched zero-contrast gratings [50,64], which support diffused field-profiles extending across the unetched high-index slab region. Furthermore, conical mounting of partially etched gratings is also found to result in reduced angle sensitivity when compared to conventional mounting [65]. One-dimensional grating structures discussed above are inherently polarization selective and polarization independence is achieved by using two-dimensional symmetric meta-atoms, such as square or hexagonal arrangement of circular features. Such polarization independent structures are particularly useful for realizing optical filters for unpolarized light [66] or for realizing resonantly enhanced fluorescence sensors [48].

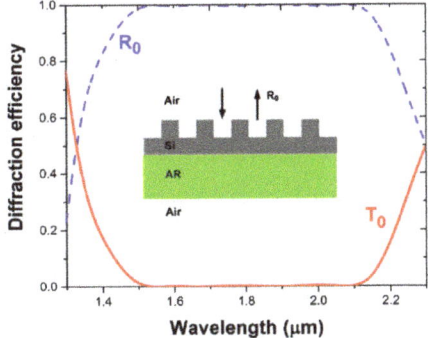

Figure 10. The simulated zeroth order reflection (R_0) and transmission (T_0) spectra for silicon zero-contrast gratings. The inset shows the cross-section of the zero-contrast gratings with the dimensions optimized using particle swarm combined with inverse-design algorithm. The optimized dimensions of the structure obtained are: etched grating height of 490 nm, unetched slab thickness of 255 nm, pitch of 827 nm, and fill factor of 0.643. (Figure reproduced with permission from ref. [60]).

3.2. Electromagnetically-Induced Transparency Analogue Resonances

Optical resonances in photonic structures can be created due to interference between coupled resonances of varying quality factor. For example, a high-quality factor and low-quality factor resonance can couple together to result in the observation of analogues of electromagnetically-induced transparency (EIT) in photonic structures [18]. The electromagnetically-induced transparency effect can be understood based on quantum interference between two different path-ways to excite an upper energy level in a three-level system, as shown in Figure 11a. This system consists of a lower level, an upper level and a meta-stable state, as denoted in the figure. A probe field can control the transition from the lower to upper state, while a control field can control the transition from the excited state to metastable state. The direct transitions from the lower state to metastable state are forbidden. In this context, the transitions induced by the probe field can interfere with the indirect transitions from lower to upper state through interaction with the metastable state, in the presence of a strong control field. This can result in destructive interference of the probe field and results in the observation of peaks at the probe field frequency in an otherwise expected dip in transmission spectrum, as shown in Figure 11b. This phenomenon is hence termed as electromagnetically induced transparency. In the context of photonic systems, the low-quality factor and high-quality factor resonant modes are equivalent to the direct and indirect lower level to upper level transition respectively, as discussed above.

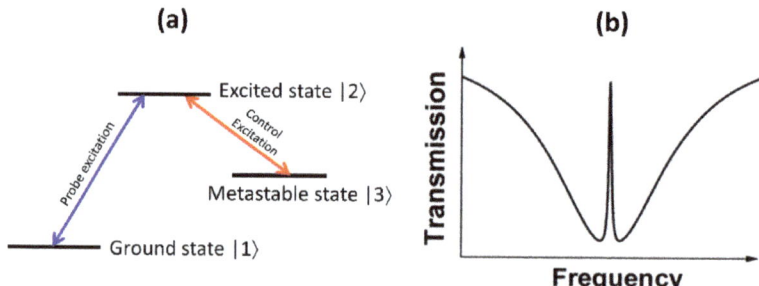

Figure 11. (**a**) The representative energy diagram for electromagnetically-induced transparency (EIT). (**b**) Schematic of the expected transmission spectrum for the EIT effect.

Photonic structures in the form of coupled Fabry-Perot cavities, coupled ring resonators, coupled photonic crystals etc. have been explored to be studied as EIT analogues [18]. In the context of arrayed metasurfaces and guided-mode resonant structures, there is also interest in coupling bright-mode and dark-mode resonances to observe EIT effects. Few examples of such implementations, supporting simulations and experimental studies are shown in Figure 12. Figure 12a–c shows an array of bar-ring structures in which the incident linearly polarized light preferentially couples to the bar and indirectly couples into the ring through the bar excited mode [67,68]. This results in excitation of dark modes in the ring, as shown in Figure 12b. The measured transmission spectra show peak in the middle of a broad transmission dip with strong field localization in the ring structures with maximum quality factor of the EIT resonance of ~300 times. Asymmetry in the array elements also results in coupling between bright and dark modes. Figure 12d–f shows one such example of rectangular bar dimer array with slight asymmetry added to one of the bars [69,70]. In this case, the coupling between the bright dipolar mode with the dark quadrapolar modes results in peaks in the transmission spectra (shown in Figure 12f) in the mid infrared wavelength region. Such structures have been used for enhanced infrared sensing [71]. Asymmetry can be introduced to the shape of a single monomeric unit as well to achieve bright-to-dark mode coupling. This is shown in Figure 12g–i, for the case of a nanocube with symmetry breaking protrusion [72]. As shown in Figure 12h, this protrusion results in the coupling of the electric dipole excitation, denoted as p_x to the magnetic dipole mode, denoted as m_z. This coupling results in characteristic dips in the measured reflection spectra, as shown in Figure 12i. Such structures have been utilized for nonlinear optical process enhancement, as discussed in Section 4. In the context of guided mode resonances, there is interest in coupling a low- and high- quality factor resonant waveguide structure. Some examples of such structures are shown in Figure 12j–l. Some of the early simulation studies of these structures consisted of a top grating-waveguide layer coupled to another bottom waveguide layer, as shown in the inset of Figure 12j. In this case, the direct coupling of the incident light to the bottom waveguide and indirect coupling through interaction with the top-layer guided-mode resonant structure can occur. This results in a sharp transmission peak, which is different from the transmission spectra of an equivalent refractive index homogenous medium. Some of the early work [73] was not even called an EIT analogue at that time. The EIT analogue ideas have been extended recently to one-dimensional and two-dimensional structures [74,75]. The spectral width of the EIT resonance and hence its overall quality factor can be engineered by choosing the separation between the top-layer grating-waveguide structure and the bottom waveguide layer. An example of such an engineered resonance shape is shown in Figure 12k,l. Such resonance shapes can potentially be used as narrow band-pass filters when compared to the more complex multi-layer dichroic filters.

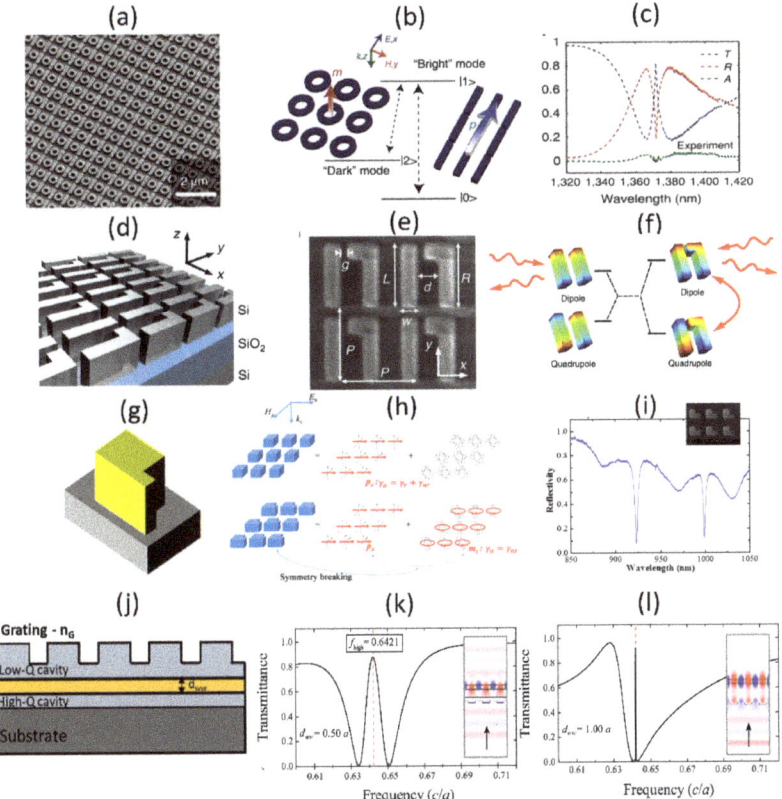

Figure 12. Different implementations of EIT-like resonance. (**a**) SEM image of bars-ring array. (**b**) Schematic of the coupling between the bar excited by the incident light and coupling to the ring structure creating magnetic dipolar type mode. (**c**) Experimentally measured transmission spectrum for the bar-split ring array showing EIT-like resonance. (**d**) Schematic of the achiral bar-dimer structures. (**e**) SEM image of asymmetric dimer achiral structures. (**f**) Coupling between the dipole and quadrapolar modes in the asymmetric dimer structures. (**g**) Perspective view of the asymmetric nanocube unit-cell. (**h**) Schematic showing the coupling of electric and magnetic dipole modes in asymmetric nanocubes. (**i**) Measured reflection spectrum for the asymmetric nanocube array. (**j**) Cross-sectional view of the structure showing upper layer grating-waveguide structure coupled to lower waveguide structure. (**k**,**l**) Simulated EIT-like resonance spectra and associated field profiles from GMR structures. (Figures a–c are reproduced with permission from ref. [68], d–f are reproduced with permission from ref. [70], g–i are reproduced with permission from ref. [72], k,l are reproduced with permission from ref. [74]).

3.3. Bound-States in Continuum Resonances

Bound-states in continuum (BIC) represent another interesting class of high-quality factor resonance. BIC are bound states in an otherwise continuum of states, considered as a trapped state with embedded eigenvalue. Figure 13 shows a schematic representation of BIC states in quantum-well systems and comparison with analogous waveguide grating systems (figure adapted from ref. [76]). Figure 13a compares a conventional quantum well with a slab-waveguide. It is found that in a conventional quantum well, the allowed states within the quantum well are bounded by the steep potential walls of the quantum well. This results in discrete, bounded states within and a continuum of states outside. This can be compared with a straight, unperturbed slab waveguide with optical mode confinement achieved by the core-cladding refractive index profile. This results in discrete, bound modes within the waveguide and a continuum of radiation modes outside. When the sharp edges of the quantum well potential profile are replaced by a modulated potential profile, it is found that there can exist bound states even within the continuum, as shown in Figure 13b. In similar lines, the introduction of a periodic modulation to the waveguide using a grating structure results in the creation of bound-states within the continuum of radiation states. These states end up being forbidden from external excitation/ coupling due to symmetry considerations [77]. This lack of in- and out-coupling or non-radiating characteristic can result in ideal infinitely high-quality factor for these bound states within the continuum resonances. However, in practice one would consider quasi-BIC states which tend towards the ideal situation and can result in large, but finite quality factors with external excitation mechanism [78,79].

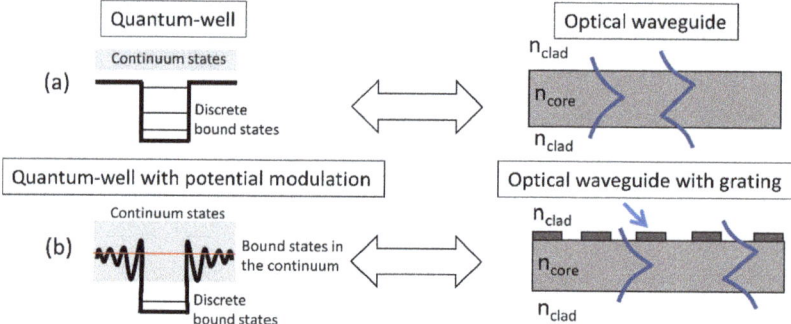

Figure 13. (a) Schematic of a quantum well with sharp potential edges and the corresponding photonic analogue showing an optical waveguide with confined modes, (b) Schematic of quantum well with modulated edges and the corresponding photonic analogue showing grating-modulated waveguide. (Figure adapted from ref. [76]).

The BIC states in the context of periodic grating structures can be broadly classified as symmetry protected BIC and accident BIC [76]. Symmetry protected BIC occurs at the high symmetry, zero wave-vector points (Γ-point in the case of photonic band structures) with direct excitation forbidden for normal incidence by the BIC resonant field symmetry. An example of such symmetry protected states in a simulated periodic dielectric constant modulated structure is shown in Figure 14a [80,81]. The schematic shows the periodic modulated structure with dielectric constant difference of $\Delta\varepsilon = \varepsilon_H - \varepsilon_L$. The structure supports multiple resonant modes (TE_0 and TE_1 modes) as shown in the figure. The bandstructure for the TE0 and TE1 resonances in the vicinity of $k_z = 0$ point shows two different types of resonances, one which is the leaky GMR and the other is the non-leaky BIC resonance [82], as shown in Figure 14b. The mode profile of the GMR and BIC resonances show odd and even order symmetry respectively and this inherently determines the ability to excite or couple into these modes through normal incidence plane wave excitation. The odd-symmetry profile can be excited with a normal incident wave, while the even-symmetry profile is forbidden from excitation. Furthermore, the GMR and BIC resonances are found to flip with change in $\Delta\varepsilon$ [80]. The same band dynamics are observed for both TE_0 and TE_1 resonant modes. The symmetry protected BIC resonances strictly remain protected only at normal incidence. With off-axis illumination, the symmetry can be broken resulting in quasi-BIC resonances with finite quality factor. BIC resonances can also be observed for non-zero k_z, which are called accidental BIC resonances [76]. Quasi-BIC resonances can also be excited at normal incidence by the introduction of asymmetry in the periodic structures [83]. Figure 14c shows schematic of such asymmetric structures. The resonant metasurfaces can be modelled by the amount of asymmetry introduced into the structure, denoted by α parameter [83]. Figure 14d shows that the asymmetry parameter can represent angular tilt, addition/ removal of material in split-ring, rectangular and bar-dimer structures in normalized units [83]. The asymmetric resonant metasurfaces discussed in Figure 12g–i which exhibit EIT-like coupling between the electric and magnetic dipolar modes can also be considered as an asymmetric structure in which quasi-BIC modes are observed. The quality factor of the BIC resonance is found to be directly related to the asymmetric parameter, with the quality factor scaling as α^{-2} [83]. In addition to resonant metasurfaces, BIC resonances are also predicted for isolated sub-wavelength particles in the form of narrow spectral features in the scattering spectra. These are termed as super-cavity modes [46]. Such high-quality factor BIC resonances in periodic grating structures, asymmetry metasurfaces and even isolated objects are finding innovative applications in BIC metasurface lasers [84,85], sensing [86], and nonlinear optics [87]. Few of the nonlinear optics applications are discussed below in Section 4.

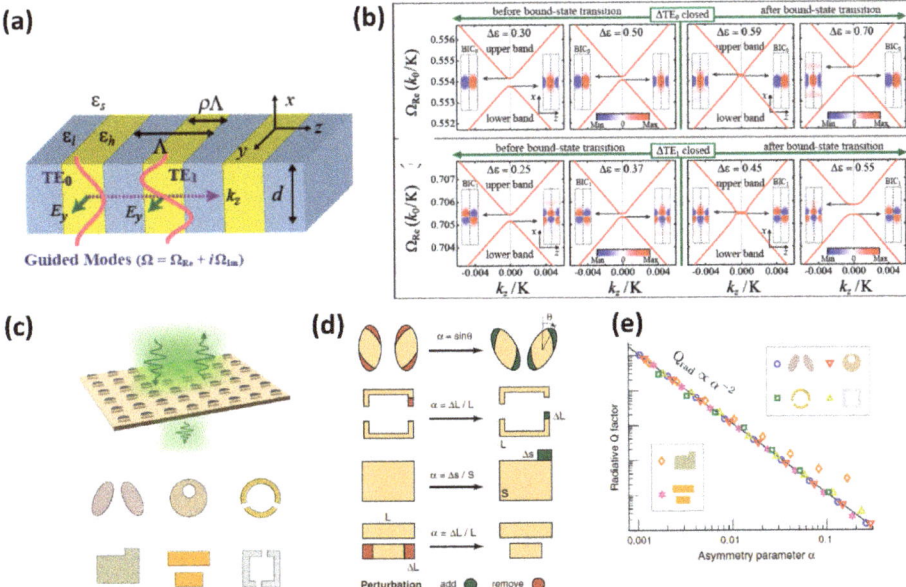

Figure 14. (a) Schematic of the periodic dielectric constant modulated grating structure with typical TE0 and TE1 modes supported by such structure. (b) Photonic band-structure calculation corresponding to the TE0 and TE1 modes showing the GMR and BIC states at either band-edge and their corresponding mode profiles. The GMR and BIC states are found to flip by changing the dielectric constant difference between the grating materials. (c) Examples of asymmetric resonant metasurfaces which support quasi-BIC resonances. (d) Modeling the asymmetricity using an asymmetry parameter, α. (e) Variation of quality factor of the quasi-BIC resonance with change in asymmetry parameter. (Figures a–b are reproduced with permission from ref. [81], c–e are reproduced with permission from ref. [83]).

4. Nonlinear Optical Studies of Resonant Dielectric Grating Structures

In this section, the various nonlinear optical processes studied in the context of guided-mode resonance structures and resonant metasurfaces are discussed. These are broadly classified based on the complexity involved in terms of the nonlinear optical processes studied or the structure being considered for this study. First, the basic nonlinear harmonic generation processes such as second and third harmonic generation are considered, following which wave-mixing processes such as four-wave mixing and sum-frequency generation are considered. This is followed by ultra-fast optical switching, photon acceleration effect, and higher harmonic generation processes. Lastly, nonlinear optical studies in hybrid metasurfaces are discussed. The discussion is aimed at outlining the salient features of the respective nonlinear processes and specific structures studied. There are few previous review articles written in the areas of nonlinear metasurfaces in plasmonic [9,11], dielectric metasurface [14] and guided-mode resonance [17] platforms. There is also interest in utilizing the metasurface to shape the wavefront of the generated nonlinear signal for beam steering or focusing applications [88–90]. These efforts are not discussed here to keep the focus solely on the resonant enhancement of nonlinear optical processes.

4.1. Second- and Third-Harmonic Generation

Second and third-order nonlinear optical processes are considered as the basic nonlinear optical processes studied in optical media under the influence of a strong incident electric field. The induced polarization or the response of the medium to the incident electric field can be expanded in a perturbative approach into various nonlinear optical processes as follows [1]:

$$\vec{P}(\omega_{out}) = \varepsilon_o \left(\chi^{(1)} . \vec{E}(\omega) + \chi^{(2)} : \vec{E}(\omega)\vec{E}(\omega) + \chi^{(3)} \vdots \vec{E}(\omega)\vec{E}(\omega)\vec{E}(\omega) + \cdots \right) \tag{5}$$

The nonlinear interactions which depend quadratically and cubically with the incident electric field gives rise to second- and third-order nonlinear processes respectively. The strength of the nonlinear optical processes can be enhanced significantly by the enhancement of the incident electric field inside the resonant metasurface. The nonlinear process typically scales as the $(Q/V)^n$ where, Q is the quality factor of the resonance under consideration and V is the cavity volume and n is the order of the nonlinearity [91]. With reduced cavity volumes in sub-wavelength metasurfaces and the ability to achieve moderately high-quality factors (few 100 s to 1000 s), the resonant nonlinear optical process can be enhanced by 10^3 to 10^5 times. This field enhancement can counteract the effect of reduced interaction length in sub-wavelength thick metasurface, which is potentially promising for realizing high efficiency nonlinear photonic devices. Second order nonlinear optical processes are observed in materials which lack inversion symmetry and in material interfaces, while third-order nonlinear optical processes are observed in all optical media [1]. This leads to the careful selection of the nonlinear media to build resonant metasurface platforms for study various nonlinear optical processes. In general, the second- and third- harmonic generation processes satisfy the frequency relationships, $\omega_{out} = \omega + \omega$ and $\omega_{out} = \omega + \omega + \omega$ respectively. The need for momentum or wave-vector matching is relaxed in sub-wavelength metasurface platforms in most implementations due to the reduced length resulting in negligible phase mismatch. Here, we broadly divide the second and third-harmonic generation studies in periodic dielectric structures into guided-mode resonance type and resonant metasurface type platforms. Few examples under each of these categories are listed in Figures 15 and 16, respectively.

Some of the early sub-wavelength periodic structures studied for nonlinear optical applications are the guided-mode resonance structures leveraging the resonances offered by the dielectric grating structures to enhance nonlinear effects from nonlinear polymer overlayers. Figure 15a,b show two such implementations using PMMA [92] and Azo-polymers [93] as the nonlinear media on top of glass and titanium oxide gratings respectively. In Figure 15a, careful attention is paid to the phase matching of the second-harmonic generation process between the counterpropagating fundamental and second-harmonic slab modes in the presence of the grating structure [92]. Experimentally measured second-harmonic signal shows enhancement corresponding to the phase matched condition when compared to the non-phase matched case. The use of higher refractive index gratings, such as periodically patterned titanium oxide layer is found to enhance the local electric field in comparison to the glass gratings and this is found to enhance second-harmonic by ~3500 times from a Azo-polymer overlayer when compared to a reference sample without the guided-mode resonance structures [93]. There has also been interest in studying nonlinear optical processes from the guided-mode resonance grating structures itself. In this context, silicon nitride gratings have been used for second- and third-harmonic generation studies [94,95]. Even though the nonlinearities in silicon nitride is weak and the index contrast with the substrate is small, the broad optical transparency window from the visible to mid infrared, makes it attractive for realizing high quality resonant structures. An example for the use of silicon nitride sub-wavelength grating structures for third-harmonic generation in the ultraviolet spectral region is shown in Figure 15c [95]. Aluminum Gallium Arsenide (AlGaAs) high-contrast grating structures with characteristic optical resonances have also been explored for second-harmonic generation studies. Schematic images of such free-standing AlGaAs high contrast grating structure and the corresponding second-harmonic microscopy images obtained for different orientations of the fundamental and second-harmonic polarization are shown in Figure 15d [96].

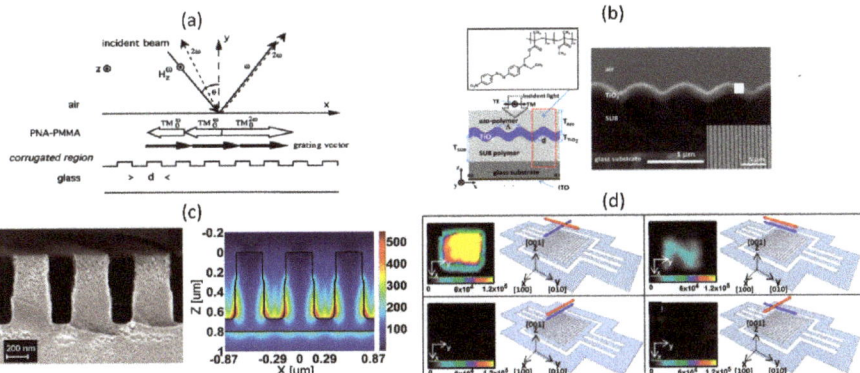

Figure 15. Various guided-mode resonance structures studied for nonlinear optical enhancement studies. (**a**) Schematic of glass-grating with PMMA layer used for phase-matched second-harmonic generation studies. (**b**) Schematic and scanning electron microscopy image of the Azo-polymer coated titanium oxide gratings used for second-harmonic generation enhancement. (**c**) Silicon nitride grating structures and simulated field profiles used for UV-third-harmonic generation. (**d**) AlGaAs high-contrast grating structures and second harmonic generation microscopy studies for different incident/ detection polarizations. (Figure a is reproduced with permission from ref. [92], b is reproduced with permission from ref. [93], c is reproduced with permission from ref. [95] and d is reproduced with permission from ref. [96]).

In the context of resonant metasurfaces for second- and third- harmonic generation studies, sub-wavelength spaced arrays of high-index semiconductors such as silicon, germanium and gallium arsenide have been studied. Fano-resonances from silicon bar-nanodisk structures, similar to the bar-ring structures shown in Figure 12a have been utilized to enhance third-harmonic generation [97]. The scanning electron microscopy image and the measured linear and third-harmonic spectra are shown in Figure 16a. Maximum third-harmonic signal enhancement of ~10^5 has been reported in this work with an overall conversion efficiency of 10^{-4}. Silicon nanodisks in ordered two-dimensional arrangement have been used to leverage magnetic dipolar resonances from the unit cell elements to enhance third-harmonic generation [34]. Figure 16b shows one such arrangement of nanodisks with the corresponding linear and nonlinear spectral measurement results. Maximum enhancement of close to two orders of magnitude with conversion efficiencies of ~10^{-7} has been reported in this work. Similar third-harmonic enhancement studies have been extended to dimer and more complex oligomeric unit cells to study the collect interaction of the individual elements in the unit cells [98,99]. There has also been interest in understanding the effect of disorder in the particle arrangement [100,101]. Figure 16c shows the arrangement of the nanodisks with controlled disorder introduced during fabrication. In this work, it has been found that the third-harmonic signal and its spatial localization are robust against disorder added to the nanodisk arrangement, making it topologically protected. Gallium Arsenide metasurfaces have been used for second-harmonic generation enhancement [102]. Asymmetric metasurfaces with high quality factor utilized for one such work with the corresponding linear and nonlinear optical spectra are shown in Figure 16d. It is found that the common [1 0 0] oriented Gallium Arsenide results in negligible second-harmonic emission along the optical axis due to the dominant longitudinally polarized nonlinear polarization, thus resulting in poor collection efficiency. One way to alter the far-field emission profile is to change the Gallium Arsenide orientation [103]. Figure 16e shows one such work on nanodisk arrays of [1 1 1] Gallium Arsenide metasurfaces. It is found from the far-field angular distribution that [1 1 1] metasurface does result in strong second-harmonic emission parallel to the optical axis when compared to [1 0 0] metasurface.

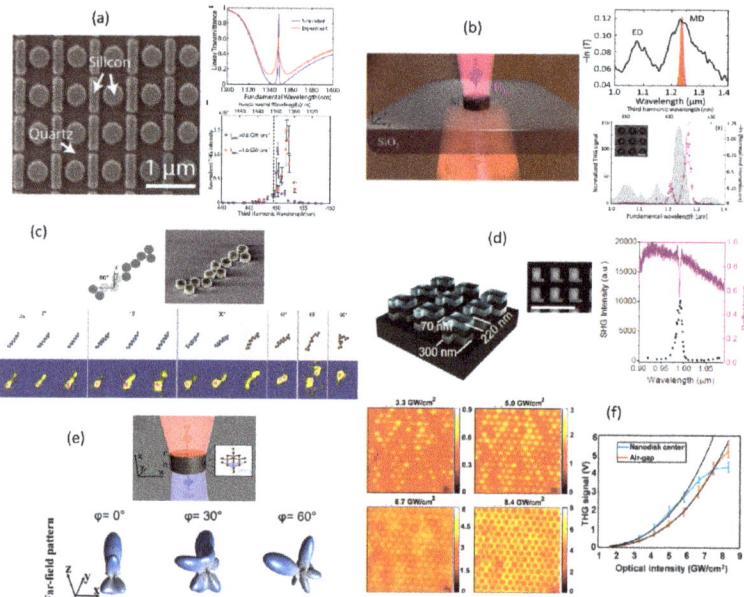

Figure 16. Various implementations of resonant metasurface for second- and third-harmonic generation studies. (**a**) Fano-resonant silicon bar-nanodisk structures for third-harmonic generation enhancement. (**b**) Silicon nanodisk array for third harmonic generation enhancement relying on magnetic dipolar modes. (**c**) Disorder robust third-harmonic generation from silicon nanodisks which are shown to be topologically protected from disorder in arrangement of the structures. (**d**) Gallium Arsenide asymmetry resonant metasurface for second-harmonic generation enhancement. (**e**) Dependence of the resonant second-harmonic far-field signal on [1 1 1] oriented Gallium Arsenide metasurface. (**f**) Spatial mapping of intensity dependent saturation of third-harmonic signal from silicon nanodisk array. (Figure a is reproduced with permission from ref. [97], b is reproduced with permission from ref. [34], c is reproduced with permission from ref. [100], d is reproduced with permission from ref. [102], e is reproduced with permission from ref. [103] and f is reproduced with permission from ref. [104]).

Spatially resolved nonlinear optical studies or nonlinear optical microscopy is also a useful tool to understand the spatial dependence of the nonlinear signal across different regions in the metasurface sample to understand the signal uniformity and can combined with spectral and intensity studies to understand spatial, spectral resonance and intensity saturation behavior of the nonlinear medium. In this context, Figure 16f shows the intensity dependence of the third-harmonic generation microscopy images across a silicon nanodisk array sample at its fundamental resonance wavelength [104]. It is found that the contrast in the third-harmonic microscopy images reverses with increasing intensity. This is attributed to the spatial position dependent onset of saturation of the third-harmonic signal as shown in the intensity dependent third-harmonic plot in Figure 16f.

4.2. Wave Mixing Processes

The wave-mixing processes can be considered as a general case of the above described harmonic generation processes. The processes of interest here are the four-wave mixing (FWM) and sum-frequency generation (SFG). In the case of FWM process, the nonlinear signal frequency is related to the incident light as follows: $\omega_4 = \omega_1 + \omega_2 - \omega_3$ [1]. Two pump frequencies (ω_1 and ω_2) being unique or identical are termed as non-degenerate and degenerate FWM processes respectively. The second-order SFG process satisfies the frequency relation: $\omega_3 = \omega_1 + \omega_2$, while degenerate third-order SFG process satisfies frequency relation of the form: $\omega_3 = 2\omega_1 + \omega_2$ or $\omega_3 = \omega_1 + 2\omega_2$ [1]. Figure 17a–c shows FWM enhancement observed for closely spaced pump-signal wavelengths in the telecom range for silicon-on-insulator based fully etched high-contrast gratings [105]. The sub-wavelength dimension high-contrast grating structures are found to support resonances with intensity enhancement of more than 8000 times and experimentally measured quality factor of ~7300. The signal and pump photons in close vicinity to this resonance results in FWM with the generation of idler with conversion efficiency of −19.5 dB as shown in Figure 17c. The use of high aspect ratio germanium (Ge) nanodisks to observe anapolar resonances [106] and the enhancement of third-order sum-frequency generation processes is shown in Figure 17d–f [107]. The higher order anapolar mode profiles are chosen with good spatial overlap to ensure enhancement of the SFG process by about two-orders of magnitude, as shown in the SFG spectrum in Figure 17f. Silicon (Si) nanodisks that support magnetic and electric-dipole resonances have also been utilized for doubly-resonant enhancement of FWM process as shown in Figure 17g–i [108]. The individual resonance spectra and the corresponding resonance for the FWM are also shown, with approximately two-orders of magnitude enhancement. Doubly-resonant structures are promising to increase the FWM efficiency using both pump and signal resonances. However, the best enhancement can be obtained only when good overlap is ensured between the interacting resonant mode profiles. Detailed spatially-resolved imaging of four-wave mixing process in singly resonant partially etched zero-contrast grating structures is shown in Figure 17j–l [109]. The structures are designed to support resonance at the signal wavelength in the 1550 to 1600 nm wavelength range. Four-wave mixing images acquired across an area of 300 × 300 microns show clear dependence of the FWM image contrast on the incident signal wavelength. A maximum FWM enhancement of 450 times has been experimentally obtained [109].

4.3. Optical Switching

The ultrafast Kerr nonlinearity and multi-photon absorption processes due to the nonlinear interaction of valence electronics in the dielectric medium with incident light can be used to perform fast optical switching at hundreds of femtosecond time scales. Such ultrafast switches have been demonstrated previously in guided-wave systems such as optical fibers and integrated waveguides utilizing self-phase and cross-phase modulation effects [2]. The optical resonances in dielectric resonant metasurfaces can be used to enhance the optical switching process through the enhancement of the nonlinear optical effect [110]. In the context of semiconductor metasurfaces, the presence of parasitic processes such as thermo-optic effects results in additional phase shift, albeit at much longer time scales. Figure 18a,b shows a schematic and scanning electron microscopy image of a silicon nanodisk array used for optical switching studies. The corresponding transmission spectra from the nanodisk array and the associated dipolar resonant modes are shown in Figure 18c. The optical switching process can be studied in a pump-probe configuration with a strong pump on-resonance leading to enhanced electric-field inside the nanodisks which results in a fast transmission dip at the probe wavelength due to enhanced multi-photon absorption. The fastest switching characteristics is obtained with a temporal response time of 65 fs with a slower extended recovery due to thermal and free-carrier recombination effects. Such resonant metasurface with ultrafast all-optical switching capability can find possible applications as fiber-connectorized photonic structures for high speed data communication and pulse-shaping [2].

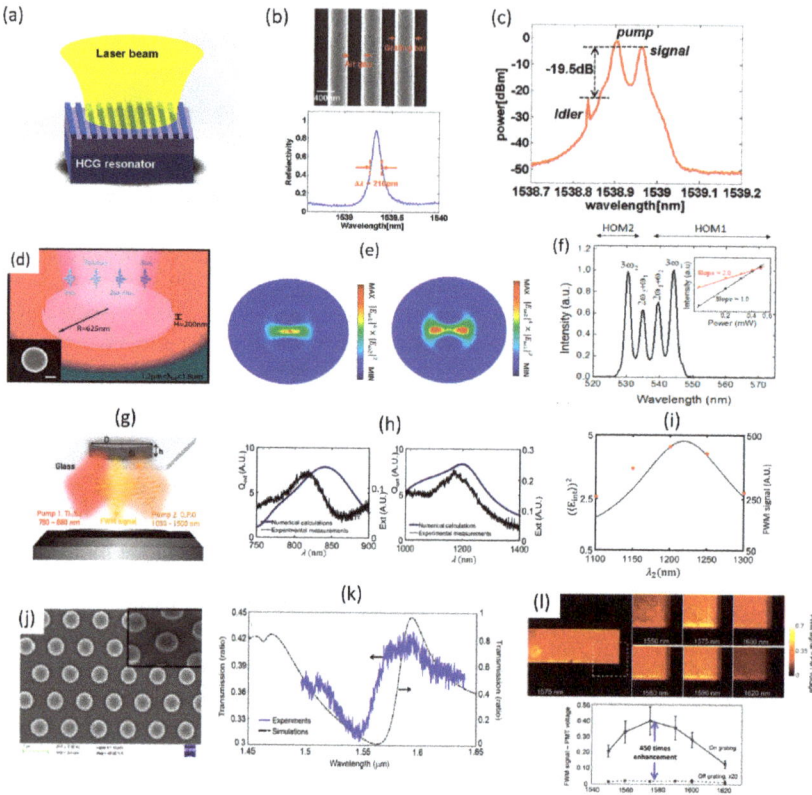

Figure 17. Various implementations of resonant FWM processes. (**a**) Schematic of high-contrast grating resonance, (**b**) Electron microscopy images and measured reflectivity spectrum of the resonance. (**c**) Measured FWM spectrum for the high-contrast grating structure. (**d**) Schematic of the high-aspect ratio Ge nanodisks for SFG studies. (**e**) Mode profiles of the nonlinear polarization for the two different SFG processes. (**f**) Measured SFG and THG spectra for the high-aspect ratio Ge nanodisks. (**g**) Schematic of Si nanodisk structures used for doubly-resonant FWM studies. (**h**) Comparison between measured and simulated scattering spectra for the two resonant modes under consideration. (**i**) Comparison of the measured FWM signal and simulated pump intensity enhancement as a function of wavelength. (**j**) Electron microscopy image of the partially etched zero-contrast grating structures used for FWM studies. (**k**) Simulated and measured transmission spectra for the zero-contrast grating structures. (**l**) FWM microscopy images for varying signal wavelength. The enhancement spectrum is also shown. (Figure a–c are reproduced with permission from ref. [105], d–f are reproduced with permission from ref. [107], g–i are reproduced with permission from ref. [108] and j–l are reproduced with permission from ref. [109]).

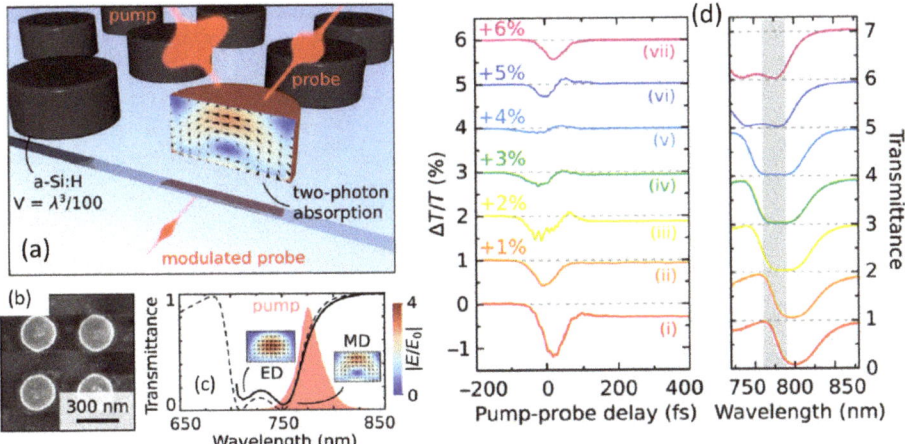

Figure 18. (**a**) Schematic of the nanodisk array used for ultrafast-optical switching studies. (**b**) Electron microscopy image and (**c**) transmission spectra for the nanodisk array with the corresponding dipolar modes marked. (**d**) Experimental results of pump-probe studied showing fast recovery or switching of probe in the presence of a resonant pump. Various pump laser wavelengths relative to the resonance are shown in the right plot. (Figures are reproduced with permission from ref. [110]).

4.4. Photon Acceleration

Time-dependent optical properties in a resonant medium can lead to noticeable spectral shift of propagating laser pulse and can manifest as wavelength shifts of corresponding nonlinear optical signals as well. Such spectral shifts of light in the presence of time-varying optical processes is termed as photon acceleration and has been studied previous in plasma media [111]. Similar effect have recently been observed in silicon based resonant metasurface due to shift in the resonant wavelength with increasing incident light fluence due to time-dependent free-carrier accumulation [112]. Schematic of the silicon-rectangular structures used to study photon acceleration is shown in Figure 19a. The corresponding transmission spectrum showing the optical resonance in the mid infrared wavelength range close to 3.6 µm and the corresponding field profile are shown in Figure 19b. With increasing incident laser fluence, the transmitted laser spectrum shifts across the metasurface resonance, as shown in Figure 19c. A comparison of the third-harmonic signal from the silicon metasurface with the un-patterned silicon film shows a significant shift of the signal to shorter wavelengths for the metasurface, while it remains unchanged for the film (shown in Figure 19d). The photon acceleration efficiency for the third-harmonic signal was measured to be ~22%. The observed blue-shift was found to be in good agreement with a time varying photon mode amplitude model considering free carrier accumulation due to four-photon absorption process [112]. Photon acceleration based on time-varying optical processes in resonant dielectric metasurfaces presents a promising platform for performing robust pulse-shaping operations [112].

4.5. Higher Order Wave-Mixing Processes

The nonlinear optical studies are not restricted to just the second and third-order nonlinear optical processes. With high enough incident light fluence and high quality factor resonant metasurface medium with strong optical nonlinearities, various higher order processes greater than third order can also be observed. This is shown in Figure 20a for a Gallium Arsenide based meta-mixer consisting of nanodisk array [113]. Such structures support magnetic and electric dipolar resonances, as shown in the inset of Figure 20a. These resonances at the incident excitation laser wavelengths can be leveraged to study various nonlinear optical processes from second- and third-harmonic to four-wave mixing to fourth-harmonic generation and even six-wave mixing processes. With a high sensitivity spectrometer, the various nonlinear processes are spectrally resolved in Figure 20b. The dependence of the various nonlinear processes on time-delay of the interacting waves is also shown in Figure 20c. Even though the overall conversion efficiency is small, this is a promising direction towards realizing complex wave-mixing processes on a small footprint platform. Further enhancement in efficiency can be achieved by improving the quality factor of the resonances [77–79] or using such resonant structures in intra-cavity configuration [114].

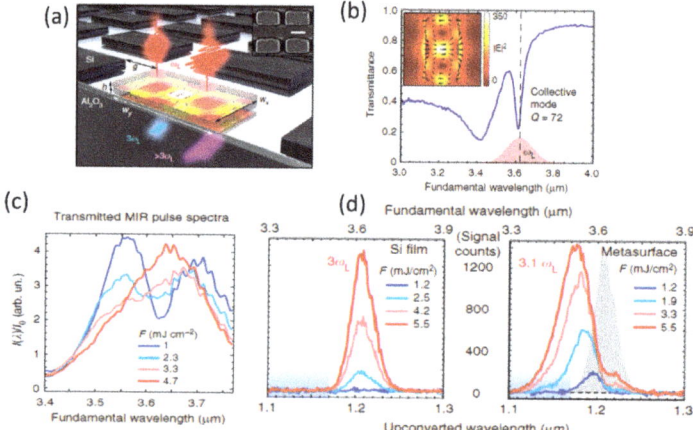

Figure 19. (a) Schematic of the silicon rectangular metasurface used for photon acceleration studies. (b) Measured transmission spectrum and field profile at resonance (inset). (c) The fundamental laser spectra transmitted through the metasurface for varying laser fluence. (d) Comparison of the third-harmonic signal generated for varying fundamental laser fluence for the un-patterned silicon film and silicon metasurface. (Figures are reproduced with permission from ref. [112]).

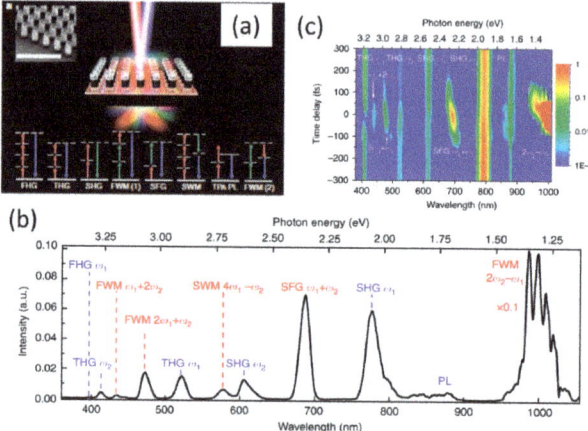

Figure 20. (a) Schematic of the Gallium Arsenide nanodisk array used for nonlinear wave mixing studies. (b) Experimentally measured spectra of various nonlinear wave mixing processes. The name of the various processes and their frequency relationship are labelled. (c) The dependence of the nonlinear wave-mixing spectra on the time-delay between the interacting excitation pulses. (Figures are reproduced with permission from ref. [113]).

4.6. Nonlinear Optics with Hybrid Metasurface

There is also interest in integration of dissimilar materials to realize hybrid metasurfaces with enhanced nonlinear optical properties when compared to the individual materials studied separately. Few examples of such hybrid integrated structures are shown in Figure 21. The hybrid integration of patterned metal metasurface with resonant quantum-well semiconductors has been studied with the objective of realizing doubly-resonant nonlinear optical metasurfaces [115,116]. The inherent resonant nonlinearities from the multi-quantum well structures are further amplified by the plasmonic resonant structures. Figure 21a shows the schematic of such hybrid plasmonic-dielectric structures. The corresponding resonant nonlinear susceptibility spectrum and the designed multi-quantum well structure is shown in Figure 21b and c respectively. The effective nonlinear susceptibility of the hybrid structure realized is close to 10^5 pm/V, one of the highest nonlinear optical susceptibility reported for any solid-state material systems. The plasmonic structures have recently been replaced by dielectric metasurfaces, as shown in Figure 21d–f [117]. The structure consists of a one-dimensional germanium guided-mode resonant structure integrated on top of the multi-quantum layer stack. The designed optical resonances of the germanium guided-mode resonance structures and the associated resonant nonlinear optical susceptibility are shown in Figure 21e. 100s of nano-watt level second harmonic generation signal has been experimentally measured from these all-dielectric hybrid metasurfaces. Such structures with the fundamental wavelength in the mid infrared wavelength range are best suited for frequency up-conversion with high conversion efficiencies from wavelength regions where there is scarcity of high efficiency detectors to wavelength regions in the near-infrared and visible region where mature, high efficiency detectors are readily available.

There is also interest in integration of dielectric metasurfaces with two-dimensional layered materials in monolayer or multi-layer form. Two-dimensional materials offer robust nonlinear optical properties in terms of strength of the nonlinear optical susceptibility, its layer number and polarization dependence [118]. Such layered materials can be readily transferred to patterned dielectric structures with simple dry-transfer or chemical-vapor deposition techniques. Figure 21g–i shows one such hybrid integration of 2D material with silicon metasurface [119]. Multi-layer Gallium selenide dry-transferred onto asymmetric silicon resonant metasurface are utilized for resonant enhancement of second-harmonic and sum-frequency generation from the 2D material layer. A schematic of such structure is shown in Figure 21g, with a red-shift in the resonance spectrum in the presence of the 2D material shown in Figure 21h. The nonlinear optical process from the hybrid 2D material- dielectric metasurface is fairly strong that continuous-wave excitation has been used to generate second-harmonic and sum-frequency signals, as shown in the experimentally measured spectrum in Figure 21i.

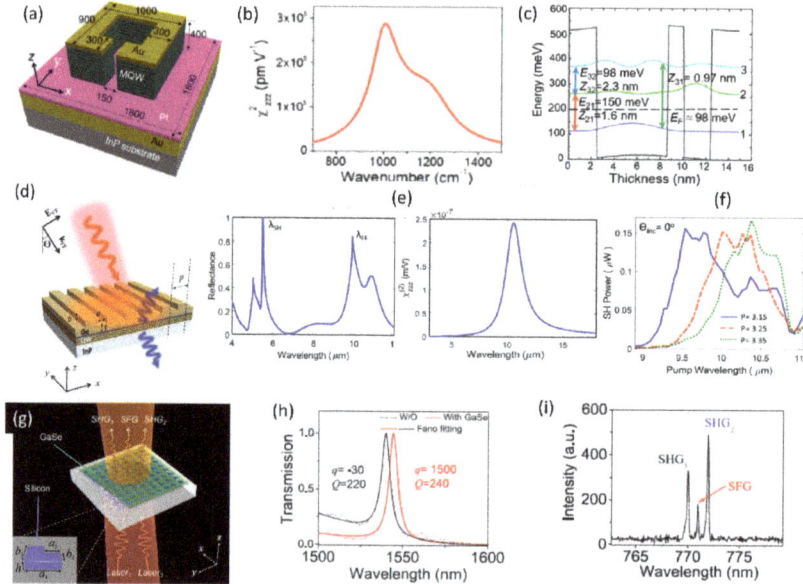

Figure 21. Various implementations of hybrid resonant metasurfaces for nonlinear optical applications. (**a–c**) Hybrid plasmonic-dielectric metasurfaces in which resonant nonlinearities in multi-quantum well structures are coupled with plasmonic resonances to study second harmonic generation. (**d–f**) All dielectric implementation of the hybrid resonant metasurface consisting of Germanium guided-mode resonance structures on top of the multi-quantum well structures. (**g–i**) Hybrid structures consisting of multi-layer Gallium Selenide on top of the asymmetric metasurface used for resonant enhancement of second-harmonic and sum-frequency generation. (Figures a–c are reproduced with permission from ref. [116], d–f are reproduced with permission from ref. [117], g–i are reproduced with permission from ref. [119]).

5. Concluding Remarks

In this review paper, an overview of various nonlinear optical processes studied in sub-wavelength periodic dielectric structures is presented. Dielectric structures are particularly attractive for nonlinear optical studies due to the ability to engineer field concentration to be located inside or outside the structure by design, high damage thresholds and ease of fabrication of complex structures. In particular, high refractive index dielectric structures are preferred for realizing such structures due to the high-quality factor of the resonances and the associated enhanced field strength. Resonant field enhancement mechanisms in such sub-wavelength structures are broadly classified here as guided-mode resonances and resonant metasurfaces. The physical mechanism behind the optical resonance phenomena in these structures are explained based on the guided-mode resonance phenomenon, EIT-like resonance and bound-state in continuum type resonances. The various nonlinear optical processes studied in these structures, which include second-/third-harmonic generation, four-wave mixing, sum-frequency generation, ultrafast optical switching based on Kerr and multiphoton absorption nonlinearities, photon acceleration of third harmonic generation signal, higher harmonic generation processes and hybrid resonant metasurfaces are discussed. Various dielectric structures in the form of one-dimensional gratings, two-dimensional arrays of nanodisks, bar-nanodisk structures, asymmetric bar dimers, asymmetric rectangular unit-cells, disordered nanodisk array, coupled GMR structures, heterogeneous structures are utilized for these nonlinear optical studies.

Looking ahead, such periodic dielectric structures are expected to find applications as miniaturized frequency converters in the form of intracavity photonic windows or as active fiber connectors [120,121].

The transition of this technology to practical, real world application does require improvements to certain aspects of this technology. In particular, scaling of the area of the metasurfaces, achieving higher conversion efficiencies, ability to extend to ultra-violet or infrared spectral windows with the exploration of novel structures and heterogenous integration capability will become essential. Optical metasurfaces are conventionally patterned using electron-beam lithography which renders it time consuming and difficult to scale to large area. The use of stepper lithography as used in standard electronic chip fabrication, interference lithography, imprint lithography or chemical synthesis techniques are promising to scale the metasurfaces to large areas [36,42]. The ability to combine multiple resonance phenomena such as quantum well based resonant nonlinearity with metasurface resonance is a promising direction to considerably improve the conversion efficiencies [116]. The use of high nonlinearity media on suitable low-index substrates, such as Gallium Phosphide on glass substrates [122] and high efficiency electro-optic polymers [123] are also possible direction to improve the efficiency. The use of multi-pass configuration, for example, in an intracavity application can also amplify the overall nonlinear process [120]. Furthermore, with emerging two-dimensional layered materials for photonic applications, the use of such materials in combination with dielectric metasurfaces is also seen as a promising direction [118]. The ability to switch or tune nonlinear optical functionality at high-speed will also be useful for fast optical modulation [124–126]. The exploration of high-quality factor resonances such as quasi bound-state in continuum is also an important direction to realize high efficiency resonant metasurfaces [83]. Overall, the emerging field of nonlinear optics in resonant metasurfaces and the resurgence it has given to guided-mode resonance platforms is creating lot of interest in nonlinear nanophotonics community. While practical applications are yet to emerge, the research efforts addressing this are in the right direction, which makes the future of this technology highly promising.

Author Contributions: Conceptualization, V.R.; writing—original draft preparation, V.R.; writing—review and editing, V.R., J.D., S.M., R.B. and L.K.A.S.; supervision, V.R.; project administration, V.R.; funding acquisition, V.R. All authors have read and agreed to the published version of the manuscript.

Funding: VR acknowledges funding support from DST SERB Early career award, ISRO Space technology cell and DST Nano mission through the NNetra program.

Conflicts of Interest: The authors declare no conflict of interest.

References

1. Boyd, R.W. *Nonlinear Optics*, 3rd ed.; Elsevier: New York, NY, USA, 2003.
2. Agrawal, G. *Applications of Nonlinear Fiber Optics*, 5th ed.; Elsevier: New York, NY, USA, 2001.
3. Shen, Y.R. *The Principles of Nonlinear Optics*, 1st ed.; Wiley-Interscience: New York, NY, USA, 1984.
4. Kivshar, Y. All-dielectric meta-optics and non-linear nanophotonics. *Natl. Sci. Rev.* **2018**, *5*, 144–158. [CrossRef]
5. Novotny, L.; Hecht, B. *Principles of Nano-Optics*, 2nd ed.; Cambridge University Press: New York, NY, USA, 2002.
6. Yu, H.; Peng, Y.; Yang, Y.; Li, Z.Y. Plasmon-enhanced light–matter interactions and applications. *npj Comput. Mater.* **2019**, *5*, 1–4. [CrossRef]
7. Tzarouchis, D.; Sihvola, A. Light scattering by a dielectric sphere: Perspectives on the Mie resonances. *Appl. Sci.* **2018**, *8*, 184. [CrossRef]
8. Kauranen, M.; Zayats, A.V. Nonlinear plasmonics. *Nat. Photonics* **2012**, *6*, 737. [CrossRef]
9. Panoiu, N.C.; Sha, W.E.; Lei, D.Y.; Li, G.C. Nonlinear optics in plasmonic nanostructures. *J. Opt.* **2018**, *20*, 083001. [CrossRef]
10. Krasnok, A.; Tymchenko, M.; Alù, A. Nonlinear metasurfaces: A paradigm shift in nonlinear optics. *Mater. Today* **2018**, *21*, 8–21. [CrossRef]
11. Liu, B.; Song, K.; Xiao, J. Two-dimensional optical metasurfaces: From plasmons to dielectrics. *Adv. Condens. Matter. Phys.* **2019**, *9*, 1–15. [CrossRef]
12. Overvig, A.C.; Shrestha, S.; Malek, S.C.; Lu, M.; Stein, A.; Zheng, C.; Yu, N. Dielectric metasurfaces for complete and independent control of the optical amplitude and phase. *Light Sci. Appl.* **2019**, *8*, 1–2. [CrossRef]

13. Kamali, S.M.; Arbabi, E.; Arbabi, A.; Faraon, A. A review of dielectric optical metasurfaces for wavefront control. *Nanophotonics* **2018**, *7*, 1041–1068. [CrossRef]
14. Sain, B.; Meier, C.; Zentgraf, T. Nonlinear optics in all-dielectric nanoantennas and metasurfaces: A review. *Adv. Photonics* **2019**, *1*, 024002. [CrossRef]
15. Lim, A.E.; Song, J.; Fang, Q.; Li, C.; Tu, X.; Duan, N.; Chen, K.K.; Tern, R.P.; Liow, T.Y. Review of silicon photonics foundry efforts. *IEEE J. Sel. Top. Quantum Electron.* **2013**, *20*, 405–416. [CrossRef]
16. Fathpour, S. Heterogeneous nonlinear integrated photonics. *IEEE J. Quantum Electron.* **2018**, *54*, 1–6. [CrossRef]
17. Quaranta, G.; Basset, G.; Martin, O.J.; Gallinet, B. Recent advances in resonant waveguide grating. *Laser Photonics Rev.* **2018**, *2*, 1800017. [CrossRef]
18. Liu, Y.C.; Li, B.B.; Xiao, Y.F. Electromagnetically induced transparency in optical microcavities. *Nanophotonics* **2017**, *6*, 789–811. [CrossRef]
19. Hsu, C.W.; Zhen, B.; Stone, A.D.; Joannopoulos, J.D.; Soljačić, M. Bound states in the continuum. *Nat. Rev. Mater.* **2016**, *1*, 1–3. [CrossRef]
20. Koshelev, K.; Kruk, S.; Melik-Gaykazyan, E.; Choi, J.H.; Bogdanov, A.; Park, H.G.; Kivshar, Y. Subwavelength dielectric resonators for nonlinear nanophotonics. *Science* **2020**, *367*, 288–292. [CrossRef]
21. Maier, S.A. *Plasmonics: Fundamentals and Applications*, 1st ed.; Springer Science & Business Media: Berlin, Germany, 2007.
22. Lakowicz, J.R.; Chowdhury, M.H.; Ray, K.; Zhang, J.; Fu, Y.; Badugu, R.; Sabanayagam, C.R.; Nowaczyk, K.; Szmacinski, H.; Aslan, K.; et al. Plasmon-controlled fluorescence: A new detection technology. Plasmonics in Biology and Medicine III. *Proc. SPIE* **2006**, *6099*, 609909.
23. Bauch, M.; Toma, K.; Toma, M.; Zhang, Q.; Dostalek, J. Plasmon-enhanced fluorescence biosensors: A review. *Plasmonics* **2014**, *9*, 781–799. [CrossRef] [PubMed]
24. Nie, S.; Emory, S.R. Probing single molecules and single nanoparticles by surface-enhanced Raman scattering. *Science* **1997**, *275*, 1102–1106. [CrossRef]
25. Kneipp, K.; Wang, Y.; Kneipp, H.; Perelman, L.T.; Itzkan, I.; Dasari, R.R.; Feld, M.S. Single molecule detection using surface-enhanced Raman scattering (SERS). *Phys. Rev. Lett.* **1997**, *78*, 1667. [CrossRef]
26. Vo-Dinh, T.; Yan, F.; Stokes, D.L. *Plasmonics-Based Nanostructures for Surface-Enhanced Raman Scattering Bioanalysis. Protein Nanotechnology*, 1st ed.; Humana Press: Totowa, NJ, USA, 2005.
27. Hulst, H.C.; van de Hulst, H.C. *Light Scattering by Small Particles*, 1st ed.; Dover Publication: Mineola, NY, USA, 1981.
28. Caldarola, M.; Albella, P.; Cortés, E.; Rahmani, M.; Roschuk, T.; Grinblat, G.; Oulton, R.F.; Bragas, A.V.; Maier, S.A. Non-plasmonic nanoantennas for surface enhanced spectroscopies with ultra-low heat conversion. *Nat. Commun.* **2015**, *6*, 1–8. [CrossRef] [PubMed]
29. Shao, M.; Ma, D.D.; Lee, S.T. Silicon nanowires–synthesis, properties, and applications. *Eur. J. Inorg. Chem.* **2010**, *27*, 4264–4278. [CrossRef]
30. Juhasz, R.; Elfström, N.; Linnros, J. Controlled fabrication of silicon nanowires by electron beam lithography and electrochemical size reduction. *Nano Lett.* **2005**, *5*, 275–280. [CrossRef] [PubMed]
31. Cao, L.; Fan, P.; Barnard, E.S.; Brown, A.M.; Brongersma, M.L. Tuning the color of silicon nanostructures. *Nano Lett.* **2010**, *10*, 2649–2654. [CrossRef]
32. Butakov, N.A.; Schuller, J.A. Designing multipolar resonances in dielectric metamaterials. *Sci. Rep.* **2016**, *6*, 38487. [CrossRef]
33. Kuznetsov, A.I.; Miroshnichenko, A.E.; Fu, Y.H.; Zhang, J.; Luk'Yanchuk, B. Magnetic light. *Sci. Rep.* **2012**, *2*, 492. [CrossRef]
34. Shcherbakov, M.R.; Neshev, D.N.; Hopkins, B.; Shorokhov, A.S.; Staude, I.; Melik-Gaykazyan, E.V.; Decker, M.; Ezhov, A.A.; Miroshnichenko, A.E.; Brener, I.; et al. Enhanced third-harmonic generation in silicon nanoparticles driven by magnetic response. *Nano Lett.* **2014**, *14*, 6488–6492. [CrossRef]
35. Noginova, N.; Barnakov, Y.; Li, H.; Noginov, M.A. Effect of metallic surface on electric dipole and magnetic dipole emission transitions in Eu 3+ doped polymeric film. *Opt. Express* **2009**, *17*, 10767–10772. [CrossRef]
36. She, A.; Zhang, S.; Shian, S.; Clarke, D.R.; Capasso, F. Large area metalenses: Design, characterization, and mass manufacturing. *Opt. Express* **2018**, *26*, 1573–1585. [CrossRef]
37. Miroshnichenko, A.E.; Kivshar, Y.S. Fano resonances in all-dielectric oligomers. *Nano Lett.* **2012**, *12*, 6459–6463. [CrossRef]
38. Wang, S.S.; Magnusson, R.J. Theory and applications of guided-mode resonance filters. *Appl. Opt.* **1993**, *32*, 2606–2613. [CrossRef] [PubMed]

39. Li, Q.; Gao, J.; Li, Z.; Yang, H.; Liu, H.; Wang, X.; Li, Y. Absorption enhancement in nanostructured silicon fabricated by self-assembled nanosphere lithography. *Opt. Mater.* **2017**, *70*, 165–170. [CrossRef]
40. Qiao, P.; Zhu, L.; Chew, W.C.; Chang-Hasnain, C.J. Theory and design of two-dimensional high-contrast-grating phased arrays. *Opt. Express* **2015**, *23*, 24508–24524. [CrossRef] [PubMed]
41. Sarabandi, K.; Behdad, N. A frequency selective surface with miniaturized elements. *IEEE Trans. Antennas Propag.* **2007**, *55*, 1239–1245. [CrossRef]
42. Madou, M.J. *Fundamentals of Microfabrication and Nanotechnology*, 3rd ed.; CRC Press: Boca Raton, FL, USA, 2018.
43. Liu, V.; Fan, S. S4: A free electromagnetic solver for layered periodic structures. *Comput. Phys. Commun.* **2012**, *183*, 2233–2244. [CrossRef]
44. Korzeniowska, B.; Nooney, R.; Wencel, D.; McDonagh, C. Silica nanoparticles for cell imaging and intracellular sensing. *Nanotechnology* **2013**, *24*, 442002. [CrossRef] [PubMed]
45. Chen, E.H.; Gaathon, O.; Trusheim, M.E.; Englund, D. Wide-field multispectral super-resolution imaging using spin-dependent fluorescence in nanodiamonds. *Nano Lett.* **2013**, *13*, 2073–2077. [CrossRef] [PubMed]
46. Rybin, M.V.; Koshelev, K.L.; Sadrieva, Z.F.; Samusev, K.B.; Bogdanov, A.A.; Limonov, M.F.; Kivshar, Y.S. High-Q supercavity modes in subwavelength dielectric resonators. *Phys. Rev. Lett.* **2017**, *119*, 243901. [CrossRef]
47. Kuznetsov, A.I.; Miroshnichenko, A.E.; Brongersma, M.L.; Kivshar, Y.S.; Luk'yanchuk, B. Optically resonant dielectric nanostructures. *Science* **2016**, *354*, aag2472. [CrossRef]
48. Menon, S.H.; Krishna, A.L.; Raghunathan, V. Silicon Nitride based Medium Contrast Gratings for Doubly Resonant Fluorescence Enhancement. *IEEE Photonics J.* **2019**, *11*, 1–11. [CrossRef]
49. Menon, S.; Lal Krishna, A.S.; Gupta, M.S.; Ameen, E.; Pesala, B.; Raghunathan, V. Silicon-nitride-based medium-contrast gratings for resonant fluorescence enhancement in the visible wavelength range. *High Contrast Metastructures VIII. Proc. SPIE* **2019**, *10928*, 1092818.
50. Chang-Hasnain, C.J.; Yang, W. High-contrast gratings for integrated optoelectronics. *Adv. Opt. Photonics* **2012**, *4*, 379–440. [CrossRef]
51. Chang, S.; Guo, X.; Ni, X. Optical metasurfaces: Progress and applications. *Annu. Rev. Mater. Res.* **2018**, *48*, 279–302. [CrossRef]
52. Magnusson, R.; Ko, Y.H. Guided-mode resonance nanophotonics: Fundamentals and applications. *Nanoengineering: Fabrication, Properties, Optics, and Devices XIII. Proc. SPIE* **2016**, *9927*, 992702.
53. Voronov, M.M. Resonant Wood's anomaly diffraction condition in dielectric and plasmonic grating structures. *arXiv* **2016**, arXiv:1612.08674.
54. Yamada, K.; Lee, K.J.; Ko, Y.H.; Inoue, J.; Kintaka, K.; Ura, S.; Magnusson, R.J. Flat-top narrowband filters enabled by guided-mode resonance in two-level waveguides. *Opt. Lett.* **2017**, *42*, 4127–4130. [CrossRef]
55. Sun, T.; Kan, S.; Marriott, G.; Chang-Hasnain, C.J. High-contrast grating resonators for label-free detection of disease biomarkers. *Sci. Rep.* **2016**, *6*, 27482. [CrossRef]
56. Karagodsky, V.; Chang-Hasnain, C.J. Physics of near-wavelength high contrast gratings. *Opt. Express* **2012**, *20*, 10888–10895. [CrossRef]
57. Karagodsky, V.; Tran, T.; Wu, M.C.; Chang-Hasnain, C.J. Double-resonant enhancement of surface enhanced Raman scattering using high contrast grating resonators. *Proc. CLEO* **2011**, *2011*, CFN1. [CrossRef]
58. Sun, T.; Chang-Hasnain, C.J. Surface-normal coupled four-wave mixing in a high contrast gratings resonator. *Proc. CLEO* **2015**, *2015*, STu2I.2. [CrossRef]
59. Tran, T.; Karagodsky, V.; Rao, Y.; Chen, R.; Chase, C.; Chuang, L.C.; Chang-Hasnain, C.J. Second harmonic generation from AlGaAs high contrast gratings. *Proc. CLEO* **2010**, *2010*, JTuD80. [CrossRef]
60. Magnusson, R. Wideband reflectors with zero-contrast gratings. *Opt. Lett.* **2014**, *39*, 4337–4340. [CrossRef] [PubMed]
61. Ko, Y.; Magnusson, R.J. Wideband dielectric metamaterial reflectors: Mie scattering or leaky Bloch mode resonance? *Optica* **2019**, *5*, 289–294. [CrossRef]
62. Molesky, S.; Lin, Z.; Piggott, A.Y.; Jin, W.; Vucković, J.; Rodriguez, A.W. Inverse design in nanophotonics. *Nat. Photonics* **2018**, *12*, 659–670. [CrossRef]
63. Jiang, J.; Fan, J.A. Global optimization of dielectric metasurfaces using a physics-driven neural network. *Nano Lett.* **2019**, *19*, 5366–5372. [CrossRef]
64. Kodali, A.K.; Schulmerich, M.; Ip, J.; Yen, G.; Cunningham, B.T.; Bhargava, R. Narrowband midinfrared reflectance filters using guided mode resonance. *Anal. Chem.* **2010**, *82*, 5697–5706. [CrossRef]
65. Ko, Y.H.; Niraula, M.; Lee, K.J.; Magnusson, R. Properties of wideband resonant reflectors under fully conical light incidence. *Opt. Express* **2016**, *24*, 4542–4551. [CrossRef]

66. Ng, R.C.; Garcia, J.C.; Greer, J.R.; Fountaine, K.T. Polarization-independent, narrowband, near-IR spectral filters via guided mode resonances in ultrathin a-Si nanopillar arrays. *ACS Photonics* **2019**, *6*, 265–271. [CrossRef]
67. Yang, Y.; Kravchenko, I.I.; Briggs, D.P.; Valentine, J. High quality factor fano-resonant all-dielectric metamaterials. *arXiv* **2014**, arXiv:1405.3901v1.
68. Yang, Y.; Kravchenko, I.I.; Briggs, D.P.; Valentine, J. All-dielectric metasurface analogue of electromagnetically induced transparency. *Nat. Commun.* **2014**, *5*, 1–7. [CrossRef]
69. Wu, C.; Arju, N.; Fan, F.; Brener, I.; Shvets, C. Spectrally Selective Chiral Silicon Metasurfaces Based on Infrared Fano Resonances. *Proc. CLEO* **2014**, *2014*, FF2C.1. [CrossRef]
70. Wu, C.; Arju, N.; Kelp, G.; Fan, J.A.; Dominguez, J.; Gonzales, E.; Tutuc, E.; Brener, I.; Shvets, G. Spectrally selective chiral silicon metasurfaces based on infrared Fano resonances. *Nat. Commun.* **2014**, *5*, 1–9. [CrossRef] [PubMed]
71. Wu, C.; Khanikaev, A.B.; Adato, R.; Arju, N.; Yanik, A.A.; Altug, H.; Shvets, G. Fano-resonant asymmetric metamaterials for ultrasensitive spectroscopy and identification of molecular monolayers. *Nat. Mater.* **2012**, *11*, 69–75. [CrossRef] [PubMed]
72. Campione, S.; Liu, S.; Basilio, L.I.; Warne, L.K.; Langston, W.L.; Luk, T.S.; Wendt, J.R.; Reno, J.L.; Keeler, G.A.; Brener, I.; et al. Broken symmetry dielectric resonators for high quality factor Fano metasurfaces. *ACS Photonics* **2016**, *3*, 2362–2367. [CrossRef]
73. Tibuleac, S.; Magnusson, R. Narrow-linewidth bandpass filters with diffractive thin-film layers. *Opt. Lett.* **2001**, *26*, 584–586. [CrossRef]
74. Lee, S.G.; Jung, S.Y.; Kim, H.S.; Lee, S.; Park, J.M. Electromagnetically induced transparency based on guided-mode resonances. *Opt. Lett.* **2015**, *40*, 4241–4244. [CrossRef]
75. Sui, C.; Han, B.; Lang, T.; Li, X.; Jing, X.; Hong, Z. Electromagnetically induced transparency in an all-dielectric metamaterial-waveguide with large group index. *IEEE Photonics J.* **2017**, *9*, 1–8. [CrossRef]
76. Koshelev, K.; Favraud, G.; Bogdanov, A.; Kivshar, Y.; Fratalocchi, A. Nonradiating photonics with resonant dielectric nanostructures. *Nanophotonics* **2019**, *8*, 725–745. [CrossRef]
77. Magnusson, R.; Lee, S.G.; Lee, K.J.; Hemmati, H.; Carney, D.J.; Bootpakdeetam, P.; Ko, Y.H. Principles of leaky-mode photonic lattices: Band flips and Bloch mode dynamics. Integrated Optics: Devices, Materials, and Technologies XXIII. *Proc. SPIE* **2019**, *10921*, 109211E.
78. Koshelev, K.; Bogdanov, A.; Kivshar, Y. Meta-optics and bound states in the continuum. *Sci. Bull.* **2019**, *64*, 836–842. [CrossRef]
79. Koshelev, K.; Bogdanov, A.; Kivshar, Y. Engineering with Bound States in the Continuum. *Opt. Photonics News* **2020**, *31*, 38–45. [CrossRef]
80. Lee, S.G.; Magnusson, R. Band dynamics of leaky-mode photonic lattices. *Opt. Express* **2019**, *27*, 18180–18189. [CrossRef] [PubMed]
81. Lee, S.G.; Magnusson, R. Band flips and bound-state transitions in leaky-mode photonic lattices. *Phys. Rev. B* **2019**, *99*, 045304. [CrossRef]
82. Ding, Y.; Magnusson, R. Band gaps and leaky-wave effects in resonant photonic-crystal waveguides. *Opt. Express* **2007**, *15*, 680–694. [CrossRef]
83. Koshelev, K.; Lepeshov, S.; Liu, M.; Bogdanov, A.; Kivshar, Y. Asymmetric metasurfaces with high-Q resonances governed by bound states in the continuum. *Phys. Rev. Lett.* **2018**, *121*, 193903. [CrossRef]
84. Kodigala, A.; Lepetit, T.; Gu, Q.; Bahari, B.; Fainman, Y.; Kanté, B. Lasing action from photonic bound states in continuum. *Nature* **2017**, *541*, 196. [CrossRef]
85. Ha, S.T.; Fu, Y.H.; Emani, N.K.; Pan, Z.; Bakker, R.M.; Paniagua-Domínguez, R.; Kuznetsov, A.I. Directional lasing in resonant semiconductor nanoantenna arrays. *Nat. Nanotechnol.* **2018**, *13*, 1042–1047. [CrossRef]
86. Yesilkoy, F.; Arvelo, E.R.; Jahani, Y.; Liu, M.; Tittl, A.; Cevher, V.; Kivshar, Y.; Altug, H. Ultrasensitive hyperspectral imaging and biodetection enabled by dielectric metasurfaces. *Nat. Photonics* **2019**, *13*, 390–396. [CrossRef]
87. Bulgakov, E.N.; Maksimov, D.N. Nonlinear response from optical bound states in the continuum. *Sci. Rep.* **2019**, *9*, 1–8. [CrossRef]
88. Li, G.; Wu, L.; Li, K.F.; Chen, S.; Schlickriede, C.; Xu, Z.; Huang, S.; Li, W.; Liu, Y.; Pun, E.Y.; et al. Nonlinear metasurface for simultaneous control of spin and orbital angular momentum in second harmonic generation. *Nano Lett.* **2017**, *17*, 7974–7979. [CrossRef]
89. Liu, B.; Sain, B.; Reineke, B.; Zhao, R.; Meier, C.; Huang, L.; Jiang, Y.; Zentgraf, T. Nonlinear Wavefront Control by Geometric-Phase Dielectric Metasurfaces: Influence of Mode Field and Rotational Symmetry. *Adv. Opt. Mater.* **2020**, *2020*, 1–11. [CrossRef]

90. Wang, L.; Kruk, S.; Koshelev, K.; Kravchenko, I.; Luther-Davies, B.; Kivshar, Y. Nonlinear wavefront control with all-dielectric metasurfaces. *Nano Lett.* **2018**, *18*, 3978–3984. [CrossRef]
91. Vahala, K. *Optical Microcavities. Advanced Applied Physics*, 1st ed.; World Scientific Pub Co Inc.: Hackensack, NJ, USA, 2004; Volume 5.
92. Blau, G.; Popov, E.; Kajzar, F.; Raimond, A.; Roux, J.F.; Coutaz, J.L. Grating-assisted phase-matched second-harmonic generation from a polymer waveguide. *Opt. Lett.* **1995**, *20*, 1101–1103. [CrossRef] [PubMed]
93. Lin, J.H.; Tseng, C.Y.; Lee, C.T.; Young, J.F.; Kan, H.C.; Hsu, C.C. Strong guided mode resonant local field enhanced visible harmonic generation in an azo-polymer resonant waveguide grating. *Opt. Express* **2014**, *22*, 2790–2797. [CrossRef] [PubMed]
94. Ning, T.; Pietarinen, H.; Hyvärinen, O.; Kumar, R.; Kaplas, T.; Kauranen, M.; Genty, G. Efficient second-harmonic generation in silicon nitride resonant waveguide gratings. *Opt. Lett.* **2012**, *37*, 4269–4271. [CrossRef]
95. Ning, T.; Hyvärinen, O.; Pietarinen, H.; Kaplas, T.; Kauranen, M.; Genty, G. Third-harmonic UV generation in silicon nitride nanostructures. *Opt. Express* **2013**, *21*, 2012–2017. [CrossRef] [PubMed]
96. Tran, T.; Karagodsky, V.; Rao, Y.; Chen, R.; Chase, C.; Chuang, L.C.; Chang-Hasnain, C. Surface-normal second harmonic emission from AlGaAs high-contrast gratings. *Appl. Phys. Lett.* **2013**, *102*, 021102. [CrossRef]
97. Yang, Y.; Wang, W.; Boulesbaa, A.; Kravchenko, I.I.; Briggs, D.P.; Puretzky, A.; Geohegan, D.; Valentine, J. Nonlinear Fano-resonant dielectric metasurfaces. *Nano Lett.* **2015**, *15*, 7388–7393. [CrossRef]
98. Shcherbakov, M.R.; Shorokhov, A.S.; Neshev, D.N.; Hopkins, B.; Staude, I.; Melik-Gaykazyan, E.V.; Ezhov, A.A.; Miroshnichenko, A.E.; Brener, I.; Fedyanin, A.A.; et al. Nonlinear interference and tailorable third-harmonic generation from dielectric oligomers. *ACS Photonics* **2015**, *2*, 578–582. [CrossRef]
99. Shorokhov, A.S.; Melik-Gaykazyan, E.V.; Smirnova, D.A.; Hopkins, B.; Chong, K.E.; Choi, D.Y.; Shcherbakov, M.R.; Miroshnichenko, A.E.; Neshev, D.N.; Fedyanin, A.A.; et al. Multifold enhancement of third-harmonic generation in dielectric nanoparticles driven by magnetic Fano resonances. *Nano Lett.* **2016**, *16*, 4857–4861. [CrossRef]
100. Kruk, S.; Poddubny, A.; Smirnova, D.; Wang, L.; Slobozhanyuk, A.; Shorokhov, A.; Kravchenko, I.; Luther-Davies, B.; Kivshar, Y. Nonlinear light generation in topological nanostructures. *Nat. Nanotechnol.* **2019**, *14*, 126–130. [CrossRef] [PubMed]
101. Kruk, S.; Poddubny, A.; Smirnova, D.; Kravchenko, I.; Luther-Davies, B.; Kivshar, Y. Disorder-Robust Nonlinear Light Generation in Topological Nanostructures. *Proc. CLEO 2019*, *2019*, FW4B-4. [CrossRef]
102. Vabishchevich, P.P.; Liu, S.; Sinclair, M.B.; Keeler, G.A.; Peake, G.M.; Brener, I. Enhanced second-harmonic generation using broken symmetry III–V semiconductor fano metasurfaces. *ACS Photonics* **2018**, *5*, 1685–1690. [CrossRef]
103. Sautter, J.D.; Xu, L.; Miroshnichenko, A.E.; Lysevych, M.; Volkovskaya, I.; Smirnova, D.A.; Camacho-Morales, R.; Zangeneh Kamali, K.; Karouta, F.; Vora, K.; et al. Tailoring second-harmonic emission from (111)-GaAs nanoantennas. *Nano Lett.* **2019**, *19*, 3905–3911. [CrossRef]
104. Deka, J.; Jha, K.K.; Menon, S.; Krishna, A.L.; Biswas, R.; Raghunathan, V. Microscopic study of resonant third-harmonic generation from amorphous silicon nanodisk arrays. *Opt. Lett.* **2018**, *43*, 5242–5245. [CrossRef] [PubMed]
105. Sun, T.; Yang, W.; Chang-Hasnain, C. Surface-normal coupled four-wave mixing in a high contrast gratings resonator. *Opt. Express* **2015**, *23*, 29565–29572. [CrossRef]
106. Miroshnichenko, A.E.; Evlyukhin, A.B.; Yu, Y.F.; Bakker, R.M.; Chipouline, A.; Kuznetsov, A.I.; Luk'yanchuk, B.; Chichkov, B.N.; Kivshar, Y.S. Nonradiating anapole modes in dielectric nanoparticles. *Nat. Commun.* **2015**, *6*, 1–8. [CrossRef]
107. Grinblat, G.; Li, Y.; Nielsen, M.P.; Oulton, R.F.; Maier, S.A. Degenerate four-wave mixing in a multiresonant germanium nanodisk. *ACS Photonics* **2017**, *4*, 2144–2149. [CrossRef]
108. Colom, R.; Xu, L.; Marini, L.; Bedu, F.; Ozerov, I.; Begou, T.; Lumeau, J.; Miroshnishenko, A.E.; Neshev, D.; Kuhlmey, B.T.; et al. Enhanced four-wave mixing in doubly resonant silicon nanoresonators. *ACS Photonics* **2019**, *6*, 1295–12301. [CrossRef]
109. Biswas, R.; Deka, J.; Jha, K.K.; Praveen, A.V.; Lal Krishna, A.S.; Menon, S.; Raghunathan, V. Resonant four-wave mixing microscopy on silicon-on-insulator based zero-contrast gratings. *OSA Contin.* **2019**, *2*, 2864–2874. [CrossRef]
110. Shcherbakov, M.R.; Vabishchevich, P.P.; Shorokhov, A.S.; Chong, K.E.; Choi, D.Y.; Staude, I.; Miroshnichenko, A.E.; Neshev, D.N.; Fedyanin, A.A.; Kivshar, Y.S. Ultrafast all-optical switching with magnetic resonances in nonlinear dielectric nanostructures. *Nano Lett.* **2015**, *15*, 6985–6990. [CrossRef] [PubMed]

111. Murphy, C.D.; Trines, R.M.; Vieira, J.; Reitsma, A.J.; Bingham, R.; Collier, J.L.; Divall, E.J.; Foster, P.S.; Hooker, C.J.; Langley, A.J.; et al. Evidence of photon acceleration by laser wake fields. *Phys. Plasmas* **2006**, *13*, 033108. [CrossRef]
112. Shcherbakov, M.R.; Werner, K.; Fan, Z.; Talisa, N.; Chowdhury, E.; Shvets, G. Photon acceleration and tunable broadband harmonics generation in nonlinear time-dependent metasurfaces. *Nat. Commun.* **2019**, *10*, 1345. [CrossRef]
113. Liu, S.; Vabishchevich, P.P.; Vaskin, A.; Reno, J.L.; Keeler, G.A.; Sinclair, M.B.; Staude, I.; Brener, I. An all-dielectric metasurface as a broadband optical frequency mixer. *Nat. Commun.* **2018**, *9*, 2507. [CrossRef]
114. Meyn, J.P.; Huber, G. Intracavity frequency doubling of a continuous-wave, diode-laser-pumped neodymium lanthanum scandium borate laser. *Opt. Lett.* **1994**, *19*, 1436–1438. [CrossRef]
115. Gomez-Diaz, J.S.; Tymchenko, M.; Lee, J.; Belkin, M.A.; Alù, A. Nonlinear processes in multi-quantum-well plasmonic metasurfaces: Electromagnetic response, saturation effects, limits, and potentials. *Phys. Rev. B* **2015**, *92*, 125429. [CrossRef]
116. Nookala, N.; Lee, J.; Tymchenko, M.; Gomez-Diaz, J.S.; Demmerle, F.; Boehm, G.; Lai, K.; Shvets, G.; Amann, M.C.; Alu, A.; et al. Ultrathin gradient nonlinear metasurface with a giant nonlinear response. *Optica* **2016**, *3*, 283–288. [CrossRef]
117. Sarma, R.; de Ceglia, D.; Nookala, N.; Vincenti, M.A.; Campione, S.; Wolf, O.; Scalora, M.; Sinclair, M.B.; Belkin, M.A.; Brener, I. Broadband and Efficient Second-Harmonic Generation from a Hybrid Dielectric Metasurface/Semiconductor Quantum-Well Structure. *ACS Photonics* **2019**, *6*, 1458–1465. [CrossRef]
118. Autere, A.; Jussila, H.; Dai, Y.; Wang, Y.; Lipsanen, H.; Sun, Z. Nonlinear optics with 2D layered materials. *Adv. Mater.* **2018**, *30*, 1705963. [CrossRef] [PubMed]
119. Yuan, Q.; Fang, L.; Fang, H.; Li, J.; Wang, T.; Jie, W.; Zhao, J.; Gan, X. Second harmonic and sum-frequency generations from a silicon metasurface integrated with a two-dimensional material. *ACS Photonics* **2019**, *6*, 2252–2259. [CrossRef]
120. Maguid, E.; Chriki, R.; Yannai, M.; Tradonsky, C.; Kleiner, V.; Hasman, E.; Friesem, A.A.; Davidson, D. Intra-cavity metasurfaces for topologically spin-controlled laser modes. *Proc. CLEO* **2018**, *2018*, SF1J.6. [CrossRef]
121. Capasso, F. The future and promise of flat optics: A personal perspective. *Nanophotonics* **2018**, *7*, 953–957. [CrossRef]
122. Emmer, H.; Chen, C.T.; Saive, R.; Friedrich, D.; Horie, Y.; Arbabi, A.; Faraon, A.; Atwater, H.A. Fabrication of single crystal gallium phosphide thin films on glass. *Sci. Rep.* **2017**, *7*, 1–6. [CrossRef] [PubMed]
123. Bloembergen, N. Nonlinear optics of polymers: Fundamentals and applications. *J. Nonlinear Opt. Phys. Mater.* **1996**, *5*, 1–8. [CrossRef]
124. Sun, M.; Xu, X.; Sun, X.W.; Valuckas, V.; Zheng, Y.; Paniagua-Domínguez, R.; Kuznetsov, A.I. Efficient visible light modulation based on electrically tunable all dielectric metasurfaces embedded in thin-layer nematic liquid crystals. *Sci. Rep.* **2019**, *9*, 1–9. [CrossRef]
125. Kim, Y.; Wu, P.C.; Sokhoyan, R.; Mauser, K.; Glaudell, R.; Kafaie Shirmanesh, G.; Atwater, H.A. Phase modulation with electrically tunable vanadium dioxide phase-change metasurfaces. *Nano Lett.* **2019**, *19*, 3961–3968. [CrossRef]
126. Seyler, K.L.; Schaibley, J.R.; Gong, P.; Rivera, P.; Jones, A.M.; Wu, S.; Yan, J.; Mandrus, D.G.; Yao, W.; Xu, X. Electrical control of second-harmonic generation in a WSe 2 monolayer transistor. *Nat. Nanotechnol.* **2015**, *10*, 407–411. [CrossRef]

© 2020 by the authors. Licensee MDPI, Basel, Switzerland. This article is an open access article distributed under the terms and conditions of the Creative Commons Attribution (CC BY) license (http://creativecommons.org/licenses/by/4.0/).

Article

Dynamical Control of Broadband Coherent Absorption in ENZ Films

Vincenzo Bruno [1], Stefano Vezzoli [2], Clayton DeVault [3,4], Thomas Roger [5], Marcello Ferrera [5], Alexandra Boltasseva [3,4], Vladimir M. Shalaev [3,4] and Daniele Faccio [1,*]

1. School of Physics and Astronomy, University of Glasgow, Glasgow G12 8QQ, UK; v.bruno.1@research.gla.ac.uk
2. The Blackett Laboratory, Department of Physics, Imperial College London, London SW7 2BW, UK; s.vezzoli@imperial.ac.uk
3. Purdue Quantum Science and Engineering Institute, Purdue University, 1205 West State Street, West Lafayette, IN 47907, USA; devaultx@gmail.com (C.D.); aeb@purdue.edu (A.B.); shalaev@purdue.edu (V.M.S.)
4. School of Electrical and Computer Engineering and Birck Nanotechnology Center, Purdue University, 1205 West State Street, West Lafayette, IN 47907, USA
5. Institute of Photonics and Quantum Sciences, Heriot-Watt University, Edinburgh EH14 4AS, UK; thomas.roger@gmail.com (T.R.); M.Ferrera@hw.ac.uk (M.F.)
* Correspondence: daniele.faccio@glasgow.ac.uk

Received: 12 December 2019; Accepted: 16 January 2020; Published: 20 January 2020

Abstract: Interferometric effects between two counter-propagating beams incident on an optical system can lead to a coherent modulation of the absorption of the total electromagnetic radiation with 100% efficiency even in deeply subwavelength structures. Coherent perfect absorption (CPA) rises from a resonant solution of the scattering matrix and often requires engineered optical properties. For instance, thin film CPA benefits from complex nanostructures with suitable resonance, albeit at a loss of operational bandwidth. In this work, we theoretically and experimentally demonstrate a broadband CPA based on light-with-light modulation in epsilon-near-zero (ENZ) subwavelength films. We show that unpatterned ENZ films with different thicknesses exhibit broadband CPA with a near-unity maximum value located at the ENZ wavelength. By using Kerr optical nonlinearities, we dynamically tune the visibility and peak wavelength of the total energy modulation. Our results based on homogeneous thick ENZ media open a route towards on-chip devices that require efficient light absorption and dynamical tunability.

Keywords: transparent conductive oxide; coherent perfect absorption; epsilon-near-zero media; light-with-light modulation; refractive index change

1. Introduction

Coherent perfect absorption (CPA) was first proposed as a time-reversed version of a laser [1]. Similar to a laser cavity, CPA occurs when light is resonant at specific wavelengths in a high-Q Fabry–Perot optical resonator. However, for CPA, the active gain material is replaced with a moderately lossy medium. Because the system's single-pass losses are typically low, perfect absorption for a given input intensity is extremely sensitive to the Q-factor and resonance wavelength [2,3].

An alternative scheme utilizes deeply subwavelength and highly absorbing materials [4–6]. Here, two counter-propagating coherent beams interfere at the film's surface and create a standing wave. Absorption in the film is then modulated by changing the relative phase of the two beams, or equivalently by scanning the film along the nodes (peak transmittance) and antinodes (peak absorption) of the interference pattern. This approach has been demonstrated in the ultrafast [7] and quantum regime [8–11], as well as in integrated photonic systems [12–16]. While a resonant cavity is not required, single-pass absorption should be 50% to achieve perfect absorption [17,18]. This is difficult to obtain in conventional dielectrics (too little losses) or metals (too high reflectivity). To circumnavigate this challenge, metasurfaces—nanostructured subwavelength films—with ideal absorptive optical properties have been used to achieve CPA [4,19]. Ideal absorption can be achieved in extremely subwavelength films over a broad range of wavelengths, making metasurface-based CPA advantageous over bulk cavity structures and compatible with integrated photonic platforms [3].

While metasurfaces and other engineered structures can exhibit CPA over large wavelength ranges, the necessary nanofabrication can be a limitation for practical CPA applications. Thin films of epsilon-near-zero (ENZ) materials, such as transparent conductive oxides (TCOs) like aluminum-doped zinc oxide (AZO) or indium tin oxide (ITO), have been proposed as a particularly suitable platform for broadband CPA [20]. ENZ materials exhibit a real part of the dielectric permittivity which crosses zero for wavelengths of practical interest in the near-infrared or visible regions [21,22]. Due to the continuity of the transverse component of the electric field at the interface, the electric field within the ENZ material can be very large and can lead to perfect absorption (PA) when illuminated at a critical angle of incidence [23,24]. In the limit of deeply subwavelength ENZ film, PA is provided by critical coupling the incident light to a fast wave propagating along the ENZ layer [24]. The proposed systems for ENZ PA are multilayer structures where the ENZ thin layer is sandwiched between two dielectrics or a dielectric and a metal structure [25]. At the critical angle where CPA happens (this is often referred to as directional PA), the loss follows a linear relationship with the ENZ film thickness which implies that CPA can occur in an arbitrarily thin ENZ film (with arbitrary small single-pass absorption) [26]. For instance, PA has been demonstrated for films of ITO film thickness as low as $0.02\ \lambda_0$ (free-space wavelength) and with only 5% single-pass absorption [27]. Electrical tuning of one port directional PA have also been shown in plasmonic strip cavity based on a ENZ thin layer, with a modulation in reflectance of the 15% [28]. Finally, broadband coherent modulation of directional PA in ENZ deeply subwavelength film have been proved by using ITO multylayer structures sandwiched between two ZnSe prisms [20]. The control of nonlinear processes by two port illumination was also theorized for deeply sub-wavelength ENZ slab [29]. Applications of CPA in deeply subwavelength ENZ films could be found in photovoltaic energy conversion or devices such as bolometers which require large absorption with small masses. However, other applications, such as in nonlinear or quantum optics, may benefit from thicker films where the efficiency of the nonlinear process and the parametric gain generally scale with thickness.

Here, we study CPA in films of TCOs near their ENZ wavelength where the film's refractive index exhibits large anomalous dispersion and a near-zero refractive index. Such films can be treated as deeply subwavelength because the effective wavelength will increase drastically for wavelength approaching the ENZ wavelength. We theoretically and experimentally explore the role of this transition region in order to achieve CPA in homogeneous AZO optically thick films and then show how this can be controlled with intense optical pump fields. It was recently shown that the combination of low refractive index and the high damage threshold of these materials allows TCOs to exhibit large and ultrafast Kerr-type optical nonlinearities in the ENZ region [30–36] and behave as efficient time-varying medium [37,38].

We perform CPA experiments in a Sagnac-like interferometer where two counter propagating light pulses are incident normal to the sample. We achieve coherent control of absorption in AZO films with different thicknesses. For all samples the total energy modulation exhibits a maximum value near the ENZ wavelength. We then demonstrate dynamical control of CPA using its strong intensity-dependent refractive index change. Our demonstration of broadband and tunable CPA in homogeneous ENZ films is relevant for practical nanoscale optical-switches and modulators where alternative nano-pattered metasurfaces would suffer from low switching efficiencies and detriments of nanofabrication processes.

2. Theoretical Investigation

Our optical system consists in two counter-propagating continuous waves (CW) impinging on a homogeneous ENZ film, E_{in}^A and E_{in}^B, respectively, at normal incidence (Figure 1a). From the transfer matrix method (TMM), we calculate the electric field at the two outputs of our symmetric system, E_{out}^C and E_{out}^D respectively. By changing the relative phase ϕ between the two counter-propagating beams, we simulate the scenario in which the film is shifted along the propagation direction and calculate the intensity of the two outputs, C and D (Figure 1b). By summing the intensity at the two outputs ($I_{Tot} = I_C + I_D$), we define the modulation visibility of the total energy as

$$V_{tot} = \frac{(I_{Tot}^{max} - I_{Tot}^{min})}{(I_{Tot}^{max} + I_{Tot}^{min})} \quad (1)$$

where I_{Tot}^{max} and I_{Tot}^{min} are the maximum and minimum of the total output energy of the system. In principle, for CPA to occur in a thin film, the transmission and reflection coefficients from both sides of the film should be equal ($|r| = |t|$) with a phase difference of $\varphi_{rt} = 0$ or π in order to achieve 100% light absorption. In this situation, the value of the total visibility is 1.

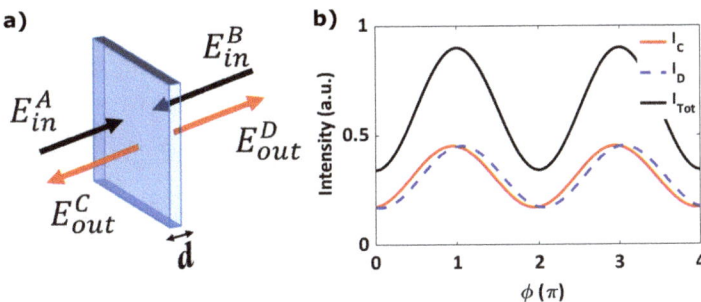

Figure 1. (a) Bi-directional coherent perfect absorption (CPA) scheme. (b) Intensity of the two output beams, C and D, and its sum as we scan the sample position in the propagation direction. This is equivalent to changing the relative phase between the two input fields ϕ.

We consider three different cases by fixing the zero crossing of the real part of the dielectric permittivity at $\lambda_{ENZ} \approx 1350$ nm, but vary the dispersion across the ENZ region as shown in Figure 2a–c where we plot the refractive index profiles (real, n, and imaginary, k, parts) for three different cases studied. These are calculated from a Drude model

$$\epsilon = \epsilon_\infty - \frac{\omega_p^2}{(\omega^2 + i\gamma\omega)} \quad (2)$$

where ϵ_∞ is the high frequency permittivity, ω_p is the plasma frequency and γ is the damping coefficient. It has been shown that this model correctly reproduces the ENZ refractive index for a variety of materials, as ITO and AZO [33,39–41]. In our case, we use $\epsilon_\infty = 3.18$ and $\omega_p = 2.4745 \times 10^{15}$ rad/s. We vary the Drude model damping coefficient, thus increasing losses and reducing the dispersion gradient in n, from (a) to (c) $\gamma = 1.0073 \times 10^{13} \rightarrow 2.4745 \times 10^{14}$ rad/s. Figure 2d–f show the visibility V_{tot} of the total energy as a function of the ENZ film's thickness for the three cases shown in Figure 2a–c. In Figure 2d (γ_a), we do not observe coherent modulation of the total energy for thickness below 1000 nm. Due to the high transmission of the thin film, the interference between the reflected and transmitted field is weak. For thicker films, r and t become more similar and stronger interference is observed. The TMM model predicts visibility with a maximum value close to one that is pinned to a wavelength slightly shorter than λ_{ENZ}. When we increase the optical losses of the ENZ slab, Figure 2e,f, the peak of the visibility becomes broader, exhibiting multiple resonances as the thickness increases but all with maximum absorption at a wavelength just below λ_{ENZ}. These results show that the system exhibits broadband coherent modulation of the energy with a maximum value close to one just below λ_{ENZ}, independently of the thickness and of the single-pass absorption. We associate this maximum to a Fabry–Perot (FP) like resonance due to interference effects in the Air/AZO/glass system. The fact that the FP resonance is 'locked' before the ENZ wavelength irrespectively of the thickness is due to the ENZ condition [42,43].

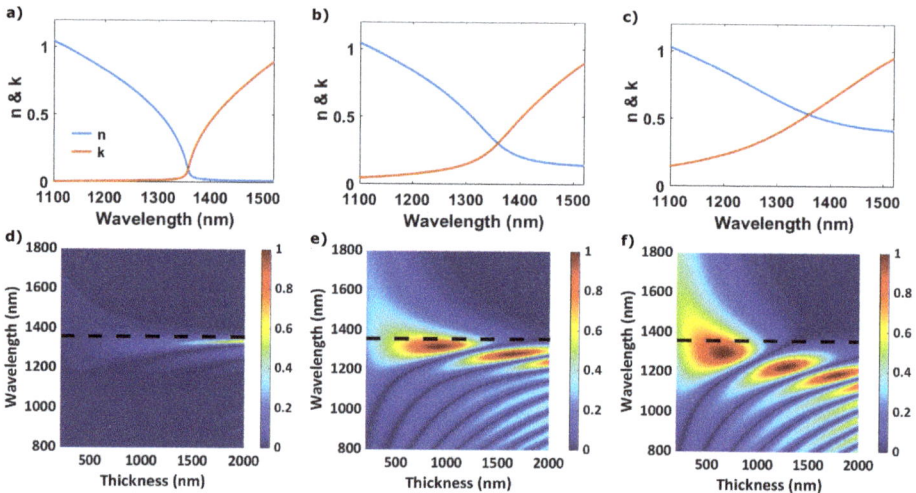

Figure 2. (a–c) Real and imaginary part of the refractive index of the three cases with $\lambda_{ENZ} \approx 1350$ nm. (d–f) Normalized visibility of the total energy as a function of the wavelength for different thicknesses. The dashed red line indicates the λ_{ENZ}. For the dispersion we use $\epsilon_\infty = 3.18$ and $\omega_p = 2.4745 \times 10^{15}$ rad/s. For the damping constant we use $\gamma_a = 1.0073 \times 10^{13}$, $\gamma_b = 0.8053 \times 10^{14}$ and $\gamma_c = 2.3614 \times 10^{14}$ rad/s.

In the ideal case without losses, the first resonance of an FP cavity is reached when the $2nd = \lambda_0$. Due to the strong gradient of the n before the ENZ region, the λ_0 at which the first resonance occurs will not scale linearly with d, but it will be locked in this spectral range with strong dispersion. Moreover, in a lossy dielectric medium, r and t become complex and their phases depend on the value of both n and k of the lossy medium. Here the first resonant order for the FP cavity is reached when $2nd = \lambda_0 (1 - \alpha/\pi)$, where α is the phase of the transmission coefficient [18]. Combining the strong dispersion of n due to the ENZ condition and the value of k, almost perfect modulation of absorption is expected at wavelength just below λ_{ENZ} even for subwavelength thickness (Figure 2f).

3. Coherent Absorption and Its Dynamical Control

We experimentally investigated the behaviour of CPA in ENZ films using AZO films illuminated by two counter-propagating laser beams in a Sagnac-like interferometer configuration. Figure 3a shows a schematic of the set-up. Laser pulses (105 fs FWHM duration, repetition rate 100 Hz) are generated by an Optical Parametric Amplifier (TOPAS) in a tunable range between 1120 nm and 1500 nm. The input power is controlled through a half wave plate and a polarizing beam splitter, which also fixes the input p-polarization (horizontal in the lab frame). The beam is split by a non-polarizing beam splitter into two beams A and B with equal energy and then recombined onto the sample at normal incidence. The AZO film (deposited on a 1 mm thick glass slide) is facing the beam A, whereas the beam B is incident on the substrate side. The two beams are focused down to 50 µm by using a pair of 125 mm lenses. By moving the sample with a piezo-electric stage, interferograms are generated at the output C and D and measured with photodiodes. We used two beam splitters to extract the light from the interferometer and send it to the photodiodes. A representative example of these interferograms are shown in Figure 3b. In order to calculate the energy visibility in the pulsed case we proceed in the same way as for the CW case, i.e. we evaluate the central portion (where the pulse intensity is maximum) of the interferogram and extrapolate the average values for maximum and minimum of the intensity.

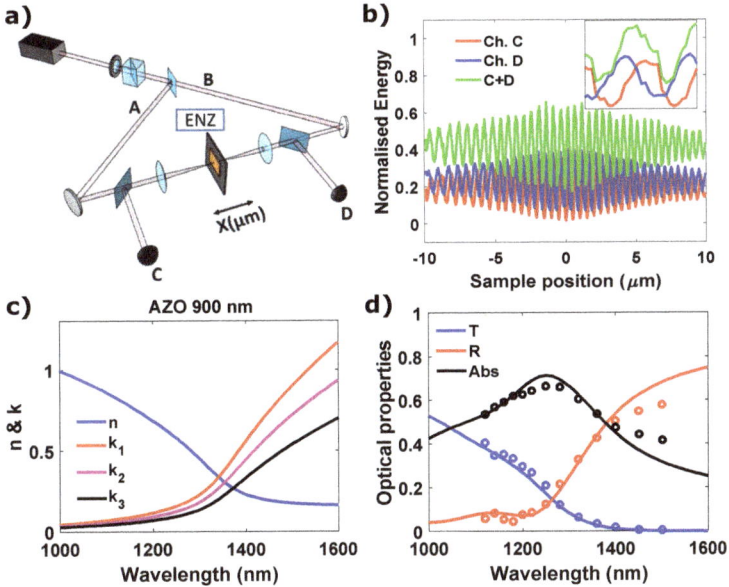

Figure 3. (a) Schematics of the Sagnac interferometer. (b) An example of measurement for $\lambda_0 = 1280$ nm, assuming energy equal to 1 at the interferometer input. The total modulation of the energy (or absorption) is given by the sum of C and D (green curve). The inset shows a zoom of the interferogram. (c) ellipsometer measurement of the index of refraction of AZO 900 nm thick film, (d) experimental (dots) and TMM simulation (solid line) of R, T and abs for the same sample.

We investigated two AZO samples of similar optical properties, i.e., n and k, in the ENZ region, but with thicknesses of 500 and 900 nm (Figure 3c). The real part of the dielectric permittivity crosses zero around 1340 nm for the 900-nm-thick film, which corresponds to where the real and imaginary part of the refractive index are equal ($n_{900} = k_{900} = 0.34$). For the 500-nm-thick sample the λ_{ENZ} is redshifted by 30 nm ($n_{500} = k_{500} = 0.52$ at the λ_{ENZ}) due to small differences in the material deposition. In the spectral range under analysis, n of the AZO 900-nm-thick film passes from close to 1 around 1100 nm to less than 0.2 for longer wavelengths. Since $\lambda_{eff} = \lambda_0/n$ inside the medium, the effective length ($L_{eff} = L/\lambda_{eff}$) of our sample is 0.8 λ_0 at 1050 nm, 0.21 λ_0 at the ($\lambda_{ENZ} = 1350$ nm, and becomes optically deeply subwavelength around 1500 nm (0.1 λ_0).

We also deposited three 900-nm-thick AZO films on a glass substrate three samples with similar n (about 20% difference), but different value of k at the crossing point ($k_1 = 0.34$, $k_2 = 0.30$ and $k_3 = 0.27$ for the 900 nm thick film). All the samples exhibit similar optical properties with an absorption close to 60% across the ENZ region (Figure 3d).

We first perform a CPA experiment for the bare glass substrate. In this case the energy modulation is almost zero for all the spectral range of interest. In Figure 4 we report the measured normalized total energy modulation visibility (red circles), together with the values predicted by the TMM (solid lines) for the AZO film. All the samples show the same trend independently from the thickness. In the case of high optical losses, for the different thicknesses the visibility is almost zero in the region where the index of refraction is close to one, whereas it increases and reaches a maximum value up to the 60% just before λ_{ENZ}. As we decrease the value of k, the trend of the visibility remains the same for all the samples, but its maximum value across the transition region decreases. Overall, the experimental results confirm the predictions that CPA can be observed in ENZ films over a broad bandwidth with thicknesses larger than the conventional subwavelength designs. The bandwidth of ~100 nm is comparable with CPA in deeply subwavelength ENZ single layer (~150 nm [20]) or white-light cavity (~100 nm [3]), whereas it is larger than metasurfaces (~40 nm [4]).

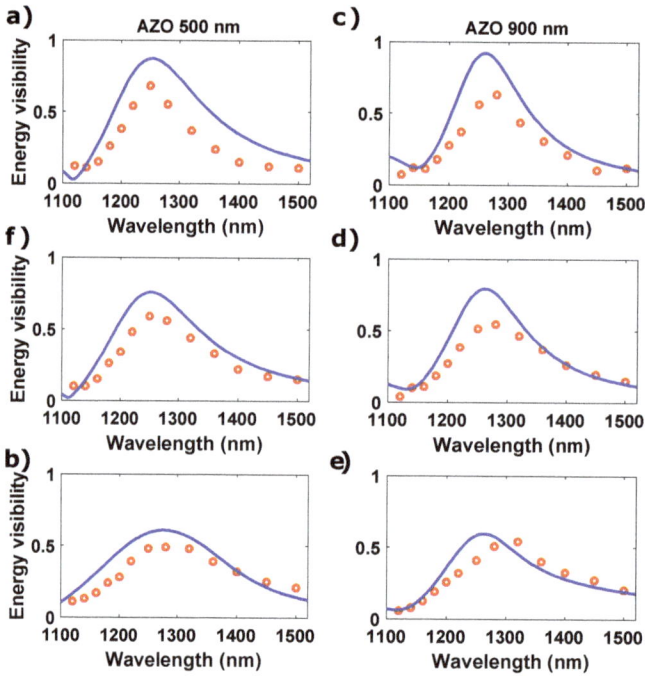

Figure 4. Experimental (circles) and transfer matrix method (TMM) simulation (solid line) of normalized visibility of the total energy for aluminum-doped zinc oxide (AZO) 500 nm and 900 nm with different values of k. (**a,b**) High losses k_1, (**c,d**) middle losses k_2 and (**e,f**) low losses k_3. For the TMM simulation we suppose $\Delta\lambda \sim 60$ nm.

We finally investigate nonlinear coherent absorption in ENZ films based on modification of the film refractive index through the nonlinear Kerr coefficient. Previously it has been demonstrated that the ENZ condition leads to the enhancement of third order nonlinearities in terms of nonlinear refractive index change for thin film of AZO [32]. This is based on the observation that when the permittivity is close to zero, any nonlinear change Δn, proportional to $\chi^{(3)}/n$, is enhanced due to the n tending to low values. In Ref. [32] a refractive index change of 400% was reported for an AZO film optically pumped with 1.3 TW/cm^2 without showing damage of the sample or saturation of the optical Kerr effect. In the same work, at $\lambda = 1310$ nm a nonlinear susceptibility of $Re[\chi^{(3)}] \sim 4.73 \times 10^{-20}$ V^2/m^2 and $Im[\chi^{(3)}] \sim 0.57 \times 10^{-20}$ V^2/m^2 was extrapolated. We therefore illuminated the AZO film in the Sagnac interferometer with two high intensity pulses at normal incidence and same wavelength. The intensities on each side are 0.8 and 0.6 TW/cm^2, respectively. By increasing the intensities from the linear regime to these maximum values, we observe that the CPA visibility passes from 68% of the linear case to 35% (Figure 5a,b). The peak of the normalized visibility also redshifts and becomes broader for both the samples, with a nearly 50 nm-shift for the 500 nm sample. Following the recent works in TCOs, this can be explained by the fact that the dielectric permittivity, and so the optical constants including λ_{ENZ}, exhibit a redshift when it is optically pumped across the ENZ wavelength [31,44]. The redshift of the λ_{ENZ} is also associated to a positive Δn and to a negative Δk [30,32]. Due to the decreasing of k, the visibility drops, as we observed for the linear case. While, the shift of the visibility peak is related to the shift of λ_{ENZ} in the same direction, and therefore to the shift of the strong dispersion which the material exhibits at wavelength shorter than the zero-crossing frequency. In Figure 5c,d we plot the experimental results together with TMM simulations. The TMM simulations are obtained considering a $\sim$60 nm shift of ω_p and a decreasing of γ ($0.15 \times 10^{15} \rightarrow 0.09 \times 10^{15}$). This correspond to a $\Delta \lambda_{ENZ} \sim 60$ nm and to a $k_{500} = 0.38$ and $k_{900} = 0.24$. These results show that enhanced nonlinearities in ENZ materials can be used to add a degree of freedom to tune the efficiency and the bandwidth of coherent absorption.

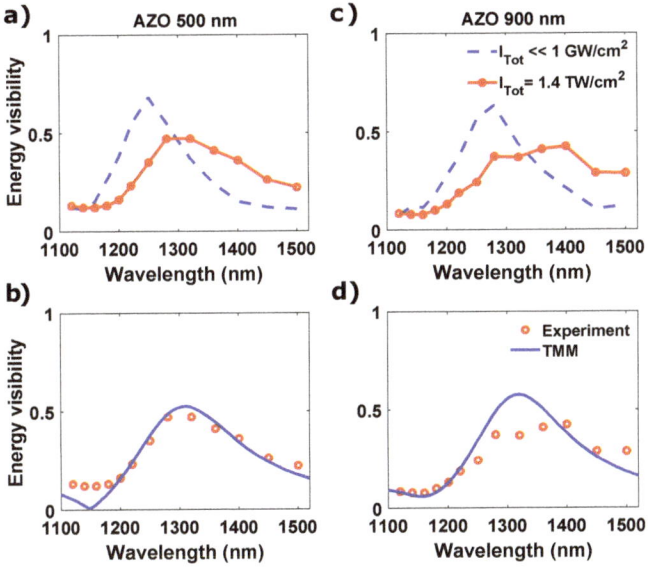

Figure 5. (**a**,**b**) Normalized visibility of the total energy for both samples, 500 nm (**a**) and 900 nm (**b**). The dashed blue curve represents the linear characterization, while the circles is the nonlinear CPA with high beam intensity. (**c**,**d**) Experimental (circles) and TMM simulation (solid line) of normalized visibility of the total energy for AZO 500 nm and 900 nm for the nonlinear CPA.

4. Conclusions

We theoretically and experimentally demonstrate coherent control of absorption in films of ENZ material. We show that it is possible to achieve a coherent absorption-mediated interferometric effect with a maximum of its effect locked just below the λ_{ENZ} wavelength. Due the strong dispersion at wavelengths below the crossing point, it can be tuned to any wavelength shorter than λ_{ENZ} by varying film thickness and the optical losses. The 60% total visibility achieved in the AZO film could be improved using a CW and collimated beam in order to achieve CPA. By using AZO's strong intensity-dependent nonlinearities, we also showed that it is possible to dynamically tune the visibility of the total energy by simply increasing the intensity of the incoming beam. The possibility to add a degree of freedom for the coherent control of the absorption in ENZ media by using intensity-dependent refractive index proposes a route towards technologies [45] such as optical data processing or devices that require efficient light absorption and dynamical tunability. All the data supporting this manuscript are available at http://researchdata.gla.ac.uk/939/.

Author Contributions: Investigation and data curation V.B. and S.V.; Sample fabrication C.D.; code for analysis S.V. and V.B., and code for measurement acquisition T.R.; Formal analysis and writing-original draft V.B., S.V. and D.F.; supervision D.F., V.M.S., A.B. and M.F.; all the author have contributed equally to the writing—review. All authors have read and agreed to the published version of the manuscript.

Funding: D. F. acknowledges financial support from EPSRC (UK, Grant No. EP/M009122/1). The Purdue team acknowledges support by the U.S. Office of Naval Research (optical characterization), U.S. Department of Energy, Office of Basic Energy Sciences, Division of Materials Sciences and Engineering under Award DE-SC0017717 (sample preparation), and Air Force Office of Scientific Research (AFOSR) award FA9550-18-1-0002.

Conflicts of Interest: The authors declare no conflict of interest.

References

1. Chong, Y.; Ge, L.; Cao, H.; Stone, A.D. Coherent perfect absorbers: Time-reversed lasers. *Phys. Rev. Lett.* **2010**, *105*, 053901. [CrossRef] [PubMed]
2. Wan, W.; Chong, Y.; Ge, L.; Noh, H.; Stone, A.D.; Cao, H. Time-reversed lasing and interferometric control of absorption. *Science* **2011**, *331*, 889–892. [CrossRef] [PubMed]
3. Baranov, D.G.; Krasnok, A.; Shegai, T.; Alù, A.; Chong, Y. Coherent perfect absorbers: Linear control of light-with-light. *Nat. Rev. Mater.* **2017**, *2*, 17064. [CrossRef]
4. Zhang, J.; MacDonald, K.F.; Zheludev, N.I. Controlling light-with-light without nonlinearity. *Light Sci. Appl.* **2012**, *1*, e18. [CrossRef]
5. Kats, M.A.; Blanchard, R.; Genevet, P.; Capasso, F. Nanometre optical coatings based on strong interference effects in highly absorbing media. *Nat. Mater.* **2013**, *12*, 20. [CrossRef] [PubMed]
6. Rao, S.M.; Heitz, J.J.; Roger, T.; Westerberg, N.; Faccio, D. Coherent control of light interaction with graphene. *Opt. Lett.* **2014**, *39*, 5345–5347. [CrossRef]
7. Fang, X.; Lun Tseng, M.; Ou, J.Y.; MacDonald, K.F.; Ping Tsai, D.; Zheludev, N.I. Ultrafast all-optical switching via coherent modulation of metamaterial absorption. *Appl. Phys. Lett.* **2014**, *104*, 141102. [CrossRef]
8. Roger, T.; Restuccia, S.; Lyons, A.; Giovannini, D.; Romero, J.; Jeffers, J.; Padgett, M.; Faccio, D. Coherent absorption of N00N states. *Phys. Rev. Lett.* **2016**, *117*, 023601. [CrossRef]
9. Altuzarra, C.; Vezzoli, S.; Valente, J.; Gao, W.; Soci, C.; Faccio, D.; Couteau, C. Coherent perfect absorption in metamaterials with entangled photons. *ACS Photon.* **2017**, *4*, 2124–2128. [CrossRef]
10. Roger, T.; Vezzoli, S.; Bolduc, E.; Valente, J.; Heitz, J.J.; Jeffers, J.; Soci, C.; Leach, J.; Couteau, C.; Zheludev, N.I.; et al. Coherent perfect absorption in deeply subwavelength films in the single-photon regime. *Nat. Commun.* **2015**, *6*, 7031. [CrossRef]
11. Lyons, A.; Oren, D.; Roger, T.; Savinov, V.; Valente, J.; Vezzoli, S.; Zheludev, N.I.; Segev, M.; Faccio, D. Coherent metamaterial absorption of two-photon states with 40% efficiency. *Phys. Rev. A* **2019**, *99*, 011801. [CrossRef]
12. Wei, P.; Croënne, C.; Tak Chu, S.; Li, J. Symmetrical and anti-symmetrical coherent perfect absorption for acoustic waves. *Appl. Phys. Lett.* **2014**, *104*, 121902. [CrossRef]
13. Akhlaghi, M.K.; Schelew, E.; Young, J.F. Waveguide integrated superconducting single-photon detectors implemented as near-perfect absorbers of coherent radiation. *Nat. Commun.* **2015**, *6*, 8233. [CrossRef] [PubMed]
14. Bruck, R.; Muskens, O.L. Plasmonic nanoantennas as integrated coherent perfect absorbers on SOI waveguides for modulators and all-optical switches. *Opt. Express* **2013**, *21*, 27652–27661. [CrossRef]
15. Xomalis, A.; Demirtzioglou, I.; Plum, E.; Jung, Y.; Nalla, V.; Lacava, C.; MacDonald, K.F.; Petropoulos, P.; Richardson, D.J.; Zheludev, N.I. Fibre-optic metadevice for all-optical signal modulation based on coherent absorption. *Nat. Commun.* **2018**, *9*, 182. [CrossRef]
16. Papaioannou, M.; Plum, E.; Valente, J.; Rogers, E.T.; Zheludev, N.I. All-optical multichannel logic based on coherent perfect absorption in a plasmonic metamaterial. *APL Photon.* **2016**, *1*, 090801. [CrossRef]
17. Dutta-Gupta, S.; Martin, O.J.; Gupta, S.D.; Agarwal, G. Controllable coherent perfect absorption in a composite film. *Opt. Express* **2012**, *20*, 1330–1336. [CrossRef]
18. Villinger, M.L.; Bayat, M.; Pye, L.N.; Abouraddy, A.F. Analytical model for coherent perfect absorption in one-dimensional photonic structures. *Opt. Lett.* **2015**, *40*, 5550–5553. [CrossRef]
19. Zhao, B.; Zhao, J.; Zhang, Z. Enhancement of near-infrared absorption in graphene with metal gratings. *Appl. Phys. Lett.* **2014**, *105*, 031905. [CrossRef]

20. Kim, T.Y.; Badsha, M.A.; Yoon, J.; Lee, S.Y.; Jun, Y.C.; Hwangbo, C.K. General strategy for broadband coherent perfect absorption and multi-wavelength all-optical switching based on epsilon-near-zero multilayer films. *Sci. Rep.* **2016**, *6*, 22941. [CrossRef]
21. Naik, G.V.; Liu, J.; Kildishev, A.V.; Shalaev, V.M.; Boltasseva, A. Demonstration of Al: ZnO as a plasmonic component for near-infrared metamaterials. *Proc. Natl. Acad. Sci. USA* **2012**, *109*, 8834–8838. [CrossRef] [PubMed]
22. Naik, G.V.; Kim, J.; Boltasseva, A. Oxides and nitrides as alternative plasmonic materials in the optical range. *Opt. Mater. Express* **2011**, *1*, 1090–1099. [CrossRef]
23. Yoon, J.; Zhou, M.; Badsha, M.A.; Kim, T.Y.; Jun, Y.C.; Hwangbo, C.K. Broadband epsilon-near-zero perfect absorption in the near-infrared. *Sci. Rep.* **2015**, *5*, 12788. [CrossRef] [PubMed]
24. Badsha, M.A.; Jun, Y.C.; Hwangbo, C.K. Admittance matching analysis of perfect absorption in unpatterned thin films. *Opt. Commun.* **2014**, *332*, 206–213. [CrossRef]
25. Jin, Y.; Xiao, S.; Mortensen, N.A.; He, S. Arbitrarily thin metamaterial structure for perfect absorption and giant magnification. *Opt. Express* **2011**, *19*, 11114–11119. [CrossRef]
26. Feng, S.; Halterman, K. Coherent perfect absorption in epsilon-near-zero metamaterials. *Phys. Rev. B* **2012**, *86*, 165103. [CrossRef]
27. Luk, T.S.; Campione, S.; Kim, I.; Feng, S.; Jun, Y.C.; Liu, S.; Wright, J.B.; Brener, I.; Catrysse, P.B.; Fan, S.; et al. Directional perfect absorption using deep subwavelength low-permittivity films. *Phys. Rev. B* **2014**, *90*, 085411. [CrossRef]
28. Park, J.; Kang, J.H.; Liu, X.; Brongersma, M.L. Electrically tunable epsilon-near-zero (ENZ) metafilm absorbers. *Sci. Rep.* **2015**, *5*, 15754. [CrossRef]
29. Vincenti, M. Non-collinear counter-propagating beams in epsilon-near-zero films: Enhancement and inhibition of nonlinear optical processes. *J. Opt.* **2017**, *19*, 124015. [CrossRef]
30. Carnemolla, E.G.; Caspani, L.; DeVault, C.; Clerici, M.; Vezzoli, S.; Bruno, V.; Shalaev, V.M.; Faccio, D.; Boltasseva, A.; Ferrera, M. Degenerate optical nonlinear enhancement in epsilon-near-zero transparent conducting oxides. *Opt. Mater. Express* **2018**, *8*, 3392–3400. [CrossRef]
31. Clerici, M.; Kinsey, N.; DeVault, C.; Kim, J.; Carnemolla, E.G.; Caspani, L.; Shaltout, A.; Faccio, D.; Shalaev, V.; Boltasseva, A.; et al. Controlling hybrid nonlinearities in transparent conducting oxides via two-colour excitation. *Nat. Commun.* **2017**, *8*, 15829. [CrossRef] [PubMed]
32. Caspani, L.; Kaipurath, R.; Clerici, M.; Ferrera, M.; Roger, T.; Kim, J.; Kinsey, N.; Pietrzyk, M.; Di Falco, A.; Shalaev, V.M.; et al. Enhanced nonlinear refractive index in ε-near-zero materials. *Phys. Rev. Lett.* **2016**, *116*, 233901. [CrossRef] [PubMed]
33. Alam, M.Z.; De Leon, I.; Boyd, R.W. Large optical nonlinearity of indium tin oxide in its epsilon-near-zero region. *Science* **2016**, *352*, 795–797. [CrossRef] [PubMed]
34. Kim, J.; Carnemolla, E.G.; DeVault, C.; Shaltout, A.M.; Faccio, D.; Shalaev, V.M.; Kildishev, A.V.; Ferrera, M.; Boltasseva, A. Dynamic control of nanocavities with tunable metal oxides. *Nano Lett.* **2018**, *18*, 740–746. [CrossRef]
35. Reshef, O.; De Leon, I.; Alam, M.Z.; Boyd, R.W. Nonlinear optical effects in epsilon-near-zero media. *Nat. Rev. Mater.* **2019**, *4*, 535–551. [CrossRef]
36. Kinsey, N.; DeVault, C.; Boltasseva, A.; Shalaev, V.M. Near-zero-index materials for photonics. *Nat. Rev. Mater.* **2019**, *4*, 742–760. [CrossRef]
37. Vezzoli, S.; Bruno, V.; DeVault, C.; Roger, T.; Shalaev, V.M.; Boltasseva, A.; Ferrera, M.; Clerici, M.; Dubietis, A.; Faccio, D. Optical Time Reversal from Time-Dependent Epsilon-Near-Zero Media. *Phys. Rev. Lett.* **2018**, *120*, 043902, doi:10.1103/PhysRevLett.120.043902. [CrossRef]
38. Bruno, V.; DeVault, C.; Vezzoli, S.; Kudyshev, Z.; Huq, T.; Mignuzzi, S.; Jacassi, A.; Saha, S.; Shah, Y.D.; Maier, S.A.; et al. Negative refraction in time-varying, strongly-coupled plasmonic antenna-ENZ systems. *arXiv* **2019**, arXiv:1908.03908.
39. Campione, S.; Brener, I.; Marquier, F. Theory of epsilon-near-zero modes in ultrathin films. *Phys. Rev. B* **2015**, *91*, 121408. [CrossRef]
40. Cleary, J.W.; Smith, E.M.; Leedy, K.D.; Grzybowski, G.; Guo, J. Optical and electrical properties of ultra-thin indium tin oxide nanofilms on silicon for infrared photonics. *Opt. Mater. Express* **2018**, *8*, 1231–1245. [CrossRef]

41. Feigenbaum, E.; Diest, K.; Atwater, H.A. Unity-order index change in transparent conducting oxides at visible frequencies. *Nano Lett.* **2010**, *10*, 2111–2116. [CrossRef] [PubMed]
42. Kim, J.; Dutta, A.; Naik, G.V.; Giles, A.J.; Bezares, F.J.; Ellis, C.T.; Tischler, J.G.; Mahmoud, A.M.; Caglayan, H.; Glembocki, O.J.; et al. Role of epsilon-near-zero substrates in the optical response of plasmonic antennas. *Optica* **2016**, *3*, 339–346. [CrossRef]
43. DeVault, C.T.; Zenin, V.A.; Pors, A.; Chaudhuri, K.; Kim, J.; Boltasseva, A.; Shalaev, V.M.; Bozhevolnyi, S.I. Suppression of near-field coupling in plasmonic antennas on epsilon-near-zero substrates. *Optica* **2018**, *5*, 1557–1563. [CrossRef]
44. Khurgin, J.B.; Clerici, M.; Bruno, V.; Caspani, L.; DeVault, C.; Kim, J.; Shaltout, A.; Boltasseva, A.; Shalaev, V.M.; Ferrera, M.; et al. Adiabatic frequency conversion in epsilon near zero materials: It is all about group velocity. *arXiv* **2019**, arXiv:1906.04849.
45. Kinsey, N.; Ferrera, M.; Shalaev, V.; Boltasseva, A. Examining nanophotonics for integrated hybrid systems: A review of plasmonic interconnects and modulators using traditional and alternative materials. *JOSA B* **2015**, *32*, 121–142. [CrossRef]

 © 2020 by the authors. Licensee MDPI, Basel, Switzerland. This article is an open access article distributed under the terms and conditions of the Creative Commons Attribution (CC BY) license (http://creativecommons.org/licenses/by/4.0/).

Article

Intersubband Optical Nonlinearity of GeSn Quantum Dots under Vertical Electric Field

Mourad Baira [1], Bassem Salem [2], Niyaz Ahamad Madhar [3] and Bouraoui Ilahi [3],*

[1] Micro-Optoelectronic and Nanostructures Laboratory, Faculty of Sciences, University of Monastir, Monastir 5019, Tunisia; mourad.baira@isimm.rnu.tn
[2] Univ. Grenoble Alpes, CNRS, CEA/LETI Minatec, LTM, F-38000 Grenoble, France; bassem.salem@cea.fr
[3] Department of Physics and Astronomy, College of Sciences, King Saud University, Riyadh 11451, Saudi Arabia; nmadhar@ksu.edu.sa
* Correspondence: bilahi@ksu.edu.sa; Tel.: +966-114676393

Received: 24 March 2019; Accepted: 10 April 2019; Published: 12 April 2019

Abstract: The impact of vertical electrical field on the electron related linear and 3rd order nonlinear optical properties are evaluated numerically for pyramidal GeSn quantum dots with different sizes. The electric field induced electron confining potential profile's modification is found to alter the transition energies and the transition dipole moment, particularly for larger dot sizes. These variations strongly influence the intersubband photoabsorption coefficients and changes in the refractive index with an increasing tendency of the 3rd order nonlinear component with increasing both quantum dot (QD) size and applied electric field. The results show that intersubband optical properties of GeSn quantum dots can be successively tuned by external polarization.

Keywords: GeSn; quantum dot; electric field; intersubband nonlinear optics; absorption coefficients; refractive index changes

1. Introduction

Self-assembled quantum dots have received an increasing interest during the past decades owing to their potentiality for novel optoelectronic devices [1,2]. Indeed, the strong carriers' confinement in these nanostructures has encouraged exploring the light emission and detection in the IR [3–6] and THz regime [7–9] using intersubband optical transitions. A particular interest has been devoted to the study of linear and nonlinear QD intersubband optical properties [7,9–17] for their importance in integrated quantum photonic technologies [18]. Despite the achieved progress, efficient light source integrable with Silicon technology has, so far, represented a challenge for Si-photonic integrated circuits [19]. Recent achievement in direct band gap GeSn material has accentuated its suitability towards comparable properties to III-V materials while being compatible with complementary metal-oxide semiconductor (CMOS) technology [20–25]. Accordingly, several reports have already demonstrated the aptness of this material for optoelectronic applications, such as light emitters [25–28] and detectors [29–31]. Furthermore, growing experimental and theoretical research activities have been developed to explore GeSn based low dimensional structures such as quantum dots [32–39]. Indeed, different synthesis roots have been reported including, colloidal QD [33], thermal diffusion [32] and self-organization [34]. Furthermore, high Tin content GeSn QD with direct band gap transition energy has recently been reported [40]. Despite the experimental and theoretical achievement, GeSn QD are still immature and a lot of works have still to be done. Recently, we have reported on the evolution of the intersubband photoabsorption coefficients (AC) and Refractive index changes (RIC) as a function of GeSn dots size and incident radiation intensity [16]. The present work treats the effect of vertical electric field on intersubband related optical properties of pyramidal GeSn QD with different sizes for CMOS compatible nonlinear optical devices.

2. Theoretical Consideration

The self-assembled GeSn QD has been considered to have a pyramidal shape with 1nm thick wetting layer (WL) embedded in Ge matrix which is one of the frequently observed shapes for semiconducting self-assembled QD [41] as illustrated by Figure 1a. Throughout this work, we set the tin composition at 30% and a QD height to base side length's (L) ratio of 1/3 (Figure 1b,c).

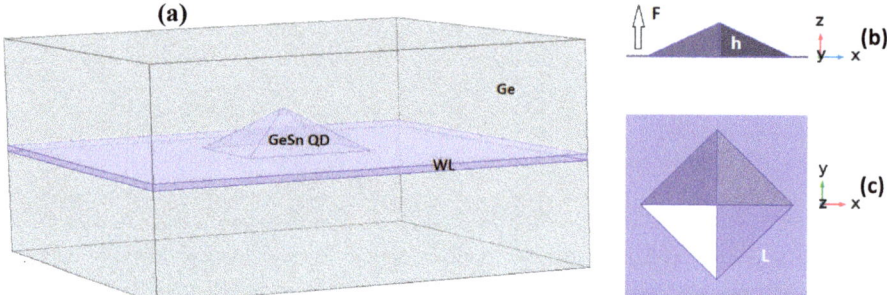

Figure 1. Schematic sketch of the pyramidal shaped self-assembled GeSn QD with 1 nm thick wetting layer (WL): (**a**) 3D projection of the pyramidal QD with wetting layer, (**b**) cross-sectional view (ZX) showing the QD height and the direction of the external electric field, (**c**) plane view (XY).

To evaluate the QD s- and p- like electrons' energy levels and associated wave functions in Γ-valley, single band 3D-Schrodinger equation (Equation (1)) is solved in Cartesian coordinates within the effective mass approximation by finite elements method offered by COMSOL multiphysics software (version 5.0, COMSOL Inc., Stockholm, Sweden) [42] for the strained pyramidal GeSn QD under vertical applied electric field (Figure 1).

$$-\frac{\hbar^2}{2}\nabla\left[\frac{1}{m^*(x,y,z)}\nabla\varnothing(x,y,z)\right] + (V(x,y,z)+eFz)\varnothing(x,y,z) = \epsilon\varnothing(x,y,z) \quad (1)$$

where ϵ, $\varnothing$, V and m^* represent the electron energy level, envelop wave function, potential barrier and effective mass, respectively. F is the external electric field and e the elementary charge. Further details can be found elsewhere [42,43]. The calculation of the transition angular frequency associated dipole moment is mandatory to evaluate the AC and RIC. Indeed, the angular frequency dependent total intersubband optical AC (α) and RIC ($\frac{\delta n}{n_r}$) are given by [10,11]:

$$\alpha(\omega) = \alpha^{(1)}(\omega) + \alpha^{(3)}(\omega, I) \quad (2)$$

$$\frac{\delta n(\omega)}{n_r} = \frac{\delta n^{(1)}(\omega)}{n_r} + \frac{\delta n^{(3)}(\omega)}{n_r} \quad (3)$$

where $\alpha^{(1)}$ and $\frac{\delta n^{(1)}}{n_r}$, denote the linear AC and RIC (Equations (4) and (6)). $\alpha^{(3)}$ and $\frac{\delta n^{(3)}}{n_r}$ represent the 3rd order nonlinear AC and RIC expressed respectively by Equations (5) and (7):

$$\alpha^{(1)}(\omega) = \frac{\omega}{\hbar}\sqrt{\frac{\mu}{\varepsilon_r}}\frac{\sigma|M_{fi}|^2\Gamma}{\left[(\omega_{fi}-\omega)^2+\Gamma^2\right]} \quad (4)$$

$$\alpha^{(3)}(\omega, I) = -\frac{\omega \sigma I}{2\varepsilon_0 n_r c \hbar^3} \sqrt{\frac{\mu}{\varepsilon_r}} \frac{|M_{fi}|^2 \Gamma}{\left[(\omega_{fi}-\omega)^2+\Gamma^2\right]^2}$$
$$\times \left[4|M_{fi}|^2 - \frac{(M_{ff}-M_{ii})^2 \left(3\omega_{fi}^2 - 4\omega_{fi}\omega + (\omega^2-\Gamma^2)\right)}{\omega_{fi}^2+\Gamma^2} \right] \quad (5)$$

$$\frac{\delta n^{(1)}(\omega)}{n_r} = \frac{\sigma |M_{fi}|^2}{2n_r^2 \varepsilon_0 \hbar} \frac{\omega_{fi}-\omega}{\left[(\omega_{fi}-\omega)^2+\Gamma^2\right]} \quad (6)$$

$$\frac{\delta n^{(3)}(\omega, I)}{n_r} = \frac{-\mu \sigma I |M_{fi}|^2}{4n_r^3 \varepsilon_0 \hbar^3 \left[(\omega_{fi}-\omega)^2+\Gamma^2\right]^2}$$
$$\times \left[4(\omega_{fi}-\omega)|M_{fi}|^2 \right.$$
$$\left. - \frac{(M_{ff}-M_{ii})^2 \{(\omega_{fi}-\omega)\times[\omega_{fi}(\omega_{fi}-\omega)-\Gamma^2]-\Gamma^2(2\omega_{fi}-\omega)\}}{\omega_{fi}^2+\Gamma^2} \right] \quad (7)$$

I is the incident in-plane polarized light intensity, σ denotes the electron density (one electron per QD) [12]. $\Gamma = 10$ ps^{-1} is the relaxation rate and n_r the GeSn material's refractive index deduced by linear interpolation [16]. ω_{fi} is the p-to-s transition frequency and $M_{fi} = \langle \varnothing_f | ex | \varnothing_i \rangle$ denotes the corresponding dipole moment for in-plane X polarized incident radiation. The subscript f and i refer to the final and initial states (QD p- and s electron states in this study). The p states are doubly degenerated (identified as px and py). A selection rule making the allowed transition to arise only from px state can be done by considering the incident radiation to be polarized along X direction [7,16,44].

3. Results and Discussion

The calculation of Γ-s and -p electron energy states and associated envelop wave functions allows to evaluate the ω_{fi} and M_{fi} required to study the electric field's impact on intersubband optical properties as a function of the QD size. The transition energy $(\epsilon_p - \epsilon_s)$ is shown Figure 2 as a function of the QD size (pyramid base side) for F = 100 kV/cm, 0 kV/cm and −100kV/cm. The dot size range is delimited to L between 25 nm to 40 nm [38] warranting efficient contribution of Γ-electrons to the intersubband transition energy.

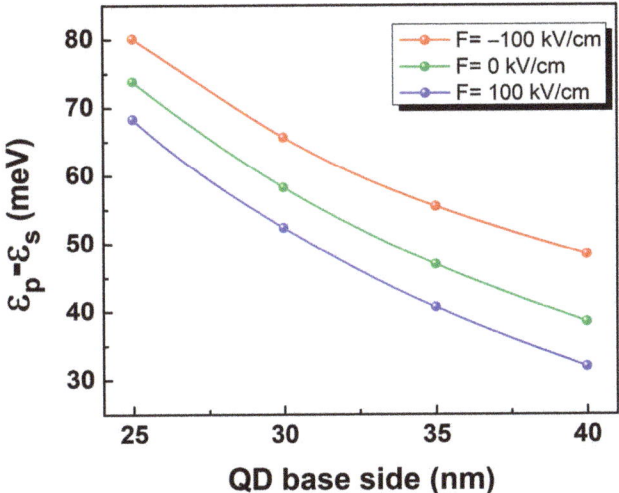

Figure 2. QD size dependent transition energy $(\epsilon_p - \epsilon_s)$ for F = 100kV/cm, 0 kV/cm and −100 kV/cm.

In absence of electric field (F = 0 kV/cm), the intraband transition decreases from 74 meV (L = 25 nm) down to 38 meV (L = 40 nm). Applying positive electric field of 100 kV/cm enhances the transition from 6 meV for the smallest QD size up to 10 meV for the largest one. Meanwhile, the energy spacing between p and s states get rather shrank by approximately 6 meV for an external electric field of −100 kV/cm. This behavior is a direct impact of the electric field driven modification of the electron confining potential's profile. To explain this trend, the electron probability density from s and p states (ZX plane) under an electric field of 100 kV/cm, 0 kV/cm and −100 kV/cm are shown by Figure 3 where a simplified band profile has also been provided for details. Indeed, the electric field has been found to induce a vertical shift of the electron probability density along z-axis. Its maximum gets vertically displaced towards the dot's tip for negative electric field and towards its base for positive one [15]. Indeed, for a QD with base side length of 40 nm and a height of 13.3 nm, the maximum ground state electron probability density is located at z = 4.5 nm for unbiased QD. Under vertical electric field, the maximum is shifted upward by approximately 2.5 nm for F = −100 kV/cm and a downward vertical shift by approximately 2 nm for F = 100 kV/cm. Consequently, in the first case, the potential minimum is created near the dot tip limiting the allowed space for electron confinement (comparable environment to a QD size reduction) enhancing the separation energy between s and p states leading to the observed blueshift (Figure 2). On the other hand, the positive electric field produces a confining potential minimum at the QD base giving rise to a lowering of the confined energy states and consequent reduction of the p-to-s transition energies.

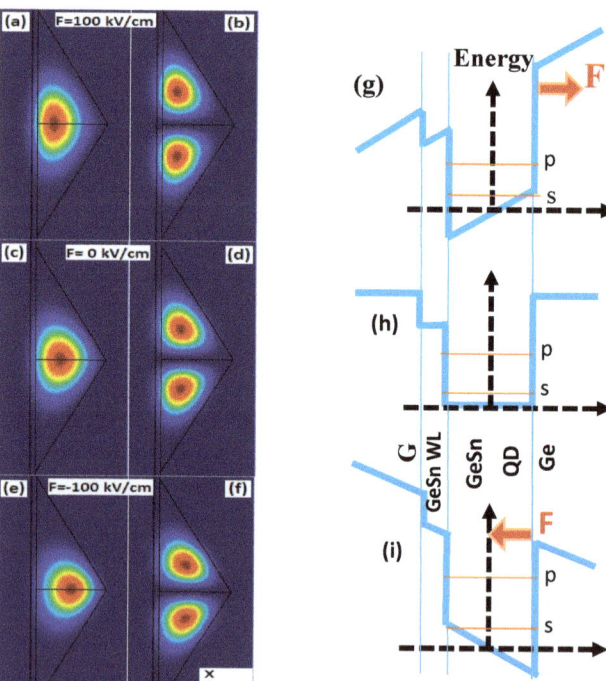

Figure 3. Probability density of s-state (**a**, **c** and **e**), p_x-state (**b**, **d** and **f**) for GeSn QD with L = 40 nm as well as a simple schematic illustration of the Γ-band electron confining profile (**g**, **h** and **i**) respectively for F = 100 kV/cm, 0 kV/cm and −100 kV/cm.

Further information can be gained through studying the evolution of the dipole moment as a function of the dot size and electric field (Figure 4). The transition dipole moment shows an increasing trend with increasing the unbiased QD size. However, it gets progressively enhanced (decreased) with increasing the QD size upon applying 100 kV/cm (−100 kV/cm) electric field. The observed relative variation traduces a high sensitivity of larger QD sizes to the applied electric field. The obtained results show that the QD intersubband optical properties can be successively adjusted by electric polarization allowing tuning not only the intersubband emission energy but also the transition dipole moment without need for QD size variation.

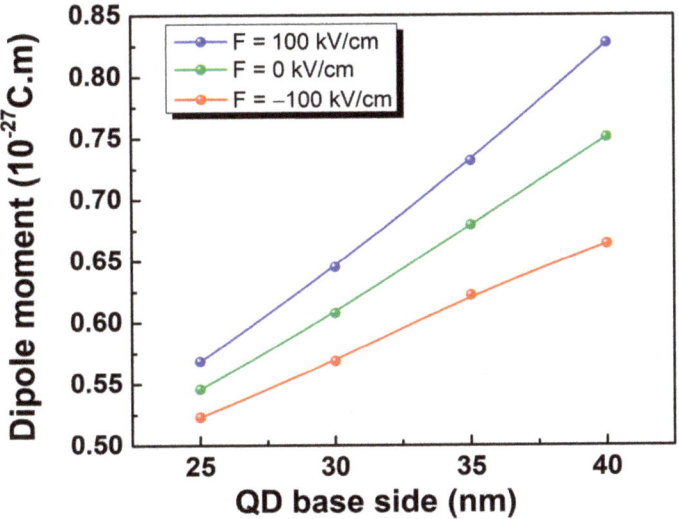

Figure 4. Intersubband dipole moment as a function of the pyramidal QD base side length for different values of the applied electric field.

Accordingly, the impact of the dot size and electric field on the AC, RIC and the corresponding linear and third order nonlinear components are shown by Figure 5, as a function of the incident photon energy, for F = 0 kV/cm, 100 kV/cm and −100 kV/cm. The results are given for the smallest and the largest dot size to illustrate the simultaneous effect of electric field and dot size. For a given applied electric field value, the observed curves shift following the decreased transition energy with the increase of the dot size. Similarly, for a given QD size, and compared to the case where no electric field is applied, the curves get blueshifted for an electric field oriented in the negative Z direction and redshifted in the opposite case following the electric field induced intersubband transition energies shift.

The resonance peak of the linear AC (Figure 5a–c) considerably quenches with increasing the dot size while no noticeable change is shown to occur upon the variation of the applied electric field. In the meantime, the peak's intensity of the third-order nonlinear AC shows an increasing trend in absolute value with increasing the applied electric field for larger QD size. Consequently, the resultant total AC exhibits strong dependence on the applied electric field. When the nonlinear part of the AC becomes comparable in magnitude to the linear one, the effect of bleaching occurs inducing a splitting of the total AC into two peaks. This saturation effect observed for the unbiased larger QD size is smoothed for F = −100 kV/cm and accentuated for F = 100 kV/cm. This behavior is analogous to that perceived upon increasing the QD size and consequent variation of the absorption threshold energy [16].

Furthermore, the linear RIC (Figure 5 d–f) shows an overall increase with increasing the applied electric field with a pronounced sensitivity for larger dot size. Meanwhile, a similar and more accentuated variation is found to occur for the third-order nonlinear RIC affecting the total changes in the refractive index curve. The observed behavior is mainly due to the simultaneous increase of the dipole moment and decrease of the intersubband transition energy.

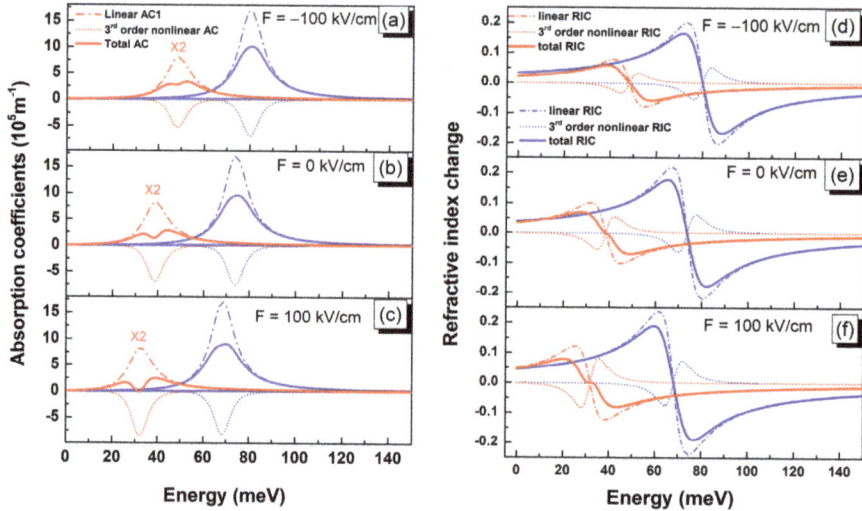

Figure 5. Absorption coefficients (**a**)–(**c**) and Refractive index change (**d**)–(**f**) as a function of the photon energy evaluated for F = −100 kV/cm (**a**) and (**d**), F = 0 kV/cm (**b**) and (**e**) and F = 100 kV/cm with an incident light intensity of 1 MW·cm^{-2}. Linear contribution (dash-dot lines), 3rd order nonlinear component (dotted lines) as well as total AC and RIC (solid lines) for QD base side length: L = 25 nm (blue), L = 40 nm (red). The AC curves for L = 40 nm are multiplied by factor 2 for better visibility.

Our calculations clearly reveal that the intersubband optical nonlinearity can be conveniently tuned by applying an external electric field for a given QD size and incident light intensity. Accordingly, the nonlinear effects can be tuned. This investigation has been conducted on GeSn QD with the available materials parameters remain a subject to experimental validation. Nonetheless, this comprehensive study could also be useful to understand the impact of the applied electric field on the intersubband optical properties of similar QD.

4. Conclusions

We have evaluated the effect of applied electric field on the intersubband optical transition, dipole moment, AC, and RIC for various GeSn QD size. The transition energy and dipole moment are found to be strongly affected by the electric field-induced confining potential profile changes. Larger size QD are found to be more sensitive to the effects of applied electric field. The intersubband-related AC and RIC can be widely tuned by employing external electric field. This comprehensive study could help future realization of CMOS compatible nonlinear optical devices.

Author Contributions: Conceptualization and Formal analysis M.B. and B.I.; Writing-original draft M.B., Investigation M.B. and B.I. Validation B.S. and N.A.M.; Supervision B.I.; Writing-review & editing B.S., N.A.M., B.I.

Funding: This research received no external funding.

Acknowledgments: The authors would like to thank the Deanship of Scientific Research at King Saud University for funding this work through the Research Group No: RG-1436-014.

Conflicts of Interest: The authors declare no conflict of interest.

References

1. Horiuchi, N. Strain-tunable dots. *Nat. Photon.* **2019**, *13*, 72. [CrossRef]
2. Su, X.; Wang, Y.; Zhang, B.; Zhang, H.; Yang, K.; Wang, R.; He, J. Bismuth quantum dots as an optical saturable absorber for a 1.3 μm Q-switched solid-state laser. *Appl. Opt.* **2019**, *58*, 1621–1625. [CrossRef]

3. Phillips, J.; Bhattacharya, P.; Kennerly, S.W.; Beekman, D.W.; Dutta, M. Self-assembled InAs-GaAs quantum-dot intersubband detectors. *IEEE J. Quantum Electron.* **1999**, *35*, 936–943. [CrossRef]
4. Chen, W.; Deng, Z.; Guo, D.; Chen, Y.; Mazur, Y.I.; Maidaniuk, Y.; Benamara, M.; Salamo, G.J.; Liu, H.; Wu, J.; et al. Demonstration of InAs/InGaAs/GaAs quantum dots-in-a-well mid-wave infrared photodetectors grown on silicon substrate. *J. Lightwave Technol.* **2018**, *13*, 2572–2581. [CrossRef]
5. Wu, J.; Jiang, Q.; Chen, S.; Tang, M.; Mazur, Y.I.; Maidaniuk, Y.; Benamara, M.; Semtsiv, M.P.; Masselink, W.T.; Sablon, K.A.; et al. Monolithically integrated InAs/GaAs quantum dot mid-infrared photodetectors on silicon substrates. *ACS Photon.* **2016**, *3*, 749–753. [CrossRef]
6. Zhuo, N.; Zhang, J.C.; Wang, F.J.; Liu, Y.H.; Zhai, S.Q.; Zhao, Y.; Wang, D.B.; Jia, Z.W.; Zhou, Y.H.; Wang, L.J.; et al. Room temperature continuous wave quantum dot cascade laser emitting at 7.2 μm. *Opt. Expr.* **2017**, *25*, 13807–13815. [CrossRef]
7. Sabaeian, M.; Riyahi, M. Truncated pyramidal-shaped InAs/GaAs quantum dots in the presence of a vertical magnetic field: An investigation of THz wave emission and absorption. *Phys. E Low-Dimensional Syst. Nanostruct.* **2017**, *89*, 105–114. [CrossRef]
8. Zibik, E.A.; Grange, T.; Carpenter, B.A.; Porter, N.E.; Ferreira, R.; Bastard, G.; Stehr, D.; Winnerl, S.; Helm, M.; Liu, H.Y.; et al. Long lifetimes of quantum-dot intersublevel transitions in the terahertz range. *Nat. Mater.* **2009**, *8*, 803–807. [CrossRef]
9. Burnett, B.A.; Williams, B.S. Density matrix model for polarons in a terahertz quantum dot cascade laser. *Phys. Rev. B* **2014**, *90*, 15530. [CrossRef]
10. Ünlü, S.; Karabulut, İ.; Şafak, H. Linear and nonlinear intersubband optical absorption coefficients and refractive index changes in a quantum box with finite confining potential. *Phys. E Low-Dimensional Syst. Nanostruct.* **2006**, *33*, 319–324. [CrossRef]
11. Vahdani, M.R.K.; Rezaei, G. Linear and nonlinear optical properties of a hydrogenic donor in lens-shaped quantum dots. *Phys. Lett. A* **2009**, *373*, 3079–3084. [CrossRef]
12. Şahin, M. Third-order nonlinear optical properties of a one- and two-electron spherical quantum dot with and without a hydrogenic impurity. *J. App. Phys.* **2009**, *106*, 063710. [CrossRef]
13. Karabulut, İ.; Baskoutas, S. Linear and nonlinear optical absorption coefficients and refractive index changes in spherical quantum dots: Effects of impurities, electric field, size, and optical intensity. *J. Appl. Phys.* **2008**, *103*, 073512. [CrossRef]
14. Sabaeian, M.; Khaledi-Nasab, A. Size-dependent intersubband optical properties of dome-shaped InAs/GaAs quantum dots with wetting layer. *Appl. Opt.* **2012**, *51*, 4176–4185. [CrossRef]
15. Sabaeian, M.; Shahzadeh, M.; Farbod, M. Electric field-induced nonlinearity enhancement in strained semi-spheroid-shaped quantum dots coupled to wetting layer. *AIP Adv.* **2014**, *4*, 127105. [CrossRef]
16. Baira, M.; Salem, B.; Madhar, N.A.; Ilahi, B. Linear and nonlinear intersubband optical properties of direct band gap GeSn quantum dots. *Nanomaterials* **2019**, *9*, 124. [CrossRef]
17. Tian, S.-C.; Lu, H.-Y.; Zhang, H.; Wang, L.-J.; Shu, S.-L.; Zhang, X.; Hou, G.-Y.; Wang, Z.-Y.; Tong, C.-Z.; Wang, L.-J. Enhancing third- and fifth-order nonlinearity via tunneling in multiple quantum dots. *Nanomaterials* **2019**, *9*, 423. [CrossRef]
18. Androvitsaneas, P.; Young, A.B.; Lennon, J.M.; Schneider, C.; Maier, S.; Hinchliff, J.J.; Atkinson, G.S.; Harbord, E.; Kamp, M.; Hofling, S.; et al. Efficient quantum photonic phase shift in a low Q-factor regime. *ACS Photon.* **2019**, *6*, 429–435. [CrossRef]
19. Marris-Morini, D.; Vakarin, V.; Ramirez, J.M.; Liu, Q.; Ballabio, A.; Frigerio, J.; Montesinos, M.; Alonso-Ramos, C.; Le Roux, X.; Serna, S.; et al. Germanium-based integrated photonics from near- to mid-infrared applications. *Nanophotonics* **2018**, *7*, 1781–1793. [CrossRef]
20. He, G.; Atwater, H. Interband transitions in Sn_xGe_{1-x} alloys. *Phys. Rev. Lett.* **1997**, *79*, 1937–1940. [CrossRef]
21. Chen, R.; Lin, H.; Huo, Y.; Hitzman, C.; Kamins, T.I.; Harris, J.S. Increased photoluminescence of strain-reduced, high-Sn composition $Ge_{1-x}Sn_x$ alloys grown by molecular beam epitaxy. *Appl. Phys. Lett.* **2011**, *99*, 181125. [CrossRef]
22. Jiang, L.; Gallagher, J.D.; Senaratne, C.L.; Aoki, T.; Mathews, J.; Kouvetakis, J.; Menéndez, J. Compositional dependence of the direct and indirect band gaps in $Ge_{1-y}Sn_y$ alloys from room temperature photoluminescence: implications for the indirect to direct gap crossover in intrinsic and n-type materials. *Semicond. Sci. Technol.* **2014**, *29*, 115028. [CrossRef]

23. Toko, K.; Oya, N.; Saitoh, N.; Yoshizawa, N.; Suemasu, T. 70 °C synthesis of high-Sn content (25%) GeSn on insulator by Sn-induced crystallization of amorphous Ge. *Appl. Phys. Lett.* **2015**, *106*, 082109. [CrossRef]
24. Taoka, N.; Capellini, G.; Schlykow, V.; Montanari, M.; Zaumseil, P.; Nakatsuka, O.; Zaima, S.; Schroeder, T. Electrical and optical properties improvement of GeSn layers formed at high temperature under well-controlled Sn migration. *Mater. Sci. Semiconduct. Proc.* **2017**, *57*, 48–53. [CrossRef]
25. Chang, C.; Chang, T.W.; Li, H.; Cheng, H.H.; Soref, R.; Sun, G.; Hendrickson, J.R. Room-temperature 2-µm GeSn PIN homojunction light-emitting diode for inplane coupling to group-IV waveguides. *Appl. Phys. Lett.* **2017**, *111*, 141105. [CrossRef]
26. Wirths, S.; Geiger, R.; Von Den Driesch, N.; Mussler, G.; Stoica, T.; Mantl, S.; Ikonic, Z.; Luysberg, M.; Chiussi, S.; Hartmann, J.M.; et al. Lasing in direct-bandgap GeSn alloy grown on Si. *Nat. Photon.* **2015**, *9*, 88–92. [CrossRef]
27. Dou, W.; Zhou, Y.; Margetis, J.; Ghetmiri, S.A.; Al-Kabi, S.; Du, W.; Liu, J.; Sun, G.; Soref, R.A.; Tolle, J. Optically pumped lasing at 3 µm from compositionally graded GeSn with tin up to 22.3%. *Opt. Lett.* **2018**, *43*, 4558–4561. [CrossRef]
28. Stange, D.; Wirths, S.; Geiger, R.; Schulte-Braucks, C.; Marzban, B.; von den Driesch, N.; Mussler, G.; Zabel, T.; Stoica, T.; Hartmann, J.M.; et al. Optically pumped GeSn microdisk lasers on Si. *ACS Photon.* **2016**, *3*, 1279–1285. [CrossRef]
29. Huang, B.J.; Lin, J.H.; Cheng, H.H.; Chang, G.E. GeSn resonant-cavity-enhanced photodetectors on silicon-on-insulator platforms. *Opt. Lett.* **2018**, *43*, 1215–1218. [CrossRef]
30. Pandey, A.K.; Basu, R.; Kumar, H.; Chang, G.E. Comprehensive analysis and optimal design of Ge/GeSn/Ge PNP infrared heterojunction phototransistors. *IEEE J. Electron Devices Soc.* **2019**, *7*, 118–126. [CrossRef]
31. Abdel-Rahman, M.; Alduraibi, M.; Hezam, M.; Ilahi, B. Sputter deposited GeSn alloy: A candidate material for temperature sensing layers in uncooled microbolometers. *Infrared Phys. Technol.* **2019**, *97*, 376–380. [CrossRef]
32. Seifner, M.S.; Hernandez, S.; Bernardi, J.; Romano-Rodriguez, A.; Barth, S. Pushing the composition limit of anisotropic $Ge_{1-x}Sn_x$ nanostructures and determination of their thermal stability. *Chem. Mater.* **2017**, *29*, 9802–9813. [CrossRef]
33. Esteves, R.J.A.; Ho, M.Q.; Arachchige, I.U. Nanocrystalline group IV alloy semiconductors: Synthesis and characterization of $Ge_{1-x}Sn_x$ quantum dots for tunable bandgaps. *Chem. Mater.* **2015**, *27*, 1559–1568. [CrossRef]
34. Lozovoy, K.A.; Kokhanenko, A.P.; Voitsekhovskii, A.V. Critical thickness of transition from 2D to 3D growth and peculiarities of quantum dots formation in $Ge_xSi_{1-x}/Sn/Si$ and $Ge_{1-y}Sn_y/Si$ systems. *Surf. Sci.* **2018**, *669*, 45–49. [CrossRef]
35. Nakamura, Y.; Masada, A.; Ichikawa, M. Quantum-confinement effect in individual $Ge_{1-x}Sn_x$ quantum dots on Si(111) substrates covered with ultrathin SiO_2 films using scanning tunneling spectroscopy. *Appl. Phys. Lett.* **2007**, *91*, 013109. [CrossRef]
36. Moontragoon, P.; Vukmirović, N.; Ikonić, Z.; Harrison, P. Electronic structure and optical properties of Sn and SnGe quantum dots. *J. Appl. Phys.* **2008**, *103*, 103712. [CrossRef]
37. Ilahi, B. Design of direct band gap type I GeSn/Ge quantum dots for mid-IR light emitters on Si substrate. *Phys. Status Solidi RRL* **2017**, *11*, 1700047. [CrossRef]
38. Baira, M.; Salem, B.; Madhar, N.A.; Ilahi, B. Tuning direct bandgap GeSn/Ge quantum dots' interband and intraband useful emission wavelength: towards CMOS compatible infrared optical devices. *Superlattice. Microstruct.* **2018**, *117*, 31–35. [CrossRef]
39. Ilahi, B.; Al-Saigh, R.; Salem, B. Impact of the wetting layer thickness on the emission wavelength of direct band gap GeSn/Ge quantum dots. *Mater. Res. Express* **2017**, *4*, 075026. [CrossRef]
40. Zhang, L.; Hong, H.; Li, C.; Chen, S.; Huang, W.; Wang, J.; Wang, H. High-Sn fraction GeSn quantum dots for Si-based light source at 1.55 µm. *Appl. Phys. Express* **2019**. [CrossRef]
41. Berbezier, I.; Ronda, A.; Portavoce, A. SiGe nanostructures: new insights into growth processes. *J. Phys. Condens. Matter* **2002**, *14*, 8283. [CrossRef]
42. Melnik, R.V.N.; Willatzen, M. Bandstructures of conical quantum dots with wetting layers. *Nanotechnology* **2014**, *15*, 1. [CrossRef]

43. Souaf, M.; Baira, M.; Nasr, O.; Alouane, M.; Maaref, H.; Sfaxi, L.; Ilahi, B. Investigation of the InAs/GaAs quantum dots' size: dependence on the strain reducing layer's position. *Materials* **2015**, *8*, 4699–4709. [CrossRef]
44. Narvaez, G.A.; Zunger, A. Calculation of conduction-to-conduction and valence-to-valence transitions between bound states in InGaAs/GaAs quantum dots. *Phys. Rev. B* **2007**, *75*, 085306. [CrossRef]

© 2019 by the authors. Licensee MDPI, Basel, Switzerland. This article is an open access article distributed under the terms and conditions of the Creative Commons Attribution (CC BY) license (http://creativecommons.org/licenses/by/4.0/).

Article

Circular Dichroism in the Second Harmonic Field Evidenced by Asymmetric Au Coated GaAs Nanowires

Alessandro Belardini [1,*], Grigore Leahu [1], Emilija Petronijevic [1], Teemu Hakkarainen [2], Eero Koivusalo [2], Marcelo Rizzo Piton [2], Soile Talmila [2], Mircea Guina [2] and Concita Sibilia [1]

1. SBAI Department, Sapienza University of Rome, 00161 Rome, Italy; grigore.leahu@uniroma1.it (G.L.); emilija.petronijevic@uniroma1.it (E.P.); concita.sibilia@uniroma1.it (C.S.)
2. Optoelectronics Research Centre, Tampere University, 33720 Tampere, Finland; teemu.hakkarainen@tuni.fi (T.H.); eero.koivusalo@tuni.fi (E.K.); marcelo.rizzopiton@tut.fi (M.R.P.); Soile.Talmila@tuni.fi (S.T.); mircea.guina@tuni.fi (M.G.)
* Correspondence: alessandro.belardini@uniroma1.it

Received: 22 January 2020; Accepted: 20 February 2020; Published: 23 February 2020

Abstract: Optical circular dichroism (CD) is an important phenomenon in nanophotonics, that addresses top level applications such as circular polarized photon generation in optics, enantiomeric recognition in biophotonics and so on. Chiral nanostructures can lead to high CD, but the fabrication process usually requires a large effort, and extrinsic chiral samples can be produced by simpler techniques. Glancing angle deposition of gold on GaAs nanowires can (NWs) induces a symmetry breaking that leads to an optical CD response that mimics chiral behavior. The GaAs NWs have been fabricated by a self-catalyzed, bottom-up approach, leading to large surfaces and high-quality samples at a relatively low cost. Here, we investigate the second harmonic generation circular dichroism (SHG-CD) signal on GaAs nanowires partially covered with Au. SHG is a nonlinear process of even order, and thus extremely sensitive to symmetry breaking. Therefore, the visibility of the signal is very high when the fabricated samples present resonances at first and second harmonic frequencies (i.e., 800 and 400 nm, in our case).

Keywords: extrinsic chirality; second harmonic generation; GaAs nanowires; plasmonic coating

1. Introduction

Three to five compounds have been utilized efficiently in photonic applications [1] as a result of their direct band gaps. Among these, the GaN semiconductor is important for its high transparency and good nonlinear properties in the visible range thanks to its high energy gap (3.4 eV) [2]. GaAs, for which the bandgap (1.42 eV) lies in the infrared, only recently have been utilized in the visible range. Indeed, a new category of applications has exploited their very high refractive index (around 4 in the visible range) to guide light in an effective way in nanostructures like nanowires (NWs) by using leaky waves [3,4], leading to different applications as emitters or even as laser sources [5].

By breaking the symmetry of the nanostructure–light interaction, it is possible to observe a circular dichroism (CD) due to the so-called extrinsic chirality or pseudo chirality [6–8]. Chirality is the lack of mirror symmetry [9], and can be probed using photoacoustic techniques that are sensitive to the differential absorptions of opposite-handed light [10–12] or by techniques sensitive to symmetry breaking such as second harmonic generation–circular dichroism (SHG-CD) [13,14]. In the case of extrinsic chirality, the high sensitivity of SHG is related to the fact that SHG can only occur in systems with broken inversion symmetry, enabling background-free measurements and leading to higher CD responses with respect other measurement systems [15]. The chiroptical responses of nanostructures have recently generated interest because most biomolecules are chiral, and their enantiomeric discrimination is relevant to industries such as pharmacology, agrochemicals and biotechnology, as well as for circularly polarized light emissions for communication and quantum optics applications [16,17]. Moreover, organic chiral molecules are used in field-effect transistor devices to detect or enhance the detection of circularly polarized light [18,19], while chiral oligothiophene thin films have shown interesting chiroptical properties that are useful to optoelectronic devices for imaging [20,21]. Thus, there is an evident need for a deep study of chiroptical effects at a nanoscale.

Recently, we observed that GaAs nanowires (NWs) offered interesting waveguiding properties even for energies above the bandgap, thanks to the high refractive index of GaAs (in particular at 800 and 400 nm) [3]. We further verified, using photoacoustic spectroscopy, that when such GaAs NWs were partially covered in gold they exhibited strong extrinsic chirality due to the breaking of the symmetry induced by the asymmetric metal coating [11,12]. We also numerically investigated near-field chiral effects in high-refractive-index nanowires with [22] and without [23] an asymmetric plasmonic layer.

Here, we present SHG-CD measurements of gold coated GaAs NWs, confirming the strong presence of extrinsic chirality and leading to potential applications in chiral light emissions and manipulation.

2. Materials and Methods

The structures under examination are nanowires of GaAs with a hexagonal cross section. They have a core of GaAs surrounded by a thin shell of AlGaAs to passivate the GaAs surface, around which there is a thin supershell of GaAs in order to prevent the oxidation of Al, as described in the scheme in Figure 1a. The geometric parameters of the four samples that were fabricated are depicted in Table 1 (the NW length L, the overall diameter D, AlGaAs shell thickness t_{AlGaAs}, and GaAs supershell thickness t_{GaAs}).

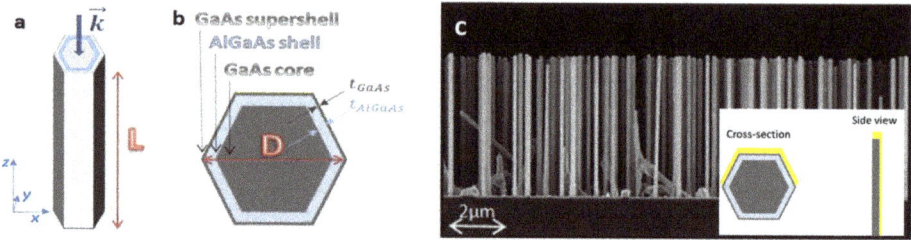

Figure 1. (a) Scheme of the NWs from [3]; (b) cross-section from [3]; (c) SEM image of Sample A (side view). The inset shows the scheme of the gold coating layer.

Table 1. Fabrication parameters of the nanowire (NW) samples. Data from [3].

Sample	L (nm)	D (nm)	t_{AlGaAs} (nm)	t_{GaAs} (nm)
A	4750 ± 34	138 ± 5	3.5	0.7
B	5190 ± 64	151 ± 5	8.6	1.7
C	4600 ± 52	165 ± 6	11.7	5.8
D	4690 ± 47	197 ± 9	27.7	5.5

The NWs were grown using molecular beam epitaxy on p-Si(111) wafers with lithography-free Si/SiOx patterns for defining the nucleation sites, as described in [24]. The lengths of the wires were about 5 microns while the diameters were in the 140–200 nm range (details in Table 1).

The as-fabricated NWs (before the gold coating) presented clear resonant modes in absorption [3] at 800 nm (close to the band edge), with the exception of Sample D (Figure 2a), and a second-order resonance at 400 nm, thus matching the first and second harmonic frequency of a standard Ti:–sapphire laser. This is due to the evidently larger overall diameter that red-shifts the modes into the transparent region in Sample D.

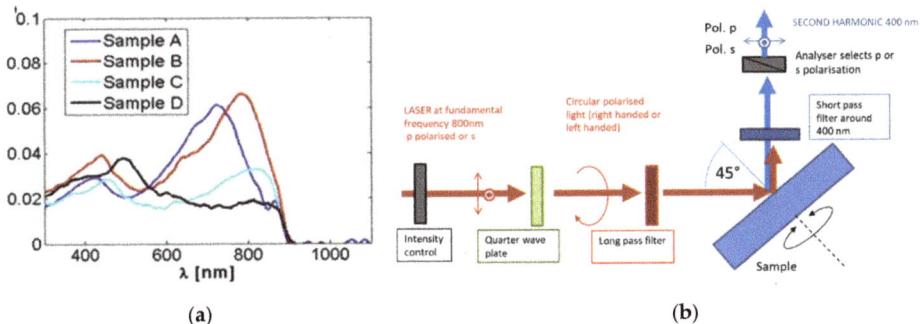

Figure 2. (a) Absorption spectra of the four samples without Au. Figure from [3]; (b) scheme of the second harmonic generation setup.

Half of each sample was asymmetrically coated with gold by glancing evaporation, as explained in more detail in [10,24]. The average Au thickness deposited on the sidewalls was around 15 nm, and the glanced evaporation resulted in the Au presence only on three out of six sidewalls, as described in the inset of Figure 1c. The absorption spectra of the samples coated with gold were similar to the ones without gold, except for a slight broadening of the resonant features, as shown in [10]. The Au-free NWs were used as reference samples for the optical measurements.

The samples were then measured by a SHG-CD setup shown in Figure 2b.

At 800 nm, a linearly polarized Ti–sapphire fs laser with a pulse duration of 100 fs and a repetition rate of 80 MHz was used on the sample at an incidence angle of 45°. The average power was attenuated with a chopper below 1 GW/cm^2 to avoid sample damage or multiphoton processes. A quarter waveplate was used to obtain either left circularly (LCP) or right circularly (RCP) polarized light. A long-pass filter removed any spurious signals at the second harmonic wavelength (400 nm).

The sample itself was mounted on an automatic azimuthal rotation stage whose axis was aligned with the incidence point. The SHG signal produced in the reflection was then detected after passing through a short pass filter that removed the first harmonic pump. The SHG was analyzed in the s (vertical) or p (horizontal) polarization state by an analyzer (linear polarizer). Since the output SHG signal in p state was larger than the one in s state, we report explicitly only the p signal in this manuscript. The signal was further filtered by a narrow bandpass filter centered at 400 nm (FWHM 10 nm) and finally detected by a photomultiplier tube in a gated photon counting regime.

As blank references, we also measured the bare p-Si(111) wafer (thickness of around 400 microns) and a sonicated sample where the wires were removed, leading to a flat GaAs layer of about 50 nm on p-Si(111) substrate.

All the SHG measurements were performed with the same intensity level of the laser.

3. Results

In Figure 3a, we show the measured p-polarized SHG signal from the blank reference sample of the bare p-Si(111) as a function of azimuthal rotation of the sample for two orthogonal circular polarization states of the laser pump (RCP and LCP), while in Figure 3b we show the measured p-polarized SHG signal of the flat GaAs sonicated substrate under the same experimental conditions.

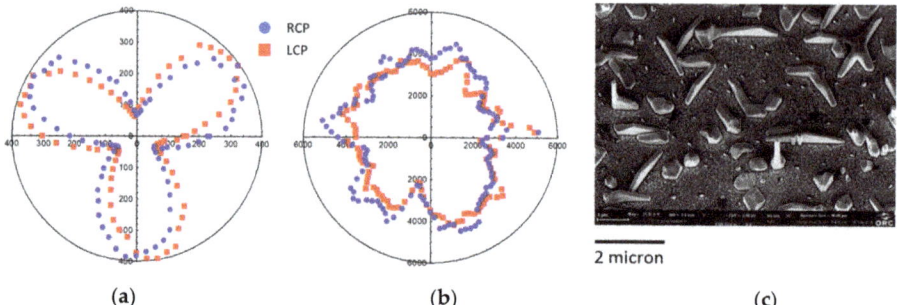

Figure 3. (a) p-polarized SHG signal from the bare p-Si(111) sample for RCP and LCP light with a maximum signal of 400 counts; (b) p-polarized SHG signal from the flat GaAs sonicated sample for RCP and LCP light with a maximum signal of 5500 counts; (c) SEM image of the sonicated GaAs sample. Horizontal residual GaAs crystallites are evident. In the measurements the azimuthal rotation angle is relative.

The Si substrate response showed a clear, but low, SHG signal with three-fold symmetry, as expected from the 111 crystallographic orientation. Si is a third order nonlinear material, and thus the SHG signal is due to surface contribution. There was a small CD due to a normal incidence on the asymmetric 111 surface.

Meanwhile, the flat GaAs sample showed a larger SHG signal (×14 times the one of Si) due to its bulk nonlinear coefficient [25]. In Figure 3c, the SEM image of the reference flat GaAs sample is shown, which was obtained by Sample C after sonication in order to remove the NWs. In the figure, the largest objects are the parasitic crystallites. The orientation of these crystallites correlated with the silicon substrate, and their microstructures showed a three-fold geometrical symmetry with two possible orientations for the crystallites, one being rotated by 180 degrees with respect to the other, leading to a six-fold microstructure symmetry (also evidenced by the hexagonal shaper of the SHG measurements). The roughness on the Si surface of the substrate in Figure 3c is parasitic polycrystalline AlGaAs/GaAs, which formed during the shell growth. Since this layer grew on the oxide-covered areas of the Si substrate, the orientation of these small crystallites was random and gave rise to the main isotropic SHG signal in Figure 3b.

In Figure 4, we show the measured SHG-CD in p-pol light for both Si(111) and the sonicated GaAs samples, defined as

$$\text{SHG-CD} = \frac{I^{(2\omega)}_{LCP} - I^{(2\omega)}_{RCP}}{I^{(2\omega)}_{LCP} + I^{(2\omega)}_{RCP}} \quad (1)$$

where $I^{(2\omega)}_{LCP}$ is the intensity of SHG signal when the fundamental pumping light is circularly left polarized, and $I^{(2\omega)}_{RCP}$ is the intensity of SHG signal when the fundamental pumping light is circularly right polarized.

In the case of Si(111), the SHG-CD was regular even when it was lower than 0.2, while in the case of GaAs it was randomly distributed at values lower than 0.1.

On the left side of each panel of Figure 5, the measured p-polarized SHG signal from Samples A,B,C,D are shown without gold as a function of the azimuthal rotation of each sample for two orthogonal circular polarization states of the laser pump (RCP and LCP), while on the right side of each panel of Figure 5, the measured p-polarized SHG signal of Samples A,B,C,D are shown with asymmetric gold coating. It is evident that the asymmetry in the structures of Samples A,B,C (with Au) led to a strong difference in the SHG signal as a function of the handedness of the circular polarized light, while the SHG response of Sample D with Au was very similar to its uncoated counterpart.

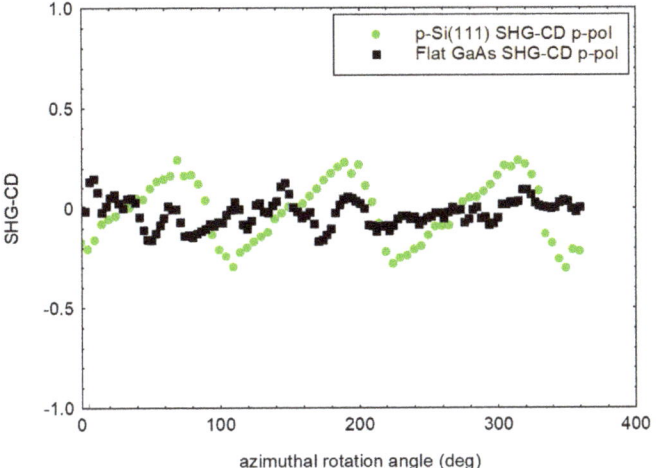

Figure 4. SHG-CD signal of the p-Si(111) reference and of the flat GaAs substrate. In the measurements the azimuthal rotation angle is relative.

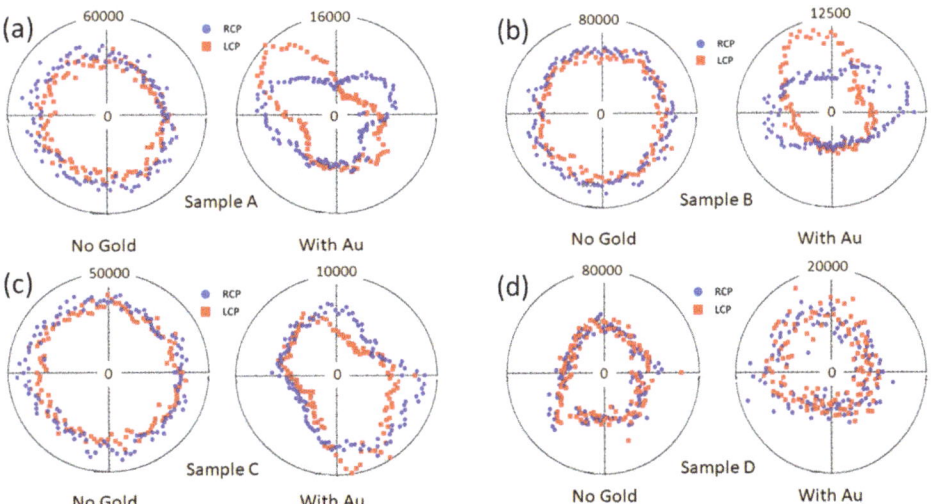

Figure 5. p-polarized SHG signal from: (**a**) Sample A; (**b**) Sample B; (**c**) Sample C; (**d**) Sample D. On the left side of each panel the samples without Au coating for RCP and LCP light are shown, while on the right side of each panel, the samples with Au coating are shown. Adapted from [26]. In the measurements the azimuthal rotation angle is relative.

By concerning the magnitude of the SHG signal, the maximum signal was 50,000 counts for Sample A without Au, while the Au coating decreased the SHG signal to 16,000 counts. The magnitude of SHG for sample A (no Au) was nine times larger than the magnitude of the flat GaAs sample, while the magnitude of Sample A (Au) was three times larger. This indicates that the SHG was enhanced by the geometrical resonances of the GaAs NWs, and that the Au layer did not increase the SHG signal but hindered it by selective absorption of one handedness of circular polarized light respect to the other, leading to a lower signal, but with a higher CD.

Samples B,C,D without Au showed maximum signals of 60,000, 40,000, and 40,000 counts, respectively, which decreased to 10,000, 10,000, and 14,000, counts, respectively, when coated with Au [26].

In Figure 6, we show the SHG-CD of the Samples calculated by Equation (1).

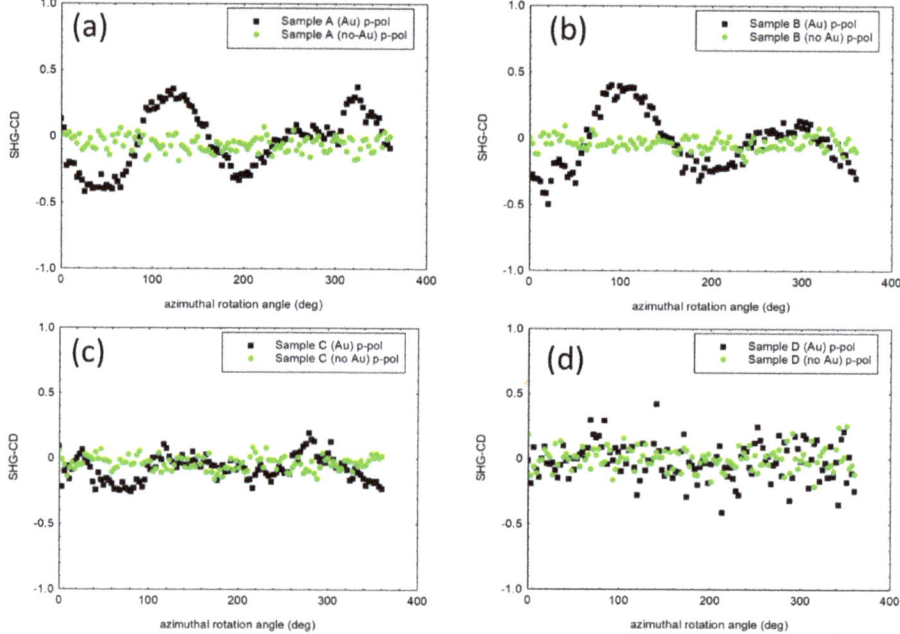

Figure 6. SHG-CD signal of the Samples without Au and with Au. SHC-CD for: (**a**) Sample A; (**b**) Sample B; (**c**) Sample C; (**d**) Sample D. In the measurements the azimuthal rotation angle is relative.

For all the four samples, the symmetric samples provided negligible SHG-CD, despite the large magnitude of SHG signals. Sample A with Au, showed a SHG-CD as high as 0.5. Similarly, the asymmetric Sample B with Au led to a SHG-CD of 0.5.

By considering Sample C, despite a SHG magnitude comparable with previous cases, the SHG-CD was dramatically decreased at a level of about 0.3-0.25 due to the resonant behavior of Sample C around 800 nm (see Figure 2a). This is because the diameter of the wires achieved larger values and thus red-shifted the spectral position of the resonance [3,10].

In the case of Sample D, the SHG-CD was negligible for both symmetric and asymmetric samples due to the complete lack of resonance at 800 nm (see Figure 2a) [3]. Here, the diameter of the wires was so large that the modes fell in the transparent region of the GaAs spectrum at larger wavelengths with respect to the band gap.

4. Discussion

Even though the lithography-free and self-assembled growth methods used for wire growth suffered from an intrinsic degree of disorder, we nevertheless saw a good agreement in the general trend of the SHG-CD signal as a function of the wires' diameters, and we were able to quantitatively compare different samples with a reasonable degree of approximation. In these experiments, we demonstrated two main issues. The first one is the possibility for strong circular dichroic responses from asymmetric samples formed by GaAs NWs partially covered in a thin Au film. This could have applications in different fields, including the ability to generate photons in a SH field, while selective pumping with circular polarized light could be viewed as a possibility for boosting the processes of circular polarized photon generation or absorption. Secondly, we observed that geometric resonance is an essential feature in this extrinsic chiral behavior. Only when resonant leaky modes were present was the geometric-induced CD enhanced in the second harmonic field. The resonance can be finely tuned by changing the diameter of the NWs. We passed from 138-nm diameter wires that showed a strong resonance around 750 nm, to 151-nm diameter wires with a strong resonance at 810 nm. As the diameter increased to 165 nm, the resonance shifted to longer a wavelength (850 nm), decreasing its magnitude as the wavelength approached the transparent region of GaAs. In these cases, we passed from a large CD of 50% (0.5) to a CD of 25%. Finally, for larger-diameter wires (197 nm), the resonance completely disappeared in the GaAs bandgap region, and negligible CD was present, despite the strong SHG signal, destroying any information about the geometrically induced asymmetry of the sample.

Author Contributions: Conceptualization, methodology and validation A.B. and C.S.; measurements A.B., G.L.; linear investigation, G.L., E.P.; fabrication, T.H., E.K., M.R.P., S.T., supervision, C.S., M.G., T.H. All authors have read and agreed to the published version of the manuscript.

Funding: Funding from the Academy of Finland Projects NESP (decision number 294630), QuantSi (323989) and NanoLight (decision number 310985) are acknowledged. EU COST MP1403 NQO is also acknowledged.

Acknowledgments: The Authors acknowledge J. T. Collins, D. C. Hooper and V. K. Valev, for fruitful discussions and help in the measurement realizations. A.B. acknowledges LASAFEM Sapienza Università di Roma Infrastructure Project prot. n. MA31715C8215A268.

Conflicts of Interest: The authors declare no conflict of interest. The funders had no role in the design of the study; in the collection, analyses, or interpretation of data; in the writing of the manuscript, or in the decision to publish the results.

References

1. Cao, L.; White, J.S.; Park, J.-S.; Schuller, J.A.; Clemens, B.M.; Brongersma, M.L. Engineering light absorption in semiconductor nanowire devices. *Nat. Mater.* **2009**, *8*, 643–647. [CrossRef] [PubMed]
2. Fazio, E.; Passaseo, A.; Alonzo, M.; Belardini, A.; Sibilia, C.; Larciprete, M.C.; Bertolotti, M. Measurement of pure Kerr nonlinearity in GaN thin films at 800 nm by means of eclipsing Z-scan experiments. *J. Opt. A Pure Appl. Opt.* **2007**, *9*, L3–L4. [CrossRef]
3. Leahu, G.; Petronijevic, E.; Belardini, A.; Centini, M.; Voti, R.L.; Hakkarainen, T.; Koivusalo, E.; Guina, M.; Sibilia, C. Photo-acoustic spectroscopy revealing resonant absorption of self-assembled GaAs-based nanowires. *Sci. Rep.* **2017**, *7*, 2833. [CrossRef] [PubMed]
4. Petronijevic, E.; Leahu, G.; Belardini, A.; Centini, M.; Voti, R.L.; Hakkarainen, T.; Koivusalo, E.; Guina, M.; Sibilia, C. Resonant Absorption in GaAs-Based Nanowires by Means of Photo-Acoustic Spectroscopy. *Int. J. Thermophys.* **2018**, *39*, 45. [CrossRef]
5. Mayer, B.; Rudolph, D.; Schnell, J.; Morkötter, S.; Winnerl, J.; Treu, J.; Müller, K.; Bracher, G.; Abstreiter, G.; Koblmüller, G.; et al. Lasing from individual GaAs-AlGaAs core-shell nanowires up to room temperature. *Nat. Commun.* **2013**, *4*, 2931. [CrossRef]
6. Verbiest, T.; Kauranen, M.; Van Rompaey, Y.; Persoons, A. Optical Activity of Anisotropic Achiral Surfaces. *Phys. Rev. Lett.* **1996**, *77*, 1456–1459. [CrossRef]
7. Plum, E.; Liu, X.-X.; Fedotov, V.A.; Chen, Y.; Tsai, D.P.; Zheludev, N.I. Metamaterials: Optical Activity without Chirality. *Phys. Rev. Lett.* **2009**, *102*, 113902. [CrossRef]
8. Volkov, S.N.; Dolgaleva, K.; Boyd, R.W.; Jefimovs, K.; Turunen, J.; Svirko, Y.; Canfield, B.K.; Kauranen, M. Optical activity in diffraction from a planar array of achiral nanoparticles. *Phys. Rev. A* **2009**, *79*, 043819. [CrossRef]
9. Bertolotti, M.; Belardini, A.; Benedetti, A.; Mandatori, A. Second harmonic circular dichroism by self-assembled metasurfaces [Invited]. *J. Opt. Soc. Am. B* **2015**, *32*, 1287–1293. [CrossRef]
10. Leahu, G.; Petronijevic, E.; Belardini, A.; Centini, M.; Mandatori, A.; Hakkarainen, T.; Koivusalo, E.; Piton, M.R.; Suomalainen, S.; Guina, M. Evidence of Optical Circular Dichroism in GaAs-Based Nanowires Partially Covered with Gold. *Adv. Opt. Mater.* **2017**, *5*, 1601063. [CrossRef]

11. Petronijevic, E.; Leahu, G.; Voti, R.L.; Belardini, A.; Scian, C.; Michieli, N.; Cesca, T.; Mattei, G.; Sibilia, C. Photo-acoustic detection of chirality in metal-polystyrene metasurfaces. *Appl. Phys. Lett.* **2019**, *114*, 053101. [CrossRef]
12. Petronijevic, E.; Leahu, G.; Belardini, A.; Centini, M.; Voti, R.L.; Hakkarainen, T.; Koivusalo, E.; Piton, M.R.; Suomalainen, S.; Guina, M.; et al. Photo-Acoustic Spectroscopy Reveals Extrinsic Optical Chirality in GaAs-Based Nanowires Partially Covered with Gold. *Int. J. Thermophys.* **2018**, *39*, 46. [CrossRef]
13. Benedetti, A.; Alam, B.; Esposito, M.; Tasco, V.; Leahu, G.; Belardini, A.; Voti, R.L.; Passaseo, A.; Sibilia, C. Precise detection of circular dichroism in a cluster of nano-helices by photoacoustic measurements. *Sci. Rep.* **2017**, *7*, 5257. [CrossRef] [PubMed]
14. Belardini, A.; Larciprete, M.C.; Centini, M.; Fazio, E.; Mandatori, A.; Chiappe, D.; Martella, C.; Toma, A.; Giordano, M.C.; De Mongeot, F.B. Circular Dichroism in the Optical Second-Harmonic Emission of Curved Gold Metal Nanowires. *Phys. Rev. Lett.* **2011**, *107*, 257401. [CrossRef] [PubMed]
15. Belardini, A.; Centini, M.; Leahu, G.; Fazio, E.; Sibilia, C.; Haus, J.; Sarangan, A. Second harmonic generation on self-assembled tilted gold nanowires. *Faraday Discuss.* **2015**, *178*, 357–362. [CrossRef]
16. Belardini, A.; Centini, M.; Leahu, G.; Hooper, D.C.; Voti, R.L.; Fazio, E.; Haus, J.W.; Sarangan, A.; Valev, V.; Mandatori, A. Chiral light intrinsically couples to extrinsic/pseudo-chiral metasurfaces made of tilted gold nanowires. *Sci. Rep.* **2016**, *6*, 31796. [CrossRef]
17. Hakkarainen, T.; Petronijevic, E.; Piton, M.R.; Sibilia, C. Demonstration of extrinsic chirality of photoluminescence with semiconductor-metal hybrid nanowires. *Sci. Rep.* **2019**, *9*, 5040. [CrossRef]
18. Yang, Y.; Rice, B.; Shi, X.; Brandt, J.; Da Costa, R.C.; Hedley, G.; Smilgies, D.-M.; Frost, J.M.; Samuel, I.D.W.; Otero-De-La-Roza, A.; et al. Emergent Properties of an Organic Semiconductor Driven by its Molecular Chirality. *ACS Nano* **2017**, *11*, 8329–8338. [CrossRef]
19. Shang, X.; Song, I.; Ohtsu, H.; Lee, Y.H.; Zhao, T.; Kojima, T.; Jung, J.H.; Kawano, M.; Oh, J.H. Supramolecular Nanostructures of Chiral Perylene Diimides with Amplified Chirality for High-Performance Chiroptical Sensing. *Adv. Mater.* **2017**, *29*, 1605828. [CrossRef]
20. Albano, G.; Salerno, F.; Portus, L.; Porzio, W.; Aronica, L.; Di Bari, L. Outstanding Chiroptical Features of Thin Films of Chiral Oligothiophenes. *ChemNanoMat* **2018**, *4*, 1059–1070. [CrossRef]
21. Albano, G.; Górecki, M.; Pescitelli, G.; Di Bari, L.; Javorfi, T.; Hussain, R.; Siligardi, G. Electronic circular dichroism imaging (CDi) maps local aggregation modes in thin films of chiral oligothiophenes. *New J. Chem.* **2019**, *43*, 14584–14593. [CrossRef]
22. Petronijevic, E.; Centini, M.; Belardini, A.; Leahu, G.; Hakkarainen, T.; Sibilia, C. Chiral near-field manipulation in Au-GaAs hybrid hexagonal nanowires. *Opt. Express* **2017**, *25*, 14148–14157. [CrossRef] [PubMed]
23. Petronijevic, E.; Mandatori, A. Enhanced Near-Field Chirality in Periodic Arrays of Si Nanowires for Chiral Sensing. *Molecules* **2019**, *24*, 853. [CrossRef] [PubMed]
24. Hakkarainen, T.V.; Schramm, A.; Mäkelä, J.; Laukkanen, P.; Guina, M. Lithography-free oxide patterns as templates for self-catalyzed growth of highly uniform GaAs nanowires on Si (111). *Nanotechnology* **2015**, *26*, 275301. [CrossRef]
25. Malvezzi, A.; Vecchi, G.; Patrini, M.; Guizzetti, G.; Andreani, L.; Romanato, F.; Businaro, L.; Di Fabrizio, E.; Passaseo, A.; De Vittorio, M. Resonant second-harmonic generation in a GaAs photonic crystal waveguide. *Phys. Rev. B* **2003**, *68*, 161306. [CrossRef]
26. Belardini, A.; Collins, J.; Hooper, D.; Leahu, G.; Petronijevic, E.; Centini, M.; Voti, R.; Hakkarainen, T.; Koivusalo, E.; Piton, M.; et al. Second Harmonic Generation Circul Dichroism in Au Coated Gaas-based Nanowires. In Proceedings of the 20th Italian National Conference on Photonic Technologies (Fotonica 2018), Lecce, Italy , 23–25 May 2018; Institution of Engineering and Technology (IET): London, UK, 2018; p. 23.

© 2020 by the authors. Licensee MDPI, Basel, Switzerland. This article is an open access article distributed under the terms and conditions of the Creative Commons Attribution (CC BY) license (http://creativecommons.org/licenses/by/4.0/).

Article

Generation of Pure State Photon Triplets in the C-Band

Xi-Rong Su [1], Yi-Wen Huang [1], Tong Xiang [1], Yuan-Hua Li [1,2] and Xian-Feng Chen [1,*]

1. State Key Laboratory of Advanced Optical Communication Systems and Networks, Department of Physics and Astronomy, Shanghai Jiao Tong University, Shanghai 200240, China; suxirong@sjtu.edu.cn (X.-R.S.); yiwenhuang@sjtu.edu.cn (Y.-W.H.); 19900104@sjtu.edu.cn (T.X.); lyhua1984@163.com (Y.-H.L.)
2. Department of physics, Jiangxi Normal University, Nanchang 330022, China
* Correspondence: xfchen@sjtu.edu.cn; Tel.: +86-21-5474-3252

Received: 11 October 2019; Accepted: 10 November 2019; Published: 13 November 2019

Abstract: In this work, the cascaded second-order spontaneous parametric down-conversion (SPDC) is considered to produce pure state photon triplets in periodically poled lithium niobite (PPLN) doped with 5% MgO. A set of parameters are optimized through calculating the Schmidt number of two-photon states generated by each down-conversion process with different pump durations and crystal lengths. We use a Gaussian filter in part and obtain three photons with 100% purity in spectrum. We provide a feasible and unprecedented scheme to manipulate the spectrum purity of photon triplets in the communication band (C-band).

Keywords: pure state; cascaded spontaneous parametric down-conversion (SPDC); numerical simulation

1. Introduction

The scheme of generating photon pairs using cascaded second-order spontaneous parametric down-conversion (SPDC) [1,2] is an indispensable ingredient of modern quantum technology and has great potential in many applications, such as quantum cryptography [3], quantum teleportation [4] and quantum entanglement swapping [5]. Recently, a wide variety of methods have been proposed to produce photon triplets. Common methods include direct generation of photon triplets [6–8], the process of four wave mixing (FWM) [9–12] and generation of three entangled photons by cascaded second-order SPDC [13–16]. Some studies propose implementing third-order SPDC in optical fibers and bulk crystals. There are always low count rates for schemes based on the $\chi^{(3)}$ process. The FWM techniques consists of stimulated SPDC and cascaded FWM. The latter can be divided into three categories according to the different ways of cascading. The cascaded second-order SPDC is considered because of the simple model, which consists of two second order SPDC processes. The mature theory and substantial experiments make it a reliable scheme.

The research about quantum correlation among individual photons lies at the core of quantum technologies. Under different conditions, the two-photon generated by SPDC will present a state of frequency positive correlation, inverse correlation or uncorrelation. The last method is used to provide a heralded source [17,18]. Previous experiments have failed to give a specific theoretical numerical analysis to judge the spectral purity of the generated photon pairs. The full use of filters [19,20] will greatly reduce the coincidence counting rate. Recently, Zhang et al. decomposed the factor mathematically to manipulate the tripartite frequency correlation [21]. But the spectrum of photons after the first SPDC and the effect on the second order down-conversion were not taken into account. They produce photons with wavelength of ~3000 nm, which is almost unavailable. So far, there is little theoretical work about pure states photon triplets in the C-band.

Quantum interference is vital for quantum information science. It is not only the basis of quantum manipulation technology, but also an important tool to implement quantum computing and quantum communication. The realization of quantum computation [22] depends on the measurement and reading of quantum states, and quantum interference is one of the most simple and feasible methods for quantum measurement. Quantum communication [23,24] is more dependent on the transmission and acquisition of information by means of interference. Three-photon interference is critical for the exploitation of quantum information in higher dimensions [25]. The GHZ interference is observed in the experiment, which lays the foundation for the subsequent quantum secret sharing [26]. In general, the photon triplets generated by the cascaded SPDC will have correction in frequency. This allows the photon pair to be resolved in the frequency dimension, thereby reducing the visibility of the interference [27]. For instance, the interference of indistinguishable photons makes the entanglement swapping and teleportation possible, which in turn opens up prospects for distributing of entanglement between distant matter qubits. The goal of our work is to prepare three photons with hyperspectral purity, which are critical for research into quantum information processes.

In this work, the suitable pump duration and crystal length are selected to eliminate the frequency correlation between the photon pairs in each SPDC process. In terms of the theoretical analysis, spectral purity of photon pairs is mainly measured by means of Schmidt number [28,29]. The conclusion of our theoretical calculation is supported by the two photons' and three photons' joint spectrum. Relevant theories will be discussed in Section 2. The common pump source used to acquire polarization-entangled photon pairs from SPDC is narrow-band or continuous wave (CW) laser, but the subsequent photon pairs have a strong correlation in frequency [30]. A broadband pumping source is adopted in our work, and the optimal pumping duration is chosen by numerical investigation in Section 3. Finally, we obtain pure-state photon triplets with two kinds of periodically poled crystals under different parameters.

2. Tripartite State and Joint Spectrum

2.1. Model

Quasi-phase matching is adopted because of the simpler and more flexible matching condition. The theoretical model consists of two parts, which are two nondegenerate SPDC processes [13]. A pair of photons called idler photons ω_0 and signal photons ω_1 are generated from the first SPDC process. The idler photons continue to be the pump source of the second SPDC process, producing photons ω_2 and ω_3.

The phase-matching conditions of the two processes are type e → o + o and type e → e + o, respectively. As shown in Figure 1, the lengths of the two crystals are L_1 and L_2 while the periodicities are Λ_1 and Λ_2, respectively. When the pump light with center frequency of ω_0 is incident into the first crystal, the generated photon pairs will be correlated in time and frequency due to the conservation of energy and momentum. The relation between the wave vectors and the frequency of the three photons are $k_p = k_1 + k_0 + k_{g1}$ and $\hbar\omega_p = \hbar\omega_1 + \hbar\omega_0$, where $k_{g1} = 2\pi m/\Lambda_1$ is the compensated wave vector. In the second down-conversion process, photon ω_0 splits into ω_2 and ω_3 while the conservation conditions are also satisfied, which are $k_0 = k_2 + k_3 + k_{g2}$ and $\hbar\omega_0 = \hbar\omega_2 + \hbar\omega_3$, where $k_{g2} = 2\pi m/\Lambda_2$. Therefore, in the whole frequency conversion process, the energy conservation and momentum conservation are also satisfied between the initial pump photon and the resulting three photons.

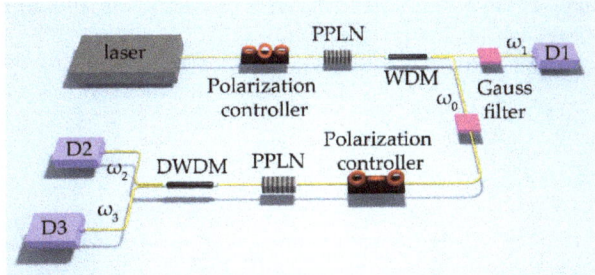

Figure 1. Theoretical model of cascaded second-order SPDC. Photons ω_0 and ω_1 are generated in the first crystal and then photons ω_2 and ω_3 are generated from the second crystal. The two parts of this model are periodically poled lithium niobite (PPLN) with lengths L_1 and L_2 respectively. Two Gauss filters are used to manipulate the joint spectrum of photonic pairs.

2.2. Hamiltonian and Probability Amplitude Function

For the convenience of calculation, our model adopts a one-dimensional collinear phase-matching structure. Since the pump field is strong, the field is treated as an electric classical field $E_p(\mathbf{r}, t) = \tilde{\alpha}(t) \exp[ik_p(\omega_p)z]$, rather than using the annihilation operator of the pump photon. A Gauss envelope is chosen as the pump function $\tilde{\alpha}_p(t) = \tilde{\alpha}_p(0) \exp(-t^2/2\tau_p^2)$. The expression corresponding to the frequency domain is

$$\alpha(\Omega_p) = \frac{\tau_p}{\sqrt{2\pi}} \exp(-\frac{\tau_p^2 \Omega_p^2}{2}) \tag{1}$$

where τ_p is the pump duration and $\Omega_p = \omega_p - \overline{\omega}_p$ is the frequency difference.

After calculating the integral of Hamiltonian [31] and simplifying the statements, the final expression of the two-photon state is

$$|\psi_2\rangle = \int_{t_0}^{t} dt' \hat{H}_I(t') = A \int d\omega_s \int d\omega_i \hat{a}_s^\dagger(\omega_s)\hat{a}_i^\dagger(\omega_i) \alpha(\omega_s, \omega_i) \varphi(\omega_s, \omega_i)|0\rangle + c.c. \tag{2}$$

where $\alpha(\omega_s, \omega_i)$ and $\varphi(\omega_s, \omega_i)$ are pump envelope function and phase-matching function, respectively. Their product is the two-photon amplitude function

$$F(\omega_s, \omega_i) = \alpha(\omega_s + \omega_i)\varphi(\omega_s, \omega_i) \tag{3}$$

The phase matching function in upper equation is

$$\varphi(\omega_s, \omega_i) = \text{sinc}\left[\frac{(k_s(\omega_s) + k_i(\omega_i) - k_p(\omega_s + \omega_i) + k_g)L}{2}\right] \tag{4}$$

For simpler operation, a coefficient $\gamma = 0.193$ is introduced to approximate the sinc function to a Gauss function to ensure that they have the same full width at half maximum (FWHM). This approximation only removes the small peak of sinc function and has no effect on the calculation of biphoton joint spectrum.

Assuming a perfect phase-matching condition, we carry out the Taylor expansion of the wave vector and preserve the first order term. That is $k_m(\omega_m) = k_{m0} + k'_m(\omega_m - \overline{\omega}_m) + \cdots$, $k'_m = \partial k_m(\omega)/\partial \omega|_{\omega=\overline{\omega}_m}$ ($m = p, s, i$). The influence of group velocity dispersion and higher order terms are not considered. The second derivative of wave vector does not change obviously with the wavelength. In addition, in the actual system, the error caused by dispersion can be overcome by compensation. The phase-matching function is described by

$$\varphi(\omega_s, \omega_i) \approx \exp\left\{-\gamma\left(\frac{\Omega_s(k'_s - k'_p) + \Omega_i(k'_i - k'_p)L}{2}\right)\right\} \quad (5)$$

where Ω_s and Ω_i are the frequency difference.

In addition to the phase-matching condition, we also consider the matching condition of group velocity [32]. But in the first SPDC, the derivative of the pump wave vector is always larger than that of the two down-converted photons. We use two Gaussian filters to remove the correlation of the two photons [33]. The two photons' amplitude function becomes

$$F(\omega_s, \omega_i) = T(\omega_0)T(\omega_1)\alpha(\omega_s + \omega_i)\varphi(\omega_s, \omega_i) \quad (6)$$

where $T(\omega_i) = \exp(-\Omega_i^2/\varsigma_i^2)$ is the corresponding filter. ς_i is the FWHM.

In the total cascaded process, the holistic Hamiltonian is the product of the Hamiltonian of two parts, that is $\hat{H} = \hat{H}_1\hat{H}_2$. The expression of the last three photon states is

$$|\psi_3\rangle = \int dt_1 dt_2 \hat{H}_1(t_1)\hat{H}_2(t_2) = B\int d\omega_1 \int d\omega_2 \int d\omega_3 \hat{a}_1^\dagger(\omega_1)\hat{a}_2^\dagger(\omega_2)\hat{a}_3^\dagger(\omega_3)F(\omega_1,\omega_2,\omega_3)|0\rangle + c.c. \quad (7)$$

When we determine the frequency distribution of the down-conversion of three photons, and there is no correlation between them, then the three photons amplitude can be equivalent to

$$F(\omega_1, \omega_2, \omega_3) = \exp(-\Omega_1^2/\sigma_1^2)\exp(-\Omega_2^2/\sigma_2^2)\exp(-\Omega_3^2/\sigma_3^2) \quad (8)$$

In practice, there are two sensitive parameters of the system that need to be strictly controlled: (1) the polarization stability of the light source and the optical path, and; (2) the temperature of the non-linear material. Both of them directly affect the refractive index of materials, thus affecting the phase-matching conditions.

2.3. Joint Spectrum and Purity

The simplest method to judge the frequency dependence of two photons produced by second order SPDC is to analyze their joint spectrum which is determined by

$$JSI(\omega_s, \omega_i) = |F(\omega_s, \omega_i)|^2 \quad (9)$$

The two photons are frequency uncorrelated if their joint spectrum is a circle or an ellipse parallel to the axis, which means the distribution of photons in frequency is independent of each other. It is impossible to obtain an optimal value simply by judging the shape or the angle with the coordinate axis because of the lack of a specific parameter to quantify the two-photon frequency correlation. Calculating the Schmidt number is the effective scheme to measure spectral correlation because it reflects the purity of correlation over frequency. It is defined as follows

$$K = \frac{1}{\text{Tr}\{\rho_1^2\}} = \frac{1}{P} \quad (10)$$

In this formula, ρ_1 is the density operator of photon ω_1 and P represents the spectral purity. There is no frequency correlation between photon pairs when the Schmidt number K reaches the minimum value of 1. After calculating the Schmidt numbers with the parameters of crystal length and pump duration in each SPDC, the results are verified and analyzed by the joint spectrum of two photons under the optimum parameters. The joint spectral intensity of the photon triplets can be written as

$$JSI(\omega_1, \omega_2, \omega_3) = |F(\omega_1, \omega_2, \omega_3)|^2 \tag{11}$$

We use the symbol quantity to carry on the maximum precision calculation. The result is converted to double type with 16 bits precision. The precision is enough that an ideal numerical simulation result can be obtained. Therefore, the error caused by the accuracy of software calculation can also be ignored.

3. Numerical Simulation Results

In this section, we discuss the generation of photon pairs from different materials, and finally obtain the pure-state photon triplets. Among the numerous nonlinear crystals, lithium niobite has a relatively higher nonlinear coefficient [34,35], which leads to a greater conversion efficiency. There is a wide range of transparency, from 420 nm to 5200 nm. In addition, lithium niobite doped with MgO has higher damage threshold, thus the periodically poled lithium niobate doped with 5% MgO (PPMgLN) will also be used as a reference for comparison. We get a set of crystal lengths which are optimum for each SPDC process through theoretical arithmetic.

In the first SPDC, the crystal and pump parameters are taken as the variables and the calculation of the Schmidt number is done. We select the appropriate pumping duration and crystal length L_1 by analyzing the obtained data. After calculation, the frequency distribution of photonic ω_0 is obtained. That is to say, the envelope information of the pump in the second SPDC is determined, which is $\exp[-(\omega_0 - \overline{\omega}_0)/\sigma_0^2]$, where σ_0 is the bandwidth of the new source ω_0. Then, the amplitude function of photon ω_2 and ω_3 is described as

$$F(\Omega_2, \Omega_3) = \exp\left[-\frac{(\Omega_2 + \Omega_3)^2}{\sigma_0^2}\right] \exp\left(-i\frac{\Delta k_2 L_2}{2}\right) \operatorname{sinc}\left[\frac{(k_0(\omega_0) - k_2(\omega_2) - k_3(\omega_3) - k_{g2})L_2}{2}\right] \tag{12}$$

In the second SPDC, the Schmidt numbers of photon state between ω_2 and ω_3 are calculated by using the bandwidth information of the generated photons ω_0 and taking the crystal length L_2 as the variable. Then we select the appropriate crystal length L_2. Each time the most appropriate parameters are determined, the two-photon joint spectrum and the final three-photon joint spectrum are given to verify the theoretical calculation.

3.1. Realization of Pure-State Photon Triplets in PPLN

The pump wavelength is 520 nm. Relevant data in the first down-conversion is shown in Figure 2. The z axis in Figure 2a describes the variation of the spectral purity with the parameters. The x-axis represents the range of the selected crystal lengths from 0 to 1 cm while the y-axis is the variation of pump duration in the range of 0–1 ps. The wavelengths of the pair of entangled photons are $\lambda_1 = 1560$ nm and $\lambda_0 = 780$ nm. The periodicity of the first PPLN is 38.47 μm. Due to the Gaussian filter with a bandwidth of 0.8 THz, the spectrum purity between photons ω_0 and ω_1 is almost 1 in the region where the crystal length and pump duration are smaller. Considering the realizability, we selected a pump duration of 100 fs and a crystal length of 0.2 cm.

Figure 2b describes the joint spectral intensity of photons ω_1 and ω_0. It is intuitive to see that there is no frequency correlation between the two photons. Figure 2c,d are the bandwidth of photons ω_1 and ω_0, respectively. Because the transmission of the filter is related to the bandwidth, the down-conversion photons of the two channels have the identical frequency distribution.

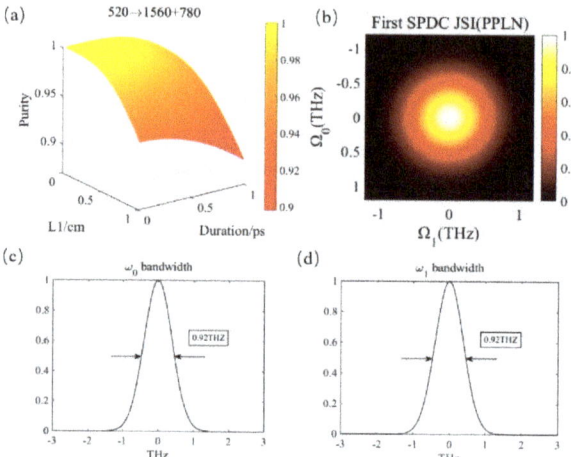

Figure 2. (a) The spectral purity in the first SPDC using PPLN. (b) The joint spectrum of photon 1 and 0. (c,d) The bandwidth of the photon pairs.

In the second SPDC process, we select the generation wavelengths of λ_2 = 1570 nm and λ_3 = 1550 nm in consideration of the matching condition of the group velocity. The polarization period of the second PPLN is 88.76 μm. Relevant data are shown in Figure 3. Figure 3a describes the calculation of the spectral purity of photons ω_2 and ω_3 with the crystal length L_2 as the independent variable. It can be seen that with the increase of crystal length, the purity increases to the maximum value of 1. We chose the best crystal length L_2 as 9.16 cm. Figure 3b is the joint spectral intensity of photon pair ω_2 and ω_3. It can be seen that the photon pairs are still frequency uncorrelated in the second SPDC process. As shown in Figure 3c,d, the bandwidth of photons ω_2 and ω_3 is different because the perfect group velocity match is not achieved, but this does not affect the correlation between them.

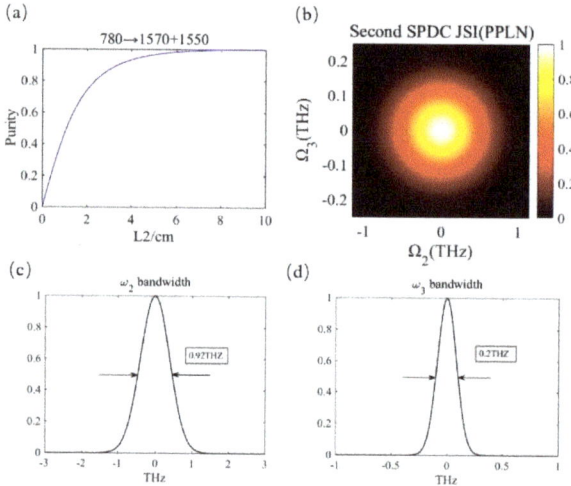

Figure 3. (a) Spectral purity data in the second SPDC. (b) The joint spectrum of photon 2 and 3. (c,d) The bandwidth of the photon pairs.

Since photons ω_1 and ω_0 are not correlated in frequency, both photon (1,2) and photon (1,3) should be irrelevant theoretically. Figure 4 shows the joint spectrum of photon triplets. From the relationship between each two-photon, as shown in three projection planes, there is no correlation between photons ω_1 and ω_2, ω_1 and ω_3. So far, we have obtained photon triplets which are not related in the frequency dimension. At the same time, all three of them are in the C-band.

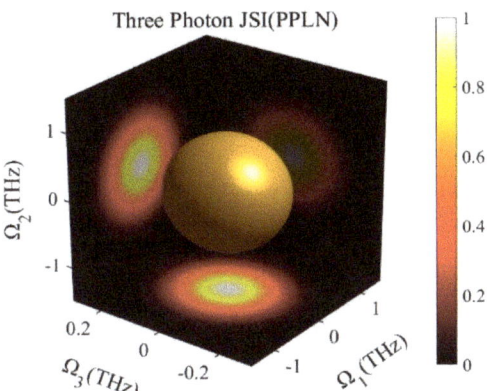

Figure 4. Joint spectrum of the three photons generated from the cascaded PPLN.

3.2. Realization of Pure-State Photon Triplets in PPMgLN

We also chose 520 nm as the pump wavelength for comparison. The group velocity matching condition of the second SPDC is not satisfied. The wavelength of photon ω_1 and ω_0 generated in the first down-conversion are 1520 nm and 790.4 nm, respectively.

The results of correlated data are given in Figure 5. The photon wavelengths generated by the second down-conversion are λ_2 = 1590 nm and λ_3 = 1571.7 nm, pumped by photon ω_0. Similar data are shown in Figure 6. The pump duration is 0.27 ps while the crystal lengths are L_1 = 0.2 cm and L_2 = 10.74 cm (corresponding Λ_1 = 34.85 µm and Λ_2 = 83.51 µm).

We also produce three photons with a purity of 100%. Due to the material differences, the center wavelengths of the photons ω_2 and ω_3 are longer than in PPLN. It takes a slightly longer crystal than PPLN to achieve the phase-matching condition. The bandwidth of the photons ω_1 and ω_2 generated in the PPMgLN is relatively wider. Figure 7 shows the joint spectral intensity of photon triplets generated by cascaded PPMgLN. The three projection planes reflect the correlation between two of the three photons.

PPLN is more suitable for weak light due to the better phase-matching conditions. According to our theoretical results, the wavelength distribution of the three-photon generated in PPLN is closer. For the same pump (λ_p = 520 nm), three photons with wavelengths of 1550 nm, 1560 nm and 1570 nm can be realized in PPLN. PPMgLN is more suitable for the pump with higher intensity, because doping MgO can increase the damage threshold of the material and obtain higher brightness photon triplets. But we can only obtain photons with wavelengths of 1520 nm, 1590 nm and 1571.7 nm. In the preparation of the light source, spectral purity is one of the core indicators. The purpose of our work is to prepare photon triplets of spectral pure-state (frequency uncorrelated), which provides a reliable scheme for the preparation of high quality sources in the field of quantum technology. There is no prior research on pure state photon triplets in the C-band before our work. Our method can also provide a heralding pure-state biphoton source with higher interference visibility. Compared with the unpredicted conditions, the heralding two-photons have superior advantages, such as avoiding the detection of noise photons, which greatly reduces the bit error rate (BER). It guarantees the realization of many of these tasks relying on qubits that are encoded in the polarization states of single photons.

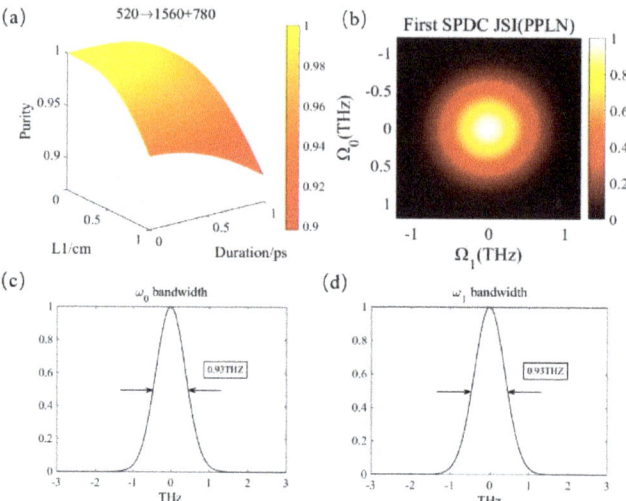

Figure 5. (a) The spectral purity in the first SPDC using MgO-doped PPLN. (b) The joint spectrum of photon 1 and 0. (c,d) The bandwidth of the photon pairs.

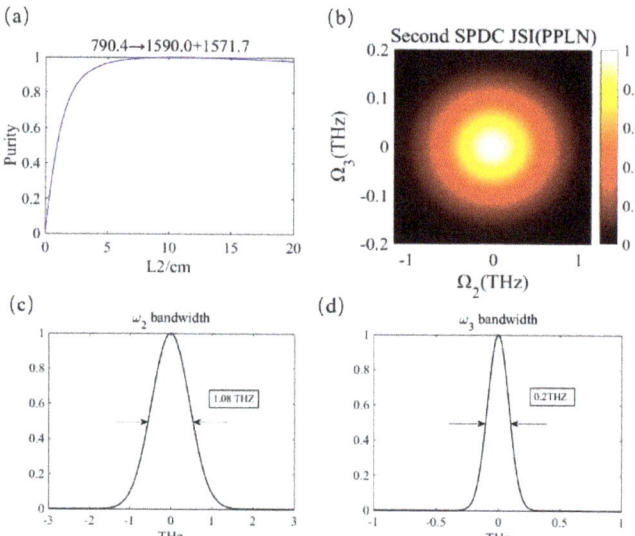

Figure 6. (a) The spectral purity data in the second SPDC. (b) The joint spectrum of photon 2 and 3. (c,d) The bandwidth of the photon pairs.

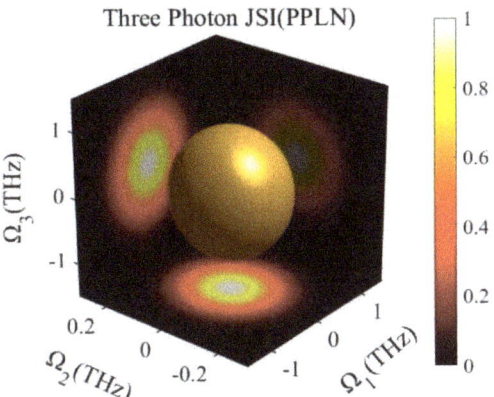

Figure 7. Joint spectrum of the three photons generated from the cascaded SPDC MgO-doped PPLN.

4. Conclusions

In summary, we discuss how to select the optimal pump and crystal parameters to obtain the pure state photon triplets by cascaded second-order SPDC using PPLN and MgO-doped PPLN. We chose 520 nm as the pump source with duration of 0.1 ps. After calculating the Schmidt number of photon pairs generated in each SPDC, we determined the crystal lengths of the two PPLN are 0.2 cm and 9.16 cm, respectively. The lengths of the two PPMgLN are 0.2 cm and 10.74 cm, respectively. According to theoretical calculation, the purity of photon pairs from each SPDC can reach 100%, that is, there is no frequency correlation. We have achieved photon triplets with a spectral purity of 100% in the C-band. We firmly believe that in the future development of quantum networks, our scheme can provide reliable pure-state photon triplet sources for various quantum information processes.

Author Contributions: Conceptualization and methodology, X.-R.S. and Y.-W.H.; validation, Y.-W.H., T.X. and Y.-H.L.; writing—original draft preparation, X.-R.S.; writing—review and editing, T.X., Y.-H.L., X.-F.C.

Funding: This work is supported by the National Key R&D Program of China (2017YFA0303701, 2018YFA0306301,); National Natural Science Foundation of China (NSFC) (11734011, 11804135); The Foundation for Development of Science and Technology of Shanghai (17JC1400400).

Conflicts of Interest: The authors declare no conflict of interest.

References

1. Giovannetti, V.; Maccone, L.; Shapiro, J.H.; Wong, F.N.C. Generating Entangled Two-Photon States with Coincident Frequencies. *Phys. Rev. Lett.* **2002**, *88*, 183602. [CrossRef] [PubMed]
2. Kitaeva, G.K.; Penin, A.N. Spontaneous Parametric Down-Conversion. *JETP Lett.* **2005**, *82*, 350. [CrossRef]
3. Jennewein, T.; Simon, C.; Weihs, G.; Weinfurter, H.; Zeilinger, A. Quantum Cryptography with Entangled Photons. *Phys. Rev. Lett.* **2000**, *84*, 4729. [CrossRef] [PubMed]
4. Bouwmeester, D.; Pan, J.W.; Mattle, K.; Eibl, M.; Weinfurter, H.; Zeilinger, A. Experimental Quantum Teleportation. *Nature* **1997**, *390*, 575. [CrossRef]
5. Pan, J.W.; Bouwmeester, D.; Weinfurter, H.; Zeilinger, A. Experimental Entanglement Swapping: Entangling Photons that Never Interacted. *Phys. Rev. Lett.* **1998**, *80*, 3891. [CrossRef]
6. Corona, M.; Garay Palmett, K.; U'Ren, A.B. Third-Order Spontaneous Parametric Down-Conversion in Thin Optical Fibers as a Photon-Triplet Source. *Phys. Rev. A* **2011**, *84*, 033823. [CrossRef]
7. Corona, M.; Garay Palmett, K.; U'Ren, A.B. Experimental Proposal for the Generation of Entangled Photon Triplets by Third-Order Spontaneous Parametric Downconversion in Optical Fibers. *Opt. Lett.* **2011**, *36*, 190–192. [CrossRef]

8. Borshchevskaya, N.A.; Katamadze, K.G.; Kulik, S.P.; Fedorov, M.V. Three-Photon Generation by Means of Third-Order Spontaneous Parametric Down-Conversion in Bulk Crystals. *Laser Phys. Lett.* **2015**, *12*, 115404. [CrossRef]
9. Douady, J.; Boulanger, B. Experimental Demonstration of a Pure Third-Order Optical Parametric Downconversion Process. *Opt. Lett.* **2004**, *29*, 2794–2796. [CrossRef]
10. Bencheikh, K.; Gravier, F.; Douady, J.; Levenson, A.; Boulanger, B. Triple Photons: A Challenge in Nonlinear and Quantum Optics. *C. R. Phys.* **2007**, *8*, 206–220. [CrossRef]
11. Wang, H.; Zheng, Z.; Wang, Y.; Jing, J. Generation of Tripartite Entanglement from Cascaded Four-Wave Mixing Processes. *Opt. Express.* **2016**, *24*, 23459–23470. [CrossRef]
12. Qin, Z.; Cao, L.; Jing, J. Experimental Characterization of Quantum Correlated Triple Beams Generated by Cascaded Four-Wave Mixing Processes. *Appl. Phys. Lett.* **2015**, *106*, 211104. [CrossRef]
13. Hamel, D.R.; Shalm, L.K.; Hubel, H.; Miller, A.J.; Marsili, F.; Verma, V.B.; Jennewein, T. Direct Generation of Three-Photon Polarization Entanglement. *Nat. Photonics* **2014**, *8*, 801. [CrossRef]
14. Shalm, L.K.; Hamel, D.R.; Yan, Z.; Simon, C.; Resch, K.J.; Jennewein, T. Three-Photon Energy–Time Entanglement. *Nat. Phys.* **2013**, *9*, 19. [CrossRef]
15. Hubel, H.; Hamel, D.R.; Resch, K.J.; Jennewein, T. Generation of Various Tri-Partite Entangled States using Cascaded Spontaneous Down-Conversion. *AIP Conf. Proc.* **2011**, *1363*, 331–334.
16. Hubel, H.; Hamel, D.R.; Fedrizzi, A.; Ramelow, S.; Resch, K.J.; Jennewein, T. Direct Generation of Photon Triplets Using Cascaded Photon-Pair Sources. *Nature* **2010**, *466*, 601. [CrossRef]
17. Mosley, P.J.; Lundeen, J.S.; Smith, B.J.; Wasylczyk, P.; U'Ren, A.B.; Silberhorn, C.; Walmsley, I.A. Heralded Generation of Ultrafast Single Photons in Pure Quantum States. *Phys. Rev. Lett.* **2008**, *100*, 133601. [CrossRef]
18. Levine, Z.H.; Fan, J.; Chen, J.; Ling, A.; Migdall, A. Heralded, Pure-State Single-Photon Source Based on a Potassium Titanyl Phosphate Waveguide. *Opt. Express* **2010**, *18*, 3708–3718. [CrossRef]
19. Grice, W.P.; Erdmann, R.; Walmsley, I.A.; Branning, D. Spectral Distinguishability in Ultrafast Parametric Down-Conversion. *Phys. Rev. A* **1998**, *57*, R2289. [CrossRef]
20. Meyer Scott, E.; Prasannan, N.; Eigner, C.; Quiring, V.; Donohue, J.M.; Barkhofen, S.; Silberhorn, C. High-Performance Source of Spectrally Pure, Polarization Entangled Photon Pairs Based on Hybrid Integrated-Bulk Optics. *Opt. Express* **2018**, *26*, 32475–32490. [CrossRef]
21. Zhang, Q.Y.; Xue, G.T.; Xu, P.; Gong, Y.X.; Xie, Z.; Zhu, S. Manipulation of Tripartite Frequency Correlation Under Extended Phase Matchings. *Phys. Rev. A* **2018**, *97*, 022327. [CrossRef]
22. Browne, D.E.; Rudolph, T. Resource-Efficient Linear Optical Quantum Computation. *Phys. Rev. Lett.* **2005**, *95*, 010501. [CrossRef]
23. Gao, T.; Yan, F.L.; Wang, Z.X. Deterministic Secure Direct Communication Using GHZ States and Swapping Quantum Entanglement. *J. Phys. A* **2005**, *38*, 5761. [CrossRef]
24. Hillery, M.; Buzek, V.; Berthiaume, A. Quantum Secret Sharing. *Phys. Rev. A* **1999**, *59*, 1829. [CrossRef]
25. Sascha, A.; Thomas, K.; Jeongwan, J.; Evan, M.S.; Jeff, Z.S.; Deny, R.H.; Kevin, J.R.; Gregor, W.; Thomas, J. Observation of Genuine Three-Photon Interference. *Phys. Rev. Lett.* **2017**, *118*, 153602.
26. Chen, Y.A.; Zhang, A.N.; Zhao, Z.; Zhou, X.Q.; Lu, C.Y.; Peng, C.Z.; Yang, T.; Pan, J.W. Experimental Quantum Secret Sharing and Third-Man Quantum Cryptography. *Phys. Rev. Lett.* **2005**, *95*, 200502. [CrossRef]
27. Humble, T.S.; Grice, W.P. Effects of Spectral Entanglement in Polarization-Entanglement Swapping and Type-I Fusion Gates. *Phys. Rev. A* **2008**, *77*, 022312. [CrossRef]
28. Gatti, A.; Corti, T.; Brambilla, E.; Horoshko, D.B. Dimensionality of the Spatiotemporal Entanglement of Parametric Down-Conversion Photon Pairs. *Phys. Rev. A* **2012**, *86*, 053803. [CrossRef]
29. Laudenbach, F.; Jin, R.B.; Greganti, C.; Hentschel, M.; Walther, P.; Hubel, H. Numerical Investigation of Photon-Pair Generation in Periodically Poled M TiO X O 4 (M=K, Rb, Cs; X=P, As). *Phys. Rev. Appl.* **2017**, *8*, 024035. [CrossRef]
30. Kim, Y.H.; Grice, W.P. Measurement of the Spectral Properties of the Two-Photon State Generated Via Type II Spontaneous Parametric Downconversion. *Opt. Lett.* **2005**, *30*, 908–910. [CrossRef]
31. Grice, W.P.; Walmsley, I.A. Spectral Information and Distinguishability in Type-II Down-Conversion with a Broadband Pump. *Phys. Rev. A* **1997**, *56*, 1627. [CrossRef]
32. Grice, W.P.; U'Ren, A.B.; Walmsley, I.A. Eliminating Frequency and Space-Time Correlations in Multiphoton States. *Phys. Rev. A* **2001**, *64*, 063815. [CrossRef]

33. Branczyk, A.M.; Ralph, T.C.; Helwig, W.; Silberhorn, C. Optimized Generation of Heralded Fock States Using Parametric Down-Conversion. *New J. Phys.* **2010**, *12*, 063001. [CrossRef]
34. Jundt, D.H. Temperature-Dependent Sellmeier Equation for the Index of Refraction, n e, in Congruent Lithium Niobate. *Opt. Lett.* **1997**, *22*, 1553–1555. [CrossRef]
35. Gayer, O.; Sacks, Z.; Galun, E.; Arie, A. Temperature and Wavelength Dependent Refractive Index Equations for MgO-Doped Congruent and Stoichiometric LiNbO 3. *Appl. Phys. B* **2009**, *94*, 367–367. [CrossRef]

© 2019 by the authors. Licensee MDPI, Basel, Switzerland. This article is an open access article distributed under the terms and conditions of the Creative Commons Attribution (CC BY) license (http://creativecommons.org/licenses/by/4.0/).

Article

Route to Intelligent Imaging Reconstruction via Terahertz Nonlinear Ghost Imaging

Juan S. Totero Gongora, Luana Olivieri, Luke Peters, Jacob Tunesi, Vittorio Cecconi, Antonio Cutrona, Robyn Tucker, Vivek Kumar, Alessia Pasquazi and Marco Peccianti *

Emergent Photonics (EP*ic*) Laboratory, Department of Physics and Astronomy, University of Sussex, Brighton BN1 9QH, UK; jt420@sussex.ac.uk (J.S.T.G.); l.olivieri@sussex.ac.uk (L.O.); l.peters@sussex.ac.uk (L.P.); jt298@sussex.ac.uk (J.T.); v.cecconi@sussex.ac.uk (V.C.); ac878@sussex.ac.uk (A.C.); rt232@sussex.ac.uk (R.T.); vk230@sussex.ac.uk (V.K.); a.pasquazi@sussex.ac.uk (A.P.)
* Correspondence: m.peccianti@sussex.ac.uk; Tel.: +44-(0)1273-873171

Received: 21 January 2020; Accepted: 17 May 2020; Published: 20 May 2020

Abstract: Terahertz (THz) imaging is a rapidly emerging field, thanks to many potential applications in diagnostics, manufacturing, medicine and material characterisation. However, the relatively coarse resolution stemming from the large wavelength limits the deployment of THz imaging in micro- and nano-technologies, keeping its potential benefits out-of-reach in many practical scenarios and devices. In this context, single-pixel techniques are a promising alternative to imaging arrays, in particular when targeting subwavelength resolutions. In this work, we discuss the key advantages and practical challenges in the implementation of time-resolved nonlinear ghost imaging (TIMING), an imaging technique combining nonlinear THz generation with time-resolved time-domain spectroscopy detection. We numerically demonstrate the high-resolution reconstruction of semi-transparent samples, and we show how the Walsh–Hadamard reconstruction scheme can be optimised to significantly reduce the reconstruction time. We also discuss how, in sharp contrast with traditional intensity-based ghost imaging, the field detection at the heart of TIMING enables high-fidelity image reconstruction via low numerical-aperture detection. Even more striking—and to the best of our knowledge, an issue never tackled before—the general concept of "resolution" of the imaging system as the "smallest feature discernible" appears to be not well suited to describing the fidelity limits of nonlinear ghost-imaging systems. Our results suggest that the drop in reconstruction accuracy stemming from non-ideal detection conditions is complex and not driven by the attenuation of high-frequency spatial components (i.e., blurring) as in standard imaging. On the technological side, we further show how achieving efficient optical-to-terahertz conversion in extremely short propagation lengths is crucial regarding imaging performance, and we propose low-bandgap semiconductors as a practical framework to obtain THz emission from quasi-2D structures, i.e., structure in which the interaction occurs on a deeply subwavelength scale. Our results establish a comprehensive theoretical and experimental framework for the development of a new generation of terahertz hyperspectral imaging devices.

Keywords: terahertz; nonlinear optical conversion; complex optical systems; adaptive imaging; single-pixel imaging; surface nonlinear photonics

1. Introduction

In recent years, there has been increasing interest in the development of imaging techniques that are capable of reconstructing the full-wave properties (amplitude and phase) of arbitrary electromagnetic field distributions [1–3]. While standard optical technologies, such as cameras and photodiodes, are usually sensitive to the field intensity, a large part of the sample information is encoded in the optical phase of the scattered field [4]. Interestingly, the direct detection of the field evolution is achievable at terahertz (THz) frequencies thanks to the availability of the time-domain spectroscopy (TDS) technique. TDS detection provides a time-resolved measurement of the electric field (e.g., via electro-optical sampling [5]), allowing researchers to retrieve the complex-valued dielectric function of a sample. Such a capability, coupled with the existence of specific and distinctive spectral fingerprints in the terahertz frequency range, are critical enabling tools for advanced applications, such as explosive detection, biological imaging, artwork conservation and medical diagnosis [6–10]. However, despite the vast body of potential applications, the development of TDS devices that are capable of high-resolution imaging is still regarded as an open challenge. A typical TDS implementation relies on complex and expensive optical components that cannot be easily integrated into high-density sensor arrays [11].

To date, THz imaging mostly relies on thermal cameras, essentially the equivalent of optical cameras, which employ arrays of micro-bolometers to measure the time-averaged intensity of the THz signal. As such, they cannot be employed for time-resolved THz detection and they are insensitive to the optical phase and temporal delay of the transmitted THz field. In an attempt to develop arrays of TDS detectors, researchers have proposed two-dimensional full-wave imaging devices that are composed of arrays of photoconductive antennas or Shack–Hartmann sensors [12,13]. However, these devices require complex and expensive technological platforms and their practicality is still a matter of debate. Furthermore, they fundamentally sample the image information in an array of single and well-separated small points. Hence, obtaining a high resolution can still require mechanical action on the sample.

A promising alternative to TDS imaging arrays is single-pixel imaging, or ghost imaging (GI). In these approaches, the sensor array is replaced by a single bucket detector, which collects the field scattered by the sample in response to a specific sequence of incident patterns. By correlating each acquired signal with its corresponding incident field distribution, it is possible to reconstruct the sample image [14–17]. However, despite its simplicity, the implementation of GI at terahertz frequencies is affected by the limited availability of wavefront-shaping devices (e.g., spatial light modulators) that are capable of impressing arbitrary patterns on an incident THz pulse. Following the initial experimental demonstrations with metallic masks and metamaterial devices [18,19], several research groups' researchers have proposed indirect patterning techniques for the generation of high-resolution THz patterns. One of the most successful approaches relies on the generation of transient photocarrier masks on semiconductor substrates [20–23]. In these experiments, a standard optical Spatial Light Modulator (SLM) impresses a spatial pattern on an ultrafast optical beam. Upon impinging on a semiconductor substrate, the latter generates a distribution of carriers matching the desired pattern profile, which acts as a transient metallic mask and can be used to pattern an external THz beam. While this technique has been successfully employed to achieve THz imaging with a deeply subwavelength resolution, it is also affected by a few limitations. In particular, recent works have shown that the maximum resolution achievable with these techniques is strongly dependent on the semiconductor substrate thickness: in Stantchev and coworkers [20,21], for example, researchers have demonstrated that deeply subwavelength resolutions are achievable only when considering patterning substrates with a thickness below 10 µm.

In a series of recent works, we have proposed a new imaging technique, time-resolved nonlinear ghost imaging (TIMING), which overcomes several of these limitations [24–26]. TIMING relies on the integration of nonlinear THz pattern generation with TDS single-pixel field detection. In this work, we discuss the main features of our approach and present our latest results on the theoretical framework underlying our image reconstruction process. Via analysis of the compression properties of the incident pattern distribution, we show how a TIMING implementation based on an optimised Walsh–Hadamard encoding scheme can significantly reduce the number of incident patterns required to obtain a high-fidelity image of the sample. Finally, we discuss how the development of ultra-thin THz emitters can provide a significant improvement to the imaging performance of TIMING.

2. Time-Resolved Nonlinear Ghost Imaging: A Conceptual Overview

A conceptual schematic of our imaging setup is shown in Figure 1a. A spatial pattern is impressed on the optical beam through a binary spatial light simulator, e.g., a digital micromirror device (DMD), obtaining the optical intensity distribution $I_n^{opt}(x,y,\omega)$. The THz patterns $E_n^0(x,y,t)$ are generated using a nonlinear conversion of $I_n^{opt}(x,y,\omega)$ in a nonlinear quadratic crystal (ZnTe) of thickness z_0. The THz pattern propagates across the crystal and interacts with the object, yielding a transmitted field, which is collected by a TDS detection setup. Different from the standard formulations in optics, which relies on the optical intensity, our object reconstruction scheme relies on the time-resolved detection of the electric field scattered by the object. More specifically, the electric field distribution is defined immediately before and after the object as $E^-(x,y,t) = E(x,y,z_0-\epsilon,t)$ and $E^+(x,y,t) = E(x,y,z_0+\epsilon,t)$, respectively, where z_0 is the object plane and $\epsilon > 0$ is an arbitrarily small distance (Figure 1a, inset). Without loss of generality, the transmission properties of the object are represented by defining the transmission function $T(x,y,t)$, which is defined on both the spatial and temporal components to account for the spectral response of the sample. To simplify our analysis, in the following, we considered two-dimensional objects, i.e., we restricted ourselves to transmission functions of the form $T(x,y,t)$. Under this position, the transmitted field is straightforwardly defined as:

$$E^+(x,y,t) = \int dt' T(x,y,t-t') E^-(x,y,t). \qquad (1)$$

The objective of a single-pixel imaging methodology is to reconstruct the transmitted field distribution $E^+(x,y,t)$ through a sequence of measurements to retrieve the transmission function of the object. In our approach, this corresponds to measuring the TDS trace of the spatially-averaged transmitted field from the object in response to a sequence of predefined patterns (a procedure known as computational ghost imaging) [27]. The nth pattern is denoted by $E_n^-(x,y,t) = P_n(x,y)f(t)$, where $P_n(x,y)$ is the deterministic spatial distribution of the pattern and $f(t)$ is the temporal profile of the THz pulse. The reconstruction process is defined as follows:

$$T(x,y,t) = C_n(t)P_n(x,y)_n - C_n(t)_n P_n(x,y)_n, \qquad (2)$$

where $\langle \cdots \rangle_n$ represents an average over the distribution patterns and the expansion coefficients $C_n(t)$ are defined as follows:

$$C_n(t) = \int dxdy\, E_n^+(x,y,t) = \int dxdydt'\, T(x,y,t-t') E_n^-(x,y,t). \qquad (3)$$

A numerical implementation of the image reconstruction process is shown in Figure 1b,c, where we employed TIMING to reconstruct the transmitted field from a semi-transparent sample (a leaf). In Figure 1b, we report the spatial average of the reconstructed field, exhibiting the characteristic temporal profile of the incident THz pulse. Since our image reconstruction operates simultaneously in time and space, it allows for not only retrieving the spatial distribution of the object but also its temporal/spectral features. The specific result of a TIMING scan is a spatiotemporal image of the transmitted field, as shown in Figure 1c.

An interesting question is whether the distance between the distribution of THz sources and the sample has any effect on the image reconstruction capability of our setup. This point is pivotal when time-resolved imaging is desired, as propagation always induces space–time coupling. This condition represents a typical challenge in mask-based ghost imaging when time-domain detection is sought. The propagation within the patterning crystal is known to lead to significant reconstruction issues when considering deeply subwavelength patterns [20–22]. These issues are related to the intrinsic space–time coupling that takes place within the crystal [28]. In essence, once the patterns are impressed on the THz wave at the surface of the crystal (at $z = 0$), they undergo diffraction. As a result, the electric field distribution $E_n^-(x, y, t)$ probing the sample is not the initial distribution $E_n^0(x, y, t)$, but rather a space-time propagated version of it. The latter is mathematically expressed as:

$$E_n^-(x, y, t) = E_n(x, y, z_0 - \epsilon, t) = \int dx dy dt' G(x - x', y - y', z_0 - \epsilon, t - t') E_n^0(x, y, t), \quad (4)$$

where $G(x, y, z_0 - \epsilon, t)$ is the dyadic Green's function propagating the field from $z = 0$ to $z = z_0 - \epsilon$. Since space–time coupling is essentially a linear process, it can be inverted by applying a Weiner filter to the reconstructed image to mitigate the effects of diffraction. In the angular spectrum coordinates (k_x, k_y, z, ω), the Weiner filter is defined as:

$$W(k_x, k_y, z, \omega) = \frac{G^\dagger(k_x, k_y, z, \omega)}{\left|G(k_x, k_y, z, \omega)\right| + \alpha NSR(k_x, k_y, \omega)}, \quad (5)$$

where $NSR(k_x, k_y, \omega)$ is the spectral noise-to-signal distribution, α is a noise-filtering fitting parameter and † stands for Hermitian conjugation [24]. As expressed by Equation (5), the Weiner filter is the equivalent of an inverse Green's function operator that is modified to take into account the presence of noise in the experimental measurements. The effect of the NSR term in the denominator, which is controlled by the parameter α, is to suppress the regions of the spectrum that are dominated by noise and could render the inversion operation an intrinsically ill-posed problem [29].

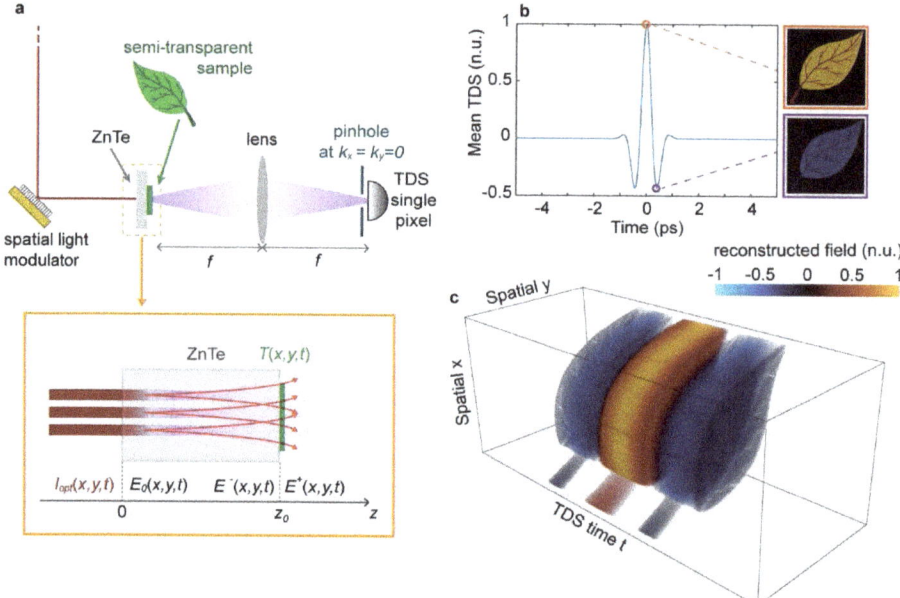

Figure 1. Conceptual description of time-resolved nonlinear ghost imaging (TIMING). (**a**) Schematic of the experimental setup. (**b**,**c**) Simulation of the TIMING reconstruction of a semi-transparent sample, including the average field transmission (panel b) and the full spatiotemporal image of the sample (panel c). The simulated object size was 10.24 cm × 10.24 cm, sampled with a spatial resolution of 512 × 512 pixels ($\Delta x = 200$ µm) and a temporal resolution of $\Delta t = 19.5$ fs. The nonlinear crystal thickness was $z_0 = 10$ µm. n.u.: normalised units, TDS: Time-domain spectroscopy.

From a physical point of view, Equations (4) and (5) can be read as follows: when performing a time-domain reconstruction of the image, the spatial distribution of $E_n^+(x, y, t)$ is acquired at a given time. However, this is not the scattered field from the object in response to the incident pattern E_n^0 at that time; there is no time in which the scattered field $E_n^0(x, y, t)$ is univocally represented in the sampling pattern E_n^-. The reason is simply that the method is slicing a fixed-time contribution of a piece of information that is warped in the space-time. This warping is introduced by the distance between sources and the object plane; therefore, it is different for any plane of the object being imaged.

Said differently, using fixed-time images to reconstruct planar features produces a fundamentally incorrect picture of the evolving scattered field, with different degrees of "distortion" introduced by the amount of propagation. It is worth noting that, although related, this is not the same concept as that of resolution degradation of incoherent near-field systems. In fact, Equation (4) shows that any space-time information retained by the field can be accessed only by accounting for near-field propagation. TIMING reconstructs the image of a scattered field from an object with fidelity by applying the backpropagation kernel from Equation (5). Another interesting aspect is whether the thickness of the nonlinear crystal accounts for an overall separation between terahertz sources and the object, affecting the achievable resolution. The difference here is that the propagation is inherently nonlinear and although the generated terahertz signal diffracts linearly, for any desired resolution, there is always a given generating crystal section that is sufficiently close to the object to illuminate it within the required near-field condition. We have recently theoretically and experimentally demonstrated that the diffraction limit does not directly apply in the nonlinear GI via the generation crystal thickness since the nonlinear conversion from optical to THz patterns is a process distributed across the crystal [25]. We argue that this general approach is particularly useful when considering samples stored in cuvettes or sample holders.

3. Compressed and Adaptive Sensing Applications

In this section, we discuss the image reconstruction performance of TIMING as a result of our particular choice of input pattern distribution. To reconstruct the sample, TIMING relies on the Walsh–Hadamard (WH) image decomposition, which constitutes the binary counterpart of standard Fourier-based image analysis [30]. In our approach, the choice of the incident pattern distribution was driven by three considerations: (i) the compatibility with the available wavefront-shaping technology impressing patterns on the optical beam, (ii) the average signal-to-noise ratio (SNR) of the signal associated with each incident pattern and (iii) the energy compaction (compressibility) properties of the image expansion base. The WH patterns can be implemented straightforwardly through a digital micromirror device (DMD) and they are known to maximise the SNR of the acquired signals in experiments [31,32]. The latter is a significant advantage when compared to standard TDS imagers, which rely on a raster-scan reconstruction approach, where either the source or receiver (or both) are sequentially moved across the sample, leading to a combination of single-pixel detection and illumination [10]. While this approach is intuitive and straightforward to implement, a single-pixel illumination usually implies a degradation of the SNR of the expansion coefficients for a fixed intensity per pixel. Furthermore, raster-scan imaging is a local reconstruction algorithm that is not suitable for compressed sensing; in mathematical terms, the raster scan corresponds to expanding the sample image in the canonical Cartesian base $E_{n,m}(x,y) = \delta(x - x_n, y - y_n)$. Trivially speaking, to reconstruct the entire image with this approach, each pixel composing it needs to be scanned.

In contrast, the WH encoding scheme is a very popular example of energy compacting (compressive) decomposition, as in the case of Fourier-based or wavelet-based image analysis [33,34]. In these approaches, the image is represented as an orthogonal basis of extended spatial functions. For example, in the case of Fourier image analysis, the sampling patterns are the basis of the two-dimensional Fourier Transform [29,35]. The choice of an expansion basis composed of extended patterns has two main advantages. First, extended patterns are generally characterised by transmitted fields with higher SNRs because distributed sources generally carry more power. In fact, for a given power limit per pixel, the Walsh–Hadamard decomposition allows for a total energy per pattern that is about N/2 higher than single-pixel illumination. Second, and more importantly, there is no one-to-one correspondence between individual image pixels and distinct measurements (as in the case of the raster scan). In fact, the incident patterns not only probe different parts of the sample in parallel but can also provide useful insights into its spatial structure, even before completing the entire set of illuminating patterns.

In practical terms, a WH pattern of size $N \times N$ is obtained by considering the tensor product between the columns (or, invariantly, rows) of the corresponding $N \times N$ Walsh–Hadamard matrix (see Figure 2a). The columns (or rows) are mutually orthogonal and form a complete tensor basis for any two-dimensional matrix. Interestingly, the columns of the Hadamard matrix can be re-arranged in different configurations, leading to matrices with different orderings [36–38]. In Figure 2, we compare two configurations: the Walsh (or sequency) order and the Hadamard (or natural) order. The Walsh ordering is particularly useful in image reconstruction as it mirrors the standard order of the discrete Fourier basis, i.e., the columns are sorted in terms of increasing spatial frequencies. This means that by using the Walsh matrix, it is possible to acquire complete lower-resolution images before completing the illumination set, which can be useful for applying decisional approaches and reducing the set dimension [39,40].

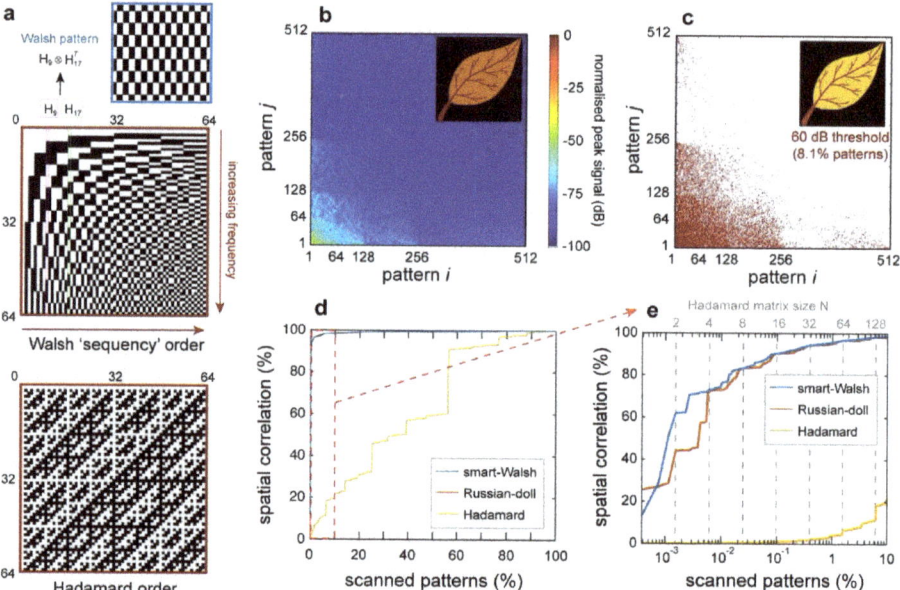

Figure 2. Walsh–Hadamard image reconstruction. (**a**) Generation of incident patterns from the Walsh–Hadamard matrix. Each pattern is defined as the tensor product between two columns of the generating matrix. The patterns can be generated from different configurations of a Hadamard matrix: we show the Walsh, or "sequency", order (top, used in TIMING) and the standard Hadamard, or "natural", order (bottom). (**b**,**c**) Reconstructed Walsh spectrum of the peak-field object transmission. Interestingly, only a fraction of the patterns (8.1%) were associated with a spectral amplitude exceeding the −60 dB threshold (with 0 dB being the energy correlation of the fittest pattern—panel c). Nevertheless, these patterns were sufficient to provide a high-fidelity reconstruction of the image (insets). (**d**,**e**) Pearson correlation coefficients between reconstructed and original images as a function of the number of patterns employed in the reconstruction. The results refer to the entire scan (panel d) and the initial 10% of patterns (panel e).

To illustrate how the image information is distributed across the basis of incident patterns, it is useful to analyse the peak-field Walsh spectrum of the reconstructed image, which is shown in Figure 2b. The WH spectrum is obtained by plotting the $C_n(t = t_{peak})$ coefficients as a function of their generating pattern indexes. As can be evinced from Figure 2b, the WH decomposition re-organises the image information into a hierarchical structure, which mirrors the spectral content of the image. Interestingly, this property is at the core of the compression properties of the WH encoding scheme, as can be exploited to significantly reduce the number of measurements required to reconstruct the image. We illustrate this result in Figure 2c, where we identify the coefficients with an amplitude exceeding a −60 dB threshold with a red marker. As shown in Figure 2c, these significant coefficients were mostly localised in correspondence with the smaller spatial frequencies of the image, and for this image, they represented 8.1% of the total number of patterns. Remarkably, this limited number of patterns was sufficient to accurately reconstruct the image (as shown in Figure 2c, inset).

For a given Walsh–Hadamard matrix, it is also critical to consider the specific order employed when selecting the sequence of columns forming the distribution of incident patterns. In our approach, we implemented an optimised ordering of the WH patterns (denoted as "smart-Walsh"), which sorts the incident patterns in terms of increasing spatial frequency (see Supplementary Video 1). In Figure 2d,e, we illustrate the fidelity of the TIMING reconstruction across the ensemble of incident patterns for different sorting schemes. The fidelity between reconstructed and original images is estimated through the Pearson correlation coefficient, which measures the spatial correlation between the two datasets and is defined as:

$$\rho(A, B) = \frac{\sum_{mn}(A_{mn} - \overline{A})(B_{mn} - \overline{B})}{\sqrt{\sum_{mn}(A_{mn} - \overline{A})^2 \cdot \sum_{mn}(B_{mn} - \overline{B})^2}}, \quad (6)$$

where $\overline{A}$ and $\overline{B}$ are the spatial averages of A and B, respectively. In our analysis, we considered the performance of our "smart-Walsh" sorting (blue line) with the natural Hadamard sorting (yellow line) and the recently proposed "Russian-doll" sorting (orange line) [38]. As shown in Figure 2d, both the smart-Walsh and the Russian-doll sorting were capable of high-fidelity reconstructions of the sample image, even just by using a fraction of patterns, especially when compared to the standard Hadamard case. Further insights on the image reconstruction performance can be obtained by analysing the image reconstruction across the first 10% of patterns (Figure 2e). Remarkably, both our approach and the Russian-doll sorting outperformed the standard Hadamard sorting, yielding a high-fidelity image (spatial correlation exceeding 90%) by considering only 0.1% of the total number of patterns. Interestingly, while the performance of our "smart-Walsh" approach matched the Russian-doll sorting as soon as each Hadamard order was completed (dashed grey lines), we observed that it outperformed it across incomplete scans.

4. Performance of Field-Based Ghost-Imaging Detection in the Fourier Plane

The possibility of performing field-sensitive detection provides TIMING with a significant advantage when compared with traditional GI. However, the typical GI correlation between detection parameters and image fidelity is broken by the nonlinear ghost imaging transformation, i.e., the need for establishing a correlation between coherent-field detection and the optical intensity patterns. More precisely, the implementation of a field average in the image extraction radically changes the way the image quality depends on the experimental parameters. Standard GI reconstruction relies on detecting the integrated scattered field to estimate the spatial correlation between the incident patterns and the sample, where:

$$C_n = \int dxdy\, I_n^+(x,y) = \int dxdydt'|T(x,y,t-t')E_n^-(x,y,t)|^2. \quad (7)$$

This corresponds to the direct acquisition of the total scattered field with a standard bucket detector, which integrates the transmitted intensity distribution. Fundamentally, it is an estimator of the total scattered power, and as such, it is directly affected by the numerical aperture of the detector and by the distance between the detector and the sample. As discussed in the literature on optical

GI, both these factors directly fix the amount of information that is available when reconstructing the image and directly affect its fidelity [15].

TIMING inherits the direct detection of the scattered THz field distribution from time-domain spectroscopy systems. By operating directly on the electric field, it allows for measuring the average THz scattered field (in a fully coherent sense) by performing a point-like detection in the Fourier plane. As defined by Equation (3), the coefficients C_n can be obtained by measuring the $(k_x, k_y) = 0$ spectral components of the THz transmitted field:

$$C_n(t) = \int dx dy\, E_n^+(x, y, t) = \mathcal{F}\left[E_n^+(x, y, t)\right]\Big|_{k_x = 0,\, k_y = 0}. \tag{8}$$

This implementation implies that the experimental measurement of the correlations C_n is not limited at all by the numerical aperture of the bucket detector. This type of measurement can be obtained by placing the object in the focal point of an arbitrary lens and by acquiring the signal in the central point of the opposite focal plane (Figure 1a). The electric field in the focal plane reads as follows:

$$E_{focal}(x', y') \propto \mathcal{F}\left[E_n^+(x, y, t)\right]\left(k_x = \frac{x'}{\lambda f}, k_y = \frac{y'}{\lambda f}\right), \tag{9}$$

where x' and y' are the physical coordinates in the Fourier plane [41].

However, in terms of implementation, the detector samples a finite small area of the Fourier plane with an area-sampling function $PH(k_x, k_y)$, obtaining the estimation $C_n{'}(t)$:

$$C_n^{'(t)} = \int PH(k_x, k_y) * \mathcal{F}\left[E_n^+(x, y, t)\right] dk_x dk_y, \tag{10}$$

where $PH(k_x, k_y)$ is physically represented by the profile of the probe beam in the electro-optical sampling (e.g., a Gaussian function), or by the shape of any aperture implemented in front of the nonlinear detection to fix its interaction area with the THz field.

The accuracy of the measurements is then directly related to how "point-like" our detection can be made. Although one could be tempted to foresee a general benefit of the high signal-to-noise ratio (SNR) resulting from large detection apertures as in the standard GI, this is also a source of artefacts, fundamentally establishing a trade-off between SNR and fidelity.

Figure 3 illustrates the effects of the size d of the sampling function $PH(x' = k_x \lambda f, y' = k_y \lambda f)$ on the image reconstruction fidelity (Figure 3e). Interestingly, the reduction of fidelity observed for increasing the sampling diameter is different from the typical limitations in standard imaging. In our case, a too-large area sampling function in the Fourier plane did not lead to a reduction in the discernible details but rather in the disappearance of entire parts of the image (see Figure 3e, insets).

Similarly, in Figure 4, we illustrate the effect of a misalignment of the sampling function PH centre with respect to the centre of the Fourier plane. Trivially, the spatial correlation between the reconstructed and original images peaks at the centre of the Fourier plane and swiftly decayed in the case of off-axis detection (Figure 4a). In these conditions, the reconstructed image showed the appearance of spurious spatial frequencies, corresponding to the (k_x, k_y) sampling position (Figure 4b,d). Interestingly, however, the overall morphology and details of the image were still present in the images, and no noticeable blurring occurred.

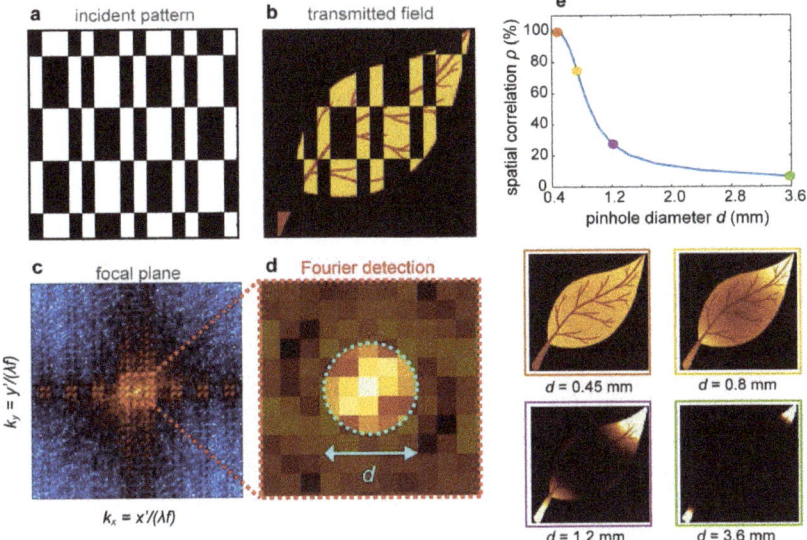

Figure 3. Influence of the pinhole size on the Fourier detection of TIMING reconstruction coefficients. (**a**–**d**) The spatial average of the transmitted field (**b**) associated with each incident pattern (**a**) could be measured by performing a point-like detection in the centre of the Fourier plane (**c**,**d**). In realistic implementations, the centre of the Fourier plane is sampled using a sampling function PH of finite diameter d. (**e**) Spatial correlation between the reconstructed and original image as a function of the sampling function diameter. A departure from the point-like approximation led to a significant corruption of the reconstructed image (insets). Interestingly, the typical image degradation did not necessarily involve the total disappearance of highly resolved details.

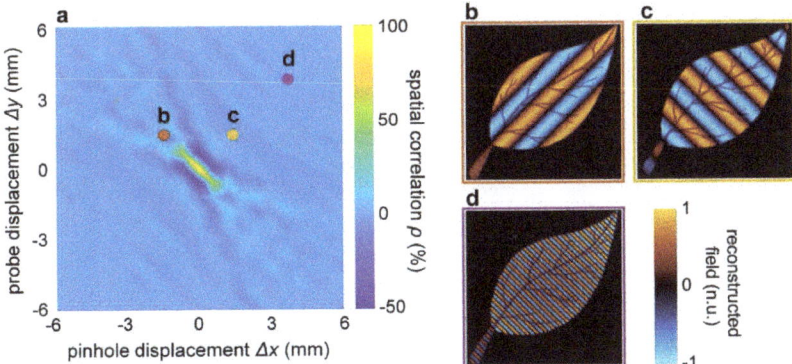

Figure 4. Influence of the pinhole displacement on the Fourier detection of TIMING reconstruction coefficients. (**a**) Spatial correlation between the reconstructed and original image as a function of the sampling function position in the focal plane. The displacement (Δx, Δy) was measured with respect to the lens axis and the sampling function diameter was set to $d = 0.36$ mm, corresponding to a spatial correlation of 100% at the centre of the Fourier plane (cf. Figure 3e). (**b**–**d**) Examples of image reconstruction with off-axis detection, illustrating the appearance of spurious spatial frequencies. Interestingly, the object morphology was still noticeable, even at a relatively large distance from the optical axis.

5. A Route towards Thinner THz Emitters: Surface Emission from Quasi-2D Semiconductor Structures

Deep near-field regimes are in general a requirement to obtain deep-subwavelength image resolutions. Here, we review this current technological solution that is under development in TIMING towards this goal.

In terms of nonlinear ghost imaging, the high resolution fundamentally results from the ability to achieve significant optical-to-terahertz conversions, keeping the sample in the proximity of the distribution of terahertz sources. This translates into the need for generating terahertz from quite thin devices (although we argued how TIMING exhibits significantly more relaxed constraints compared to previous literature [25]).

Although the technology is continuously evolving, the best-performing and most practical off-the-shelf sources are within the class of electro-optical switches. The terahertz emission is generated by a transient current that is sustained by an external electric source and is triggered by a change of conductivity induced by an ultrafast optical absorption [5]. This specific approach benefits from a virtually high optical-to-terahertz conversion efficiency since the actual source of radiation is a current sustained by the electric source. However, this technology is difficult to translate to TIMING since the integration into a single device of a dense distribution of independent electrical switches emitting terahertz signals is extremely challenging.

In terms of direct optical-to-terahertz conversion, improving the efficiency of nonlinear converters is undoubtedly a central research area with a vast spectrum of proposed solutions ranging from novel materials to the design of sophisticated propagation geometries, which allows for very long interaction lengths. However, very few alternatives are currently available for emitters with a thickness below the micrometre scale. One general issue is that the efficiency of bulk nonlinear interactions tend to be vanishingly low at this scale, whereas the ruling mechanisms of the nonlinear interactions are dominated by peculiar physical mechanisms that exist only in quasi-2D frameworks. Some very promising, recently explored solutions comprise exploiting spin-mediated current transients (spintronic emitters) in nano-hetero-metallic structures [42]. On the other hand, a significant fraction of the work in this research area focuses on achieving a very large interfacial nonlinear response or inducing carrier-mediated nonlinear dynamics at a surface.

In general, these effects are fundamentally driven by breaking the lattice symmetry, which is produced by the material discontinuity at the interface. The requirement of tightly reduced interaction lengths makes low-bandgap semiconductors, such as Indium Arsenide (InAs) and Indium Antimonide (InSb), very popular experimental frameworks. What motivated the interest in these systems is the surprisingly high conversion efficiency per interaction length [43–45]. In a traditional NIR ultrafast excitation setting, the mean absorption length for photons is very small, typically within the scale of $l_d = 140$ nm at a wavelength $\lambda = 800$ nm. At low fluences (below 100 nJ/cm^2), InAs is probably considered the benchmark surface emitter. In this case, the generation is driven by the very large difference in mobility between holes and electrons via the photo-Dember effect (Figure 5c,d): when a high density of photogenerated pairs is induced in the proximity of the surface, electrons quickly diffuse away from the surface, leaving uncompensated carriers of the opposite sign. Such a charge unbalance creates a fast stretching dipole, or equivalently, a local current transient that is the source of the terahertz emission [46].

At very high pumping energies (above 10 µJ/cm^2), this phenomenon becomes critically saturated due to the electromagnetic screening role of dense carrier densities. Conversely, the optical surface rectification (SOR) dominates the emission [43]. The optical surface rectification is a quadratic phenomenon induced by the contribution of a local static field at the surface, which is induced by surface states within the bulk cubic nonlinear response (Figure 5a,b). The DC field effectively plays the role of a field contribution in a four-wave mixing process in a mechanism commonly referred to as a field-induced quadratic response [45,47] and is described using:

$$E_{THz} \propto \chi^{(3)} E_{surf} E_\omega^* E_\omega, \tag{11}$$

where $\chi^{(3)}$ is the third-order susceptibility of InAs, E_{surf} is the intrinsic surface potential field, E_ω is the incident optical field and * stands for the complex conjugate. Quite interestingly, because the phenomenon is driven by a surface potential, it is also a measurable way to probe the dynamics of the carrier at the surface, and it has been proposed as the optical analogy of a Kelvin probe [48].

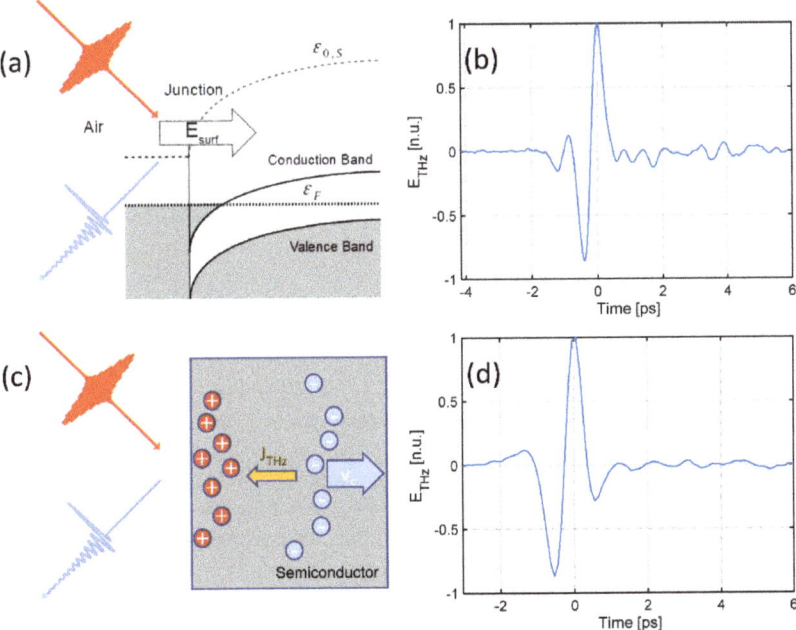

Figure 5. Surface emission driving mechanisms. (**a**) Surface optical rectification—a surface field at the air–semiconductor barrier combines with the optical field in a four-wave mixing process (cubic), generating a terahertz mixing product (see Equation (7)). (**b**) Measurement of the terahertz emission using surface optical rectification with an optical pulsed excitation fluence of 7 mJ/cm² (1 kHz repetition rate) and a pulse with a wavelength of 800 nm and a duration of 90 fs. (**c**) Simplified sketch of the photo-Dember process in InAs. The absorption of an ultrashort pulse generates a high density of photogenerated hole–electron pairs within the optical penetration depth (140 nm). The fast diffusion of the electrons induces a transient current J_{THz}, which is the source of the terahertz emission. (**d**) Measurement of the terahertz emission by photo-Dember mechanism with an optical pulsed excitation fluence of 0.28 µJ/cm² (80 MHz repetition rate) and pulse with a wavelength of 800 nm and a duration of 140 fs.

6. Discussions and Conclusions

In this work, we have provided an overview of the advantages and implementation challenges of a time-resolved nonlinear ghost-imaging approach to THz single-pixel imaging. By combining nonlinear THz generation and single-pixel TDS detection, we demonstrated the high-resolution reconstruction of a semi-transparent sample with a subwavelength resolution (512 × 512 pixels). By providing a detailed analysis of the Walsh–Hadamard reconstruction scheme, we have shown how a specific choice of patterns and the order of acquisition can play a beneficial role in speeding-up the reconstruction of the peak-field transmission from the sample. Remarkably, we have shown that less than 10% of the incident samples were required to achieve a high-fidelity reconstruction of the sample image in a general sequential reconstruction. Our approach, which is based on a lexicographical sorting of the incident patterns in terms of their spatial frequency (an approach we denoted as a "smart-Walsh" reconstruction), is general and image-independent and can be applied to reduce the overall reconstruction time for unknown samples. Interestingly, such a result could be further improved by considering that even a smaller percentage of incident patterns are required to reconstruct the sample: in our case, only 8% of the patterns were associated with an expansion coefficient exceeding 60dB. In practical terms, this would correspond to a 92% shorter acquisition time, corresponding to a 12.5× speed up of the image reconstruction process when compared to a full scan based on the Hadamard encoding scheme. These numbers suggest that the reconstruction process could be significantly sped up through the application of adaptive-basis-scan algorithms and deep-learning-enhanced imaging, which identify and predict the best set of scanning patterns in real time [40,49–51].

Interestingly, our results suggest that the nonlinear GI methodology is not limited by the numerical aperture of the optical system in a "conventional" sense. Said differently, it operates under the assumption of a very low numerical aperture to obtain a faithful spectral representation of the image. However, our results highlight that the image reconstruction is quite sensitive to the size and alignment of the pinhole function selecting the $(k_x, k_y) = 0$ components of the scattered field. Most importantly, in sharp contrast with previous literature on the topic, the reconstruction accuracy cannot simply be represented as a matter of effective "resolution". The drop in reconstruction fidelity, in fact, is not driven by the attenuation of high-frequency spatial components (i.e., blurring) as in standard imaging, but it can lead to the appearance of artefacts and spurious spatial frequencies. To the best of our knowledge, the reconstruction limits of single-pixel time-domain imaging have never been formalised elsewhere.

Finally, although thin emitters are a general requirement for this approach, TIMING exhibits relaxed constraints between the nonlinear interaction length and the image resolution. Yet, solutions for sub-micron-thick large-area terahertz generation are practically possible, enabling resolutions within the same scale or better. A promising platform to achieve this goal is narrow-bandgap semiconductor devices based on InAs or InSb platforms. These materials not only provide extremely high optical-to-terahertz conversion efficiency per unit length but they are also suitable for large-scale fabrication and deployment in real-world devices thanks to their established deployment in the electronic domain.

We believe that TIMING is a significant step forward in the development of terahertz micro-diagnostics based on hyperspectral imaging devices. Our approach also addresses fundamental criticalities in the imaging reconstruction process, which generally affect any high-resolution imaging domain where high temporal resolution is sought. As such, TIMING establishes a comprehensive theoretical and technological platform that paves the way for new generations of terahertz imaging devices satisfying the requirements for high-resolution and spectral sensitivity in real-world applications.

Supplementary Materials: The following are available online at http://www.mdpi.com/2072-666X/11/5/521/s1.

Author Contributions: Conceptualisation, J.S.T.G., A.P. and M.P.; Methodology, J.S.T.G., L.O., A.P. and M.P.; Investigation, J.S.T.G., J.T., L.P., V.C., A.C., L.O., R.T. and V.K.; Supervision, J.S.T.G., A.P. and M.P.; Writing—original draft, J.S.T.G.; Writing—review & editing, J.S.T.G. and M.P. Project administration, M.P.; Funding acquisition, M.P. All authors have read and agreed to the published version of the manuscript.

Funding: This project has received funding from the European Research Council (ERC) under the European Union's Horizon 2020 research and innovation programme (grant agreement no. 725046). J.T. acknowledges the support of the Engineering and Physical Sciences Research Council (EPSRC) through the studentship EP/N509784/1. J.S.T.G, L.P., V.C., V.K. and R.T. acknowledge the support of the European Union's "Horizon 2020" research and innovation program, grant agreement no. 725046, ERC-CoG project TIMING. J.S.T.G. acknowledges funding from the Helena Normanton Fellowship of the University of Sussex, UK. A.C. acknowledges the support from the University of Palermo, Italy through the mobility fellowship "Corso di Perfezionamento all'estero 2018".

Conflicts of Interest: The authors declare no conflict of interest.

References

1. Tikan, A.; Bielawski, S.; Szwaj, C.; Randoux, S.; Suret, P. Single-shot measurement of phase and amplitude by using a heterodyne time-lens system and ultrafast digital time-holography. *Nat. Photonics* **2018**, *12*, 228. [CrossRef]
2. Shechtman, Y.; Eldar, Y.C.; Cohen, O.; Chapman, H.N.; Miao, J.; Segev, M. Phase Retrieval with Application to Optical Imaging: A contemporary overview. *IEEE Signal Process. Mag.* **2015**, *32*, 87–109. [CrossRef]
3. Rivenson, Y.; Zhang, Y.; Günaydın, H.; Teng, D.; Ozcan, A. Phase recovery and holographic image reconstruction using deep learning in neural networks. *Light Sci. Appl.* **2018**, *7*, 17141. [CrossRef] [PubMed]
4. Hack, E.; Zolliker, P. Terahertz holography for imaging amplitude and phase objects. *Opt. Express* **2014**, *22*, 16079–16086. [CrossRef]
5. Song, H.-J.; Nagatsuma, T. *Handbook of Terahertz Technologies: Devices and Applications*; CRC Press: New York, NY, USA, 2015.
6. Kampfrath, T.; Tanaka, K.; Nelson, K.A. Resonant and nonresonant control over matter and light by intense terahertz transients. *Nat. Photonics* **2013**, *7*, 680–690. [CrossRef]
7. Frühling, U.; Wieland, M.; Gensch, M.; Gebert, T.; Schütte, B.; Krikunova, M.; Kalms, R.; Budzyn, F.; Grimm, O.; Rossbach, J.; et al. Single-shot terahertz-field-driven X-ray streak camera. *Nat. Photonics* **2009**, *3*, 523–528. [CrossRef]
8. Yu, L.; Hao, L.; Meiqiong, T.; Jiaoqi, H.; Wei, L.; Jinying, D.; Xueping, C.; Weiling, F.; Yang, Z. The medical application of terahertz technology in non-invasive detection of cells and tissues: Opportunities and challenges. *RSC Adv.* **2019**, *9*, 9354–9363. [CrossRef]
9. Yang, X.; Zhao, X.; Yang, K.; Liu, Y.; Liu, Y.; Fu, W.; Luo, Y. Biomedical Applications of Terahertz Spectroscopy and Imaging. *Trends Biotechnol.* **2016**, *34*, 810–824. [CrossRef]
10. Mittleman, D.M. Twenty years of terahertz imaging [Invited]. *Opt. ExpressOe* **2018**, *26*, 9417–9431. [CrossRef]
11. Sengupta, K.; Nagatsuma, T.; Mittleman, D.M. Terahertz integrated electronic and hybrid electronic–photonic systems. *Nat. Electron.* **2018**, *1*, 622–635. [CrossRef]
12. Lee, A.W.; Hu, Q. Real-time, continuous-wave terahertz imaging by use of a microbolometer focal-plane array. *Opt. Lett.* **2005**, *30*, 2563–2565. [CrossRef] [PubMed]
13. Abraham, E.; Cahyadi, H.; Brossard, M.; Degert, J.; Freysz, E.; Yasui, T. Development of a wavefront sensor for terahertz pulses. *Opt. Express* **2016**, *24*, 5203. [CrossRef] [PubMed]
14. Moreau, P.-A.; Toninelli, E.; Gregory, T.; Padgett, M.J. Ghost Imaging Using Optical Correlations. *Laser Photonics Rev.* **2018**, *12*, 1700143. [CrossRef]
15. Padgett, M.J.; Boyd, R.W. An introduction to ghost imaging: Quantum and classical. *Phsical Eng. Sci.* **2017**, *375*, 20160233. [CrossRef] [PubMed]
16. Shapiro, J.H.; Boyd, R.W. The physics of ghost imaging. *Quantum Inf Process* **2012**, *11*, 949–993. [CrossRef]
17. Erkmen, B.I.; Shapiro, J.H. Ghost imaging: From quantum to classical to computational. *Adv. Opt. Photon. Aop* **2010**, *2*, 405–450. [CrossRef]
18. Chan, W.L.; Charan, K.; Takhar, D.; Kelly, K.F.; Baraniuk, R.G.; Mittleman, D.M. A single-pixel terahertz imaging system based on compressed sensing. *Appl. Phys. Lett.* **2008**, *93*, 121105. [CrossRef]
19. Watts, C.M.; Shrekenhamer, D.; Montoya, J.; Lipworth, G.; Hunt, J.; Sleasman, T.; Krishna, S.; Smith, D.R.; Padilla, W.J. Terahertz compressive imaging with metamaterial spatial light modulators. *Nat. Photonics* **2014**, *8*, 605–609. [CrossRef]
20. Stantchev, R.I.; Sun, B.; Hornett, S.M.; Hobson, P.A.; Gibson, G.M.; Padgett, M.J.; Hendry, E. Noninvasive, near-field terahertz imaging of hidden objects using a single-pixel detector. *Sci. Adv.* **2016**, *2*, e1600190. [CrossRef]
21. Stantchev, R.I.; Phillips, D.B.; Hobson, P.; Hornett, S.M.; Padgett, M.J.; Hendry, E. Compressed sensing with near-field THz radiation. *Optica* **2017**, *4*, 989. [CrossRef]
22. Zhao, J.; Yiwen, E.; Williams, K.; Zhang, X.-C.; Boyd, R.W. Spatial sampling of terahertz fields with sub-wavelength accuracy via probe-beam encoding. *Light Sci. Appl.* **2019**, *8*, 55. [CrossRef] [PubMed]

23. Chen, S.-C.; Du, L.-H.; Meng, K.; Li, J.; Zhai, Z.-H.; Shi, Q.-W.; Li, Z.-R.; Zhu, L.-G. Terahertz wave near-field compressive imaging with a spatial resolution of over λ/100. *Opt. Lett.* **2019**, *44*, 21. [CrossRef] [PubMed]
24. Olivieri, L.; Totero Gongora, J.S.; Pasquazi, A.; Peccianti, M. Time-Resolved Nonlinear Ghost Imaging. *ACS Photonics* **2018**, *5*, 3379–3388. [CrossRef]
25. Olivieri, L.; Gongora, J.S.T.; Peters, L.; Cecconi, V.; Cutrona, A.; Tunesi, J.; Tucker, R.; Pasquazi, A.; Peccianti, M. Hyperspectral terahertz microscopy via nonlinear ghost imaging. *Optica* **2019**, *7*, 186. [CrossRef]
26. Olivieri, L.; Totero Gongora, J.S.; Pasquazi, A.; Peccianti, M. Time-resolved nonlinear ghost imaging: Route to hyperspectral single-pixel reconstruction of complex samples at THz frequencies. In *Society of Photo-Optical Instrumentation Engineers (SPIE) Conference Series, Proceedings of the Nonlinear Frequency Generation and Conversion: Materials and Devices XVIII, San Francisco, CA, USA, 5–9 February 2019*; Schunemann, P.G., Schepler, K.L., Eds.; SPIE: San Francisco, CA, USA, 2019; p. 44.
27. Shapiro, J.H. Computational ghost imaging. *Phys. Rev. A At. Mol. Opt. Phys.* **2008**, *78*, 061802. [CrossRef]
28. Peccianti, M.; Clerici, M.; Pasquazi, A.; Caspani, L.; Ho, S.P.; Buccheri, F.; Ali, J.; Busacca, A.; Ozaki, T.; Morandotti, R. Exact Reconstruction of THz Sub-λSource Features in Knife-Edge Measurements. *IEEE J. Sel. Top. Quantum Electron.* **2013**, *19*, 8401211. [CrossRef]
29. Khare, K. *Fourier Optics and Computational Imaging*; Wiley, John Wiley & Sons Ltd.: Chichester, West Sussex, 2016; ISBN 978-1-118-90034-5.
30. Harwit, M. *Hadamard Transform Optics*; Elsevier: New York, NY, USA, 2012; ISBN 978-0-323-15864-0.
31. Davis, D.S. Multiplexed imaging by means of optically generated Kronecker products: 1. The basic concept. *Appl. Opt.* **1995**, *34*, 1170–1176. [CrossRef]
32. Streeter, L.; Burling-Claridge, G.R.; Cree, M.J.; Künnemeyer, R. Optical full Hadamard matrix multiplexing and noise effects. *Appl. Opt.* **2009**, *48*, 2078–2085. [CrossRef]
33. Eldar, Y.C.; Kutyniok, G. *Compressed Sensing: Theory and Applications*; Cambridge University Press: Cambridge, UK, 2012; ISBN 978-1-107-00558-7.
34. Dongfeng, S.; Jian, H.; Wenwen, M.; Kaixin, Y.; Baoqing, S.; Yingjian, W.; Kee, Y.; Chenbo, X.; Dong, L.; Wenyue, Z. Radon single-pixel imaging with projective sampling. *Opt. ExpressOe* **2019**, *27*, 14594–14609. [CrossRef]
35. Wenwen, M.; Wenwen, M.; Dongfeng, S.; Dongfeng, S.; Jian, H.; Jian, H.; Kee, Y.; Kee, Y.; Kee, Y.; Yingjian, W.; et al. Sparse Fourier single-pixel imaging. *Opt. ExpressOe* **2019**, *27*, 31490–31503. [CrossRef]
36. Hedayat, A.; Wallis, W.D. Hadamard Matrices and Their Applications. *Ann. Stat.* **1978**, *6*, 1184–1238. [CrossRef]
37. Wang, L.; Zhao, S. Fast reconstructed and high-quality ghost imaging with fast Walsh–Hadamard transform. *Photonics Res.* **2016**, *4*, 240. [CrossRef]
38. Sun, M.-J.; Meng, L.-T.; Edgar, M.P.; Padgett, M.J.; Radwell, N. A Russian Dolls ordering of the Hadamard basis for compressive single-pixel imaging. *Sci. Rep.* **2017**, *7*. [CrossRef] [PubMed]
39. Rousset, F.; Ducros, N.; Farina, A.; Valentini, G.; D'Andrea, C.; Peyrin, F. Adaptive Basis Scan by Wavelet Prediction for Single-Pixel Imaging. *IEEE Trans. Comput. Imaging* **2017**, *3*, 36–46. [CrossRef]
40. Kravets, V.; Kondrashov, P.; Stern, A. Compressive ultraspectral imaging using multiscale structured illumination. *Appl. Opt. AO* **2019**, *58*, F32–F39. [CrossRef] [PubMed]
41. Born, M.; Wolf, E. *Principles of Optics: Electromagnetic Theory of Propagation, Interference and Diffraction of Light*, 7th expanded; Cambridge University Press: Cambridge, NY, USA, 1999; ISBN 978-0-521-64222-4.
42. Seifert, T.; Jaiswal, S.; Martens, U.; Hannegan, J.; Braun, L.; Maldonado, P.; Freimuth, F.; Kronenberg, A.; Henrizi, J.; Radu, I.; et al. Efficient metallic spintronic emitters of ultrabroadband terahertz radiation. *Nat. Photonics* **2016**, *10*, 483–488. [CrossRef]
43. Reid, M.; Fedosejevs, R. Terahertz emission from (100) InAs surfaces at high excitation fluences. *Appl. Phys. Lett.* **2005**, *86*, 011906. [CrossRef]
44. Reid, M.; Cravetchi, I.V.; Fedosejevs, R. Terahertz radiation and second-harmonic generation from InAs: Bulk versus surface electric-field-induced contributions. *Phys. Rev. B* **2005**, *72*, 035201. [CrossRef]
45. Peters, L.; Tunesi, J.; Pasquazi, A.; Peccianti, M. High-energy terahertz surface optical rectification. *Nano Energy* **2018**, *46*, 128–132. [CrossRef]
46. Apostolopoulos, V.; Barnes, M.E. THz emitters based on the photo-Dember effect. *J. Phys. D Appl. Phys.* **2014**, *47*, 374002. [CrossRef]

47. Cazzanelli, M.; Schilling, J. Second order optical nonlinearity in silicon by symmetry breaking. *Appl. Phys. Rev.* **2016**, *3*, 011104. [CrossRef]
48. Peters, L.; Tunesi, J.; Pasquazi, A.; Peccianti, M. Optical Pump Rectification Emission: Route to Terahertz Free-Standing Surface Potential Diagnostics. *Sci. Rep.* **2017**, *7*. [CrossRef] [PubMed]
49. Dai, H.; Gu, G.; He, W.; Liao, F.; Zhuang, J.; Liu, X.; Chen, Q. Adaptive compressed sampling based on extended wavelet trees. *Appl. Opt.* **2014**, *53*, 6619. [CrossRef] [PubMed]
50. Dai, H.; Gu, G.; He, W.; Ye, L.; Mao, T.; Chen, Q. Adaptive compressed photon counting 3D imaging based on wavelet trees and depth map sparse representation. *Opt. Express* **2016**, *24*, 26080. [CrossRef]
51. Barbastathis, G.; Ozcan, A.; Situ, G. On the use of deep learning for computational imaging. *Opt. Opt.* **2019**, *6*, 921–943. [CrossRef]

© 2020 by the authors. Licensee MDPI, Basel, Switzerland. This article is an open access article distributed under the terms and conditions of the Creative Commons Attribution (CC BY) license (http://creativecommons.org/licenses/by/4.0/).

MDPI
St. Alban-Anlage 66
4052 Basel
Switzerland
Tel. +41 61 683 77 34
Fax +41 61 302 89 18
www.mdpi.com

Micromachines Editorial Office
E-mail: micromachines@mdpi.com
www.mdpi.com/journal/micromachines

www.ingramcontent.com/pod-product-compliance
Lightning Source LLC
LaVergne TN
LVHW070742100526
838202LV00013B/1288